Cardinals

Linguistic Inquiry Monographs
Samuel Jay Keyser, general editor

A complete list of books published in the Linguistic Inquiry Monographs series appears at the back of this book

Cardinals
The Syntax and Semantics of Cardinal-Containing Expressions

Tania Ionin and Ora Matushansky

The MIT Press
Cambridge, Massachusetts
London, England

This book was set in Times LT Std by Toppan Best-set Premedia Limited. Printed and bound in the United States of America.

Library of Congress Cataloging-in-Publication Data

Names: Ionin, Tania, author. | Matushansky, Ora, author.
Title: Cardinals : the syntax and semantics of cardinal-containing expressions / Tania Ionin and Ora Matushansky.
Description: Cambridge, MA : The MIT Press, [2018] | Series: Linguistic inquiry monographs ; 79 | Includes bibliographical references and index.
Identifiers: LCCN 2018004431| ISBN 9780262038737 (hardcover : alk. paper) | ISBN 9780262535786 (paperback : alk. paper)
Subjects: LCSH: Grammar, Comparative and general--Numerals. | Cardinal numbers.
Classification: LCC P275 .I66 2018 | DDC 415--dc23 LC record available at https://lccn.loc.gov/2018004431

10 9 8 7 6 5 4 3 2 1

To our daughters:
The two who arrived before this book's inception: Dana and Elise
And the two who arrived during the writing process: Alice and Ruth

Contents

Series Foreword

We are pleased to present the seventy-ninth volume in the series *Linguistic Inquiry Monographs*. These monographs present new and original research beyond the scope of the article. We hope they will benefit our field by bringing to it perspectives that will stimulate further research and insight.

Originally published in limited edition, the *Linguistic Inquiry Monographs* are now more widely available. This change is due to the great interest engendered by the series and by the needs of a growing readership. The editors thank the readers for their support and welcome suggestions about future directions for the series.

Samuel Jay Keyser
for the Editorial Board

Acknowledgments

We gratefully acknowledge the debt we owe to three anonymous reviewers for *Linguistic Inquiry Monographs* for their detailed and insightful comments on an earlier version of this monograph, and to David Hill for his extremely thorough and insightful copyediting. We are also grateful to the anonymous reviewers for the *Journal of Semantics* for their helpful comments on our 2006 paper on this topic. We would like to thank Eddy Ruys for discussion, data, help, tea, and sympathy, and the audiences of various conferences and workshops where parts of this work were presented, as well as our department members, for their comments.

We are grateful to many informants for providing information about and examples from seventeen different languages discussed in this work. In alphabetical order by language, we would like to express our gratitude to the following linguist and nonlinguist informants. *Arabic:* Abdulkafi Albirini, Abdelaadim Bidaoui, Zainab Hermes, Sarah Ouwayda, and Elias Shakkour; *Basque:* Ikuska Ansola and Itxaso Rodriguez; *Bulgarian:* Roumyana Pancheva and Roumyana Slabakova; *Dutch:* Dana Matushansky and Eddy Ruys; *Finnish:* Elsi Kaiser; *German:* Klaus von Heusinger; *Modern Greek:* Marina Terkourafi; *Hebrew:* Nora Boneh and Yael Gertner; *Inari Sami:* Ida Toivonen; *Polish:* Barbara Citko, Ania Lubowicz, and Barbara Tomaszewicz, as well as Heidi Klockmann for helpful discussion and for pointing us to an important example; *Romanian:* Roxana Girju; *Russian:* Timur Oikhberg; *Scottish Gaelic:* Roibeard Ó Maolalaigh; *Sinhala:* Wijeratne Wijesinghe; *Spanish:* Ikuska Ansola and Karlos Arregi; *Turkish:* Gulsat Aygen; *Modern Welsh:* Peredur Webb-Davies.

Finally, we would like to thank our husbands and children for their patience during the long and torturous process of book writing, during which the number of children multiplied by two, but the number of husbands was thankfully neither added to nor subtracted from.

Additional notes: The authors are listed in alphabetical order. In chapter 2, the two figures depicting schematic representations of spatial relationships are redrawn from "Making space for measures," a paper by Ora Matushansky and Joost Zwarts presented at the forty-seventh annual meeting of the New England Linguistic Society (University of Massachusetts, Amherst, October 14–16, 2016).

Abbreviations

1	first person		FOC	focus
2	second person		FRACT	fraction
3	third person		FUT	future
ABL	ablative		GEN	genitive
ABS	absolutive		GRD	gerund
ACC	accusative		HAB	habitual
ADE	adessive		HUMAN	human
ADJ	adjectival affix		ILL	illative
ADN	adnumerative		IMPER	imperative
AGR	agreement marker		IMPF	imperfect
AOR	aorist		IMPFV	imperfective
APPROX	approximative		INDEF	indefinite
BOUND	bound		INE	inessive
C	common		INF	infinitive
CL	classifier		INFL	inflected
COLL	collective		INS	instrumental
CONJ	conjunction		LKR	linker
CONN	connective		LNK	linking vowel
COUNT	count		LOC	locative
CX	connecting vowel		M	masculine
DAT	dative		N	neuter
DEF	definite		NC	noun-concord marker
DET	determiner		NEG	negative
DIM	diminutive		NMN	nominalizer
DPNDNT	dependent		NOM	nominative
DU	dual		NHUMAN	nonhuman
EMPH	emphatic marker		NV	nonvirile
ERG	ergative		OBL	oblique
F	feminine		ORD	ordinal

PART	partitive	PSM	postverbal-subject marker
PASS	passive		
PAST	past	PTCP	participle
PAUC	paucal	REDUP	reduplicative marker
PL	plural	REFL	reflexive
POSS	possessive	SBJV	subjunctive
PRED	predicative particle	SG	singular
PRES	present	SR	subject-relativizing
PRFV	perfective	TOT	totality
PRM	previous-reference marker	V	virile
		VN	verbal noun
PROG	progressive	VOC	vocative
PRT	particle		

1 Introduction

The linguistic properties of numerals have long been of interest to theoretical linguists as well as psycholinguists, and they have been studied from a variety of perspectives. Research on the linguistic properties of numerals can be divided into three main types of approaches: (i) formal linguistic approaches to the syntax and semantics of numeral-containing expressions; (ii) experimental research on the semantic and pragmatic properties of numerals; and (iii) the study of mathematical cognition within cognitive psychology. (The last two approaches also include much work on the acquisition of numerals by children.) Research on numerals can inform our understanding both of linguistic theory and of human cognition, which makes numerals an area of interest to scholars in linguistics, psycholinguistics, and cognitive psychology.

Our own work falls within formal linguistic approaches to numerals. Before proceeding to the details of our analysis, we provide a brief overview of issues in the linguistic study of numerals; for a more detailed overview organized by topic, see the numerals bibliography of Ionin and Matushansky 2013b.

For the purposes of language, we distinguish between numerals and numbers. Numerals, for us, are linguistic objects that include the notion of cardinality in their denotation. Numbers, on the other hand, are studied by philosophy of mathematics, where natural numbers are viewed as abstract objects (Frege 1884; Zalta 1999; Moltmann 2012a,b), either preexisting or constructed, and defined in purely mathematical terms. Our work does not have anything to add to the study of numbers, beyond noting that they can be referred to by syntactically complex expressions involving numerals, which are the focus of this work.

1.1 Numerals in Theoretical Linguistics: A Brief Preview

Formal generative analyses of numerals have typically focused on their syntactic status (as lexical vs. functional, as heads vs. maximal projections) and on their semantic status (as determiners vs. predicates vs. modifiers). The relevant evidence comes from a large body of languages. In particular, numeral systems that have received much attention in formal linguistic literature include those found in Slavic languages (e.g., Mel'čuk 1985; Corbett 1993; Franks 1995; Rappaport 2002; and much work by Adam Przepiorkowski— see Przepiorkowski and Rosen 2008 for references); Welsh (Biblical as well as Modern—Hurford 1975; Mittendorf and Sadler 2005); Semitic languages (e.g., Sadler 2010; Danon 2012; Ouwayda 2014); Finno-Ugric languages (e.g., Nelson and Toivonen 2000; Karttunen 2006); Bantu languages (e.g., Zweig 2006; Krifka and Zerbian 2008); Basque (e.g., Etxeberria and Etxepare 2008); and classifier languages, such as Mandarin (e.g., Zhang 2012), Japanese (e.g., Kobuchi-Philip 2003), Korean (e.g., Jong-Bok Kim 2011), and Thai (e.g., Jenks 2011).

For comprehensive overviews of crosslinguistic variation in numeral systems, see Hurford 1975, 1987; Corbett 1978, 1993; and Comrie 2005a,b. Depending on the language, the relevant properties of numerals that are examined include word order, case assignment, and agreement for gender and number. Most literature, including our own work, focuses on cardinal numerals; for discussion of ordinals and other numeral derivatives, see Greenberg 2000; Stolz and Veselinova 2005; Stump 2010; and Bylinina et al. 2015, among others.

Theoretical accounts of the syntactic status of cardinals can be divided into those that treat cardinals as functional elements (e.g., Ritter 1991; Cardinaletti and Giusti 1992; Kayne 2003) and those that treat them as lexical, for example as nouns (Jackendoff 1977; Przepiorkowski 1997) or adjectives (Allegranza 1998). What all of these accounts have in common is that they make no distinction between simplex cardinals such as *four* and complex cardinals such as *twenty-four* or *four hundred*. Whatever analysis (as head or maximal projection, as a functional or lexical element) is given for simplex cardinals, it is assumed to also hold for complex cardinals. See chapters 2 and 3 for more discussion.

With regard to the semantics of cardinal numerals, classic approaches to cardinals have treated them as determiners or generalized quantifiers (e.g., Montague 1974; Bennett 1974; Barwise and Cooper 1981; Scha 1981; Van der Does 1992, 1993), as predicates (Partee 1986), as semantic modifiers (Link 1987; Verkuyl 1993; Carpenter 1998; Landman 2003, among others), and, more recently, as degrees (Kennedy 2013, 2015); see chapter 2 for more discussion. Many approaches concerned with modified numerals, such as *at least three*, also take a stand on the semantics of unmodified numerals (see, e.g., Krifka 1999; Hackl 2000; and Nouwen 2010, among many others); see chapter

9 for more discussion. What unites all of the above proposals is that they do not make a distinction between simplex and complex cardinals, assuming that the semantic analysis of *three* also holds for *three hundred* or *one thousand and twenty-nine*. That is, classic semantic analyses of cardinals, like classic syntactic analyses of cardinals, do not concern themselves with the structure of complex cardinals.

In Ionin and Matushansky 2004, 2006, we took up precisely this question and proposed that complex cardinals are derived compositionally and that in an NP that contains a complex cardinal, such as *twenty-four books* or *four hundred books*, the complex cardinal does not form a unit to the exclusion of the lexical NP. On our analysis, cardinals are lexical elements (nouns, adjectives, or even verbs, depending on the language) and cardinal-containing NPs are built using syntactic means (complementation, coordination, prepositions) as well as standard principles of semantic composition. Later approaches in both syntax (Zweig 2006; Danon 2012) and semantics (Rothstein 2013, 2016, 2017), while differing from our proposal, have adopted the basic view that complex cardinals have internal linguistic structure. We discuss these proposals at later points in this book (see chapter 2 for discussion of Rothstein's work, see chapter 3 for Zweig 2006, and see chapter 7 for a discussion of Danon 2012 with regard to cardinals in Hebrew).

1.2   Numerals in Experimental Semantics/Pragmatics

Much literature in semantics and pragmatics has focused on whether cardinal numerals have 'at least' or 'exactly' readings: does *three* mean 'at least three' or 'exactly three'? Even though cardinals are most commonly used with 'exactly' readings (one does not usually talk about *three books* if there are four), the traditional view of cardinals as quantifiers (e.g., Barwise and Cooper 1981) treats them as having 'at least' readings. On this view, the reason that *three* is usually interpreted as 'exactly three' is scalar implicature, generated under the Gricean Quantity Maxim (Horn 1972): if we wanted to talk about four books, we would have said *four*, not *three*; since we said *three*, we must have meant 'exactly three'. This is analogous to the *some–all* distinction: *some* is commonly taken to mean 'some and possibly all', with scalar implicature resulting in the interpretation 'some but not all'.

More recently, a number of alternative proposals have been put forth. Relevance-theoretic views such as Carston's (1998) treat cardinals as inherently underspecified, with both 'exactly' and 'at least' readings driven by relevance considerations. Another possibility is that cardinals are lexically ambiguous between the 'exactly' and 'at least' readings: for example, Geurts

(2006) derives the 'at least' reading from the 'exactly' reading via semantic type shifting. In contrast, Breheny (2008) treats the 'at least' reading as derived from the 'exactly' reading via a pragmatic process.

The claims of these competing proposals have been tested in a series of experimental studies (see Huang and Snedeker 2009 and the references cited therein). The evidence is conflicting. On the one hand, Huang and Snedeker (2009) found that there is no processing cost to interpreting *three* as 'exactly three', while there is a processing cost for interpreting *some* as 'some but not all'; on the assumption that computing a scalar implicature carries a processing cost, this finding suggests that the 'exactly' reading of cardinals, unlike the 'not all' reading of *some*, is not due to scalar implicature. On the other hand, Panizza, Chierchia, and Clifton (2009) did find evidence for a computation of scalar implicatures on cardinals, based on the processing of cardinals in upward-entailing versus downward-entailing contexts.

A different set of experimental studies have focused on the behavior of comparative versus superlative numerals (*more than three* vs. *at least four*). While the Generalized-Quantifier Theory of Barwise and Cooper (1981) gives the same quantifier analysis to both types of modified numeral, more recent literature (e.g., Krifka 1999; Hackl 2000; Geurts and Nouwen 2007; Büring 2008, 2009) has challenged this view, showing differences in the behavior of these two types of expressions. For instance, superlative but not comparative numerals have been argued to have a modal component. Studies such as Geurts et al. 2010 and Cummins and Katsos 2010 have used experimental methodology to examine the readings of comparative versus superlative numerals and the inferences that they give rise to.

In our work, we similarly reject the view that comparative and superlative numerals are generalized quantifiers, but we do so for a different reason: our analysis of complex cardinals necessitates an analysis on which sequences such as *more than three* or *at least four* are not treated as constituents, as discussed in more detail in chapter 9.

1.3 Numerals in Human Cognition

Linguists as well as psychologists have studied the relationship between human language and the mathematical number system. One reason for this interest is that both language and formal mathematics are unique to humans; indeed, many researchers have tied the two together, arguing that language is what makes formal mathematics possible. For instance, Hurford (1987) argues that the number system grew out of an interaction between human cognitive and linguistic capacities, and language is what allows us to reason about higher numbers. Wiese (2007) goes further, proposing that language provides humans not only with the ability to reason about numbers but with the numbers themselves. According to Wiese, the emergence of verbal counting systems allowed for the coevolution of a complex number system. Hurford and Wiese both assume one single cognitive system that underlies numerical capacity, but much work by Elizabeth Spelke and colleagues (see Spelke 2011 for an overview) has argued that there are actually two distinct cognitive systems underlying number cognition. One system (the exact-number system for small numbers) allows us to track small numbers, while the other (the approximate-number system) allows us to compare and combine sets. Much experimental research has shown that these systems are not unique to adult humans but are also available to preverbal infants, rats, and pigeons (see Dehaene 1997 for an overview of the research). There is a large body of research both on numerical systems in infants (e.g., Wynn 1992; Xu and Spelke 2000; and Izard et al. 2009, among others) and on numerical cognition among speakers whose language has few or no number words (e.g., Pica et al. 2004; Everett 2005; Frank et al. 2008). Both lines of research converge on the claim that precise larger numbers are not part of our innate cognitive abilities but emerge in humans as a result of language.

While the thrust of our work is quite different from the studies discussed above—we run no experiments, and we make no claims about human cognitive capacities—we fully endorse the view that human language is required for the creation of numbers. Furthermore, we show exactly how this works, by providing the linguistic tools and operations which are necessary to talk about both simplex and complex numerals. While it might seem that the creation of complex cardinals requires arithmetical operations, we demonstrate that purely linguistic means suffice for encoding multiplication, addition, and subtraction.

1.4   Goals of This Monograph

In the present work, we take our original proposal for the structure of complex cardinals in Ionin and Matushansky 2006 and explore the implications of this proposal both for a greater number and variety of languages and for a greater variety of linguistic phenomena (including concord and agreement, the semantics of coordination, and modified numerals) than in that work. The basic tenets of our proposal are summarized in (1).

(1) a. Simplex cardinals are lexical rather than functional, and they do not form a special linguistic class; depending on the language, and the type of cardinal, they may be nouns or adjectives (or more rarely,

verbs). At the same time, they do not necessarily exhibit all properties of nouns or adjectives, respectively. Rather, cardinals form a cline, from the more adjectival lower cardinals to the more nominal higher cardinals. Nominal cardinals are often defective nouns, which lack some or all of the features borne by "regular" nouns in a given language. Similarly, adjectival cardinals do not behave in every respect like "regular" adjectives.

b. Simplex cardinals have the semantics of modifiers (type $\langle\langle e,t\rangle,\langle e,t\rangle\rangle$), which allows them to combine with either a lexical NP (as in *three books*) or an NP that contains a multiplicand (as in *three hundred books*). The result is full recursivity for complex cardinals that involve multiplication. Complex cardinals do not form a unit to the exclusion of the lexical NP.

The semantics that we propose for simplex cardinals are in principle compatible with different syntactic structures. For complex cardinals involving multiplication (as in *two hundred books*), the simplest mapping between semantics and syntax is the cascading structure we proposed in Ionin and Matushansky 2006, given in (2). We show that this structure is indeed attested in many languages, as evidenced by the case assignment and agreement patterns within, and by, cardinal-containing NPs.

(2)

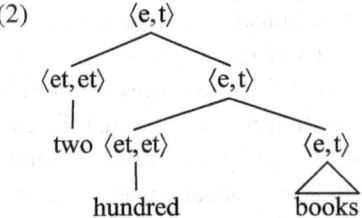

While this work builds on the basic proposal of Ionin and Matushansky 2006, it also extends and modifies it in several ways. First, in this book we provide an overview of both NP-internal and NP-external agreement with cardinals, something only briefly mentioned in our 2006 paper. Second, we provide further evidence that complex cardinals are built using morphosyntactic means that are attested elsewhere in the language. While Ionin and Matushansky 2006 makes this argument primarily on the basis of case assignment crosslinguistically, in this work we take the argument further, considering complex cardinals that are built using prepositions (e.g., in Biblical Welsh), as well as complex cardinals in the construct state in Hebrew. Third, relative to our 2006 paper, we largely revise the analysis of both the syntax and the semantics of complex cardinals involving addition (such as *twenty-four*).

Fourth, we consider the implications of our proposal for phenomena not addressed in the 2006 paper, including the composition of cardinals with relational nouns (e.g., *three sisters*), the internal composition of modified numerals (e.g., *more than three books*), and the use of cardinals inside partitives (e.g., *one of my friends*).

Fifth, and perhaps most importantly, the current work considers a much wider range of languages than we were able to include in our 2006 paper. While this work is not intended to give comprehensive crosslinguistic coverage of numeral systems (for this purpose, see Hurford 1975 and Corbett 1978), we do strive to examine numeral systems from languages that are typologically and structurally distinct. Our choice of languages is further determined by the linguistic properties of numerals in a given language, by the availability of relevant linguistic literature, and by our access to native-speaker informants. Specifically, we choose languages in which cardinals show different behaviors with regard to such properties as case assignment, agreement, and the morphosyntactic processes involved in complex cardinal formation, and languages in which cardinals belong to different lexical categories (such as noun vs. adjective). This choice of languages allows us to better illustrate the range of facts that need to be accounted for by any theory of complex cardinals, as well as to demonstrate that crosslinguistically complex cardinals use in their composition morphosyntactic processes attested elsewhere in human language. Keeping these considerations in mind, we provide in-depth coverage of numeral systems in several languages, including Russian, Polish, Hebrew, Standard Arabic, Biblical Welsh, and Dagaare, and we furthermore consider some aspects of numeral systems in a number of other languages, including Latin, Bulgarian, Western Armenian, Lebanese Arabic, Hungarian, Finnish, Inari Sami, Turkish, Basque, Miya, and Seri. While we discuss several Slavic languages—Slavic languages having some of the most complicated systems of cardinal formation—we dedicate the most attention to Russian because as native speakers we are most aware of the properties of its system of complex-cardinal formation.

One class of languages that we leave aside completely is generalized-classifier languages, such as Chinese and Japanese, except for a brief discussion in chapter 3. Given that classifier languages typically lack case assignment with or by cardinals, as well as morphologically overt agreement with cardinal-containing NPs, there is not much that we can say about the composition of complex cardinals in these languages. Other issues that we leave aside are ordinals and other numeral derivatives, though we do discuss fractions in chapter 10. The reason that we include fractions but not ordinals in our discussion is that fractions show some of the properties of the external syntax of cardinals, which is not true for ordinals.

The overall goal of this monograph is to test our original proposal with regard to phenomena and languages not considered in our prior work and to strive for a wider coverage of the facts that any theory of complex cardinals must explain.

1.5 Organization of This Book

As stated above, the primary focus of this work is the syntax and semantics of cardinals and the formation of complex cardinals. Therefore, our choice of phenomena and languages to concentrate on is primarily determined by their relevance to our main concerns.

We start by providing an overview of the original proposal from Ionin and Matushansky 2006 in chapter 2, in order to motivate the second tenet in (1) and the cascading structure in (2). Chapter 2 also considers the implications of our proposal for the semantics of relational nouns and provides a novel analysis for NPs such as *two sisters*. In chapter 3, we take up the first tenet in (1) and provide evidence from Finnish, Latin, Russian, and Seri for the status of simplex cardinals as lexical rather than functional elements. Moving on to complex cardinals involving multiplication (as in *three hundred books*), chapter 3 considers the possible syntactic structures that can map onto the semantics in (2). We argue that while the semantics in (2) is universal, the syntax–semantics mappings differ by language as well as by cardinal: for instance, lower cardinals are more likely to be adjectival and occur in adjunct position, while higher cardinals are more likely to be nominal and function as syntactic heads.

As discussed in chapters 2 and 3, our semantic analysis requires the lexical NP that combines with a cardinal to be semantically singular (atomic). This proposal receives support from languages such as Finnish, Hungarian, and Turkish, in which the lexical NP that occurs with a cardinal is morphologically singular, but runs into apparent difficulty with languages such as English and Russian, where the lexical NP is plural (e.g., *books* in *three hundred books*). In chapter 3, we discuss some ways of implementing the atomicity requirement, while chapter 4 takes up the question of why the lexical NP combining with a cardinal is (in some languages) morphologically plural. Chapter 4 provides a novel analysis, on which the plural marking on the lexical NP that occurs with a cardinal is not about inherent plurality but rather is due to agreement. In order to establish this, we show that the plural marking is conditioned by the same factors as those influencing agreement elsewhere within the same language; these factors, discussed in chapter 4, include animacy in Miya, individuation in Dutch, and specificity in Western Armenian.

While the first four chapters examine simplex cardinals as well as complex cardinals involving multiplication, chapter 5 shifts the focus to complex cardinals involving addition (as in *twenty-four books*) or both addition and multiplication (as in *thirty-three thousand four hundred and ninety-seven books*). In this chapter, we modify the analysis proposed in Ionin and Matushansky 2006 and argue that complex cardinals involving addition are derived not only via coordination followed by NP deletion but also by means of prepositional structures. Once again, we provide crosslinguistic evidence in support of our proposal, from languages as diverse as Biblical Hebrew, Russian, and Inari Sami. We also take up the question (briefly touched upon in Ionin and Matushansky 2006) of the semantics of complex cardinals involving addition. We show that the classic Boolean analysis of conjunction cannot account for complex cardinals involving addition, which necessitate two more interpretations of 'and' (additive and disjunctive), each of which is independently motivated by conjunctions not involving cardinals. Additive complex cardinals such as *twenty-two books* require the additive interpretation of 'and', which can be attributed to the set-product operation proposed by Heycock and Zamparelli (2000, 2003, 2005) or derived from the Boolean 'and', as suggested by Champollion (2016). Conversely, conjunction under a cardinal, as in *five men and women*, requires the disjunctive interpretation of 'and', also attested in examples like *every boy and girl*.

Chapters 2 through 5 also consider extralinguistic conventions constraining the construction of complex cardinals, including ordering constraints and the constraints on which numerals can serve as bases. We show that such constraints find parallels in other domains, and that they must be built into any analysis of numerals. We argue (against Wiese 2003) that the existence of extralinguistic constraints does not speak in favor of treating cardinals as extralinguistic elements.

Chapters 6 and 7 move on to consider the numeral systems of selected languages in more detail. Chapter 6 examines Slavic languages, where the facts of declensional morphology, case assignment, and agreement both provide evidence for and challenge some aspects of our proposal. The bulk of chapter 6 is dedicated to Russian, but we also address in detail the facts of Polish, with brief mention of Bulgarian, Ukrainian, and Belarusian where relevant. Chapter 7 takes up several unrelated languages whose numeral systems have certain characteristics that make them particularly important test cases for our proposal. This chapter covers Semitic languages (Modern Hebrew, Standard Arabic, and Lebanese Arabic); Welsh, both Biblical and Modern; and the Gur language Dagaare.

Chapters 8, 9, and 10 consider linguistic constructions that contain cardinals; we focus on phenomena that are most closely related to our central focus, complex cardinals. In chapter 8, we discuss the English *modified-cardinal construction*, in which the cardinal is preceded by the indefinite article and an adjective (as in *an amazing sixteen performances*). Expanding on the analysis for this construction proposed in Ionin and Matushansky 2004, we discuss its peculiar properties in light of our analysis of cardinals. Chapter 9 moves on to a consideration of so-called modified numerals, including comparative numerals (*more than five books*), superlative numerals (*at least four books*), and prepositional numerals (*around ten books, between four and eight books*). Given the vast body of research on this topic, it is imperative for us to show that our assumptions about the internal structure and composition of cardinal-containing NPs are compatible with modification. Specifically, our semantic analysis predicts that the cardinal should combine with the lexical NP before combining with the comparative/superlative/prepositional elements. Drawing on crosslinguistic data from Modern Hebrew, Basque, Russian, and Dutch, and building on the proposal in Arregi 2010 (and arguing against Corver and Zwarts 2006), we show that there is indeed much independent evidence for this syntactic configuration. Finally, chapter 10 takes up the syntax and semantics of partitives (*two of the books*) and fractions (*one-third of the cake*) and shows them to be compatible with our proposal.

Chapter 11 concludes the monograph with open questions and suggestions for further research.

2   The Semantics of Cardinals

In Ionin and Matushansky 2006, we put forth a novel proposal on which complex cardinals are derived compositionally from simplex cardinals, which have the semantic type of *modifiers* ($\langle\langle e,t\rangle,\langle e,t\rangle\rangle$). We summarize this proposal and review the motivation for it in section 2.1. In section 2.2, we continue to follow Ionin and Matushansky 2006 in explaining the source of existential force of cardinal-containing NPs. Sections 2.3 and 2.4 take up two new issues, respectively: the semantics of cardinals inside measure phrases and the semantics of cardinals inside arithmetic expressions. Then, in section 2.5, we consider a recent alternative proposal of Rothstein (2013, 2016, 2017), who, like Ionin and Matushansky (2006), adopts a compositional approach to the composition of complex cardinals. We discuss the motivation for her approach, and we point out empirical problems for it, ultimately concluding that it has no advantage over our proposal. Finally, in section 2.6, we consider the consequences of our proposal for the combination of cardinals with relational nouns.

2.1   Composition of Complex Cardinals

We start with the assumption that the semantics of *three* is the same regardless of whether it occurs inside the NP *three birds* or the NP *three hundred birds*. Our hypothesis is that complex cardinals are derived from simplex cardinals: *three hundred* should be semantically related to *three* as well as to *hundred*. In this chapter, we consider only those complex cardinals that involve multiplication, such as *three hundred, eight hundred thousand, five hundred million*, and so on. We will examine the semantics of complex cardinals that involve addition, such as *forty-two* or *two hundred and three*, in chapter 5.

2.1.1 Cardinals as Predicates or Determiners

The three main approaches to the semantics of cardinals in the literature have treated them as determiners (e.g., Montague 1974; Bennett 1974; Barwise and Cooper 1981; Scha 1981; Van der Does 1992, 1993), as predicates (Partee 1986), or as modifiers (Link 1987; Verkuyl 1993; Carpenter 1998; Landman 2003, among others). However, as shown below, none of these approaches make any distinction between simplex and complex cardinals, or explain what relationship (if any) exists between *three* and *three hundred*.

Consider first what happens if simplex cardinals are analyzed as predicates, type $\langle e,t \rangle$, with the lexical entry in (1a) for *three* and the lexical entry in (1c) for *hundred*. There is no problem in deriving the meaning of *three books*, as in (1b). It is seemingly possible to derive *three hundred books* via Predicate Modification (Heim and Kratzer 1998), as shown in (2). However, the resulting truth conditions would not be correct, as shown in (1d). The NP *three hundred books* would be self-contradictory, since nothing can simultaneously have the cardinality 100 (consist of a hundred atoms) and the cardinality 3 (consist of three atoms).

(1) a. $[\![\text{three}]\!] = \lambda x \in D_e \,.\, |x| = 3$
     b. $[\![\text{three books}]\!] = \lambda x \in D_e \,.\, [\![\text{books}]\!](x) \wedge |x| = 3$
     c. $[\![\text{hundred}]\!] = \lambda x \in D_e \,.\, |x| = 100$
     d. $[\![\text{three hundred books}]\!] = \lambda x \in D_e \,.\, [\![\text{books}]\!](x) \wedge |x| = 100 \wedge |x| = 3$

(2)

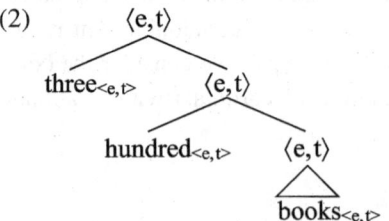

If simplex cardinals are instead determiners, type $\langle\langle e,t\rangle,\langle\langle e,t\rangle,t\rangle\rangle$, then, as shown by (3), semantic composition is impossible. If *hundred* is combined with *books* first, the resulting NP is a generalized quantifier (type $\langle\langle e,t\rangle,t\rangle$), which cannot then be combined with another cardinal of type $\langle\langle e,t\rangle,\langle\langle e,t\rangle,t\rangle\rangle$. Having *hundred* combine with *two* before it combines with a lexical NP would not solve the problem, since it would require semantic composition of two determiners; not only is there no semantic rule for combining two determiners, there is independent evidence that a sequence of two determiners is ungrammatical (cf. **no every cat, *the each child, *every that girl*, etc.).

(3)

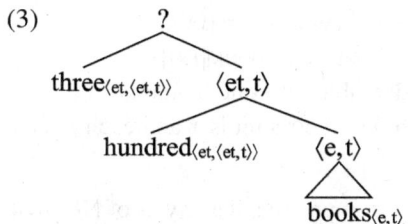

To sum up, standard approaches to cardinals cannot successfully derive the meaning of complex cardinals from that of simplex cardinals. Another approach to the semantics of cardinals that has been put forth in recent literature (Kennedy 2013, 2015 and Rothstein 2013, 2016, 2017, among others) treats cardinals as degrees; we will address this proposal in detail in section 2.5. First, however, we will present our own analysis of cardinals as semantic modifiers.

2.1.2 Proposal: Cardinals Are Modifiers

In light of the inability of the standard approaches to derive the meaning of complex cardinals, in Ionin and Matushansky 2006 we propose that simplex cardinals such as *three* or *hundred* have the semantic type of *modifiers* ($\langle\langle e,t\rangle,\langle e,t\rangle\rangle$), with the sample lexical entry in (4), where S is a partition Π of an entity x if it is a cover of x and its cells do not overlap (cf. Higginbotham 1981, 110; Gillon 1984; Verkuyl and Van der Does 1991; Schwarzschild 1994), as in (5)–(6).[1]

(4) $[\![three]\!] = \lambda P \in D_{\langle e,t\rangle} . \lambda x \in D_e . \exists S \in D_{\langle e,t\rangle} [\Pi(S)(x) \land |S| = 3$
$\land \forall s \in S \, P(s)]$

(5) *Partition*

    $\Pi(S)(x)$ is true iff

        S is a *cover* of x, and

        $\forall z,y \in S \, [z = y \lor \neg\exists a \, [a \leq_i z \land a \leq_i y]]$.

        (Forbidding that cells of the partition overlap ensures that no element is counted twice.)

(6) A set of individuals C is a *cover of a plural individual X* iff

    X is the sum of all members of C: $\sqcup C = X$.

The notion of a partition in (5) (see Higginbotham 1981, 110; Verkuyl and Van der Does 1991) corresponds to a cover of a plural individual X into the corresponding number (in this case, three) of possibly plural individuals such that they do not overlap (do not share any parts).[2] This means that an NP such as *hundred books* has the extension in (7a), stated informally in (7b).[3]

(7) a. ⟦hundred books⟧ = $\lambda x \in D_e$. $\exists S$ [$\Pi(S)(x) \wedge |S| = 100$
$\wedge \forall s \in S$ ⟦book⟧(s)]

   b. $\lambda x \in D_e$. x is a plural individual divisible into one hundred
   nonoverlapping individuals p_i such that their sum is x and each p_i is a
   book

Ionin and Matushansky (2006) provide a compositional analysis of NPs that
contain complex cardinals involving multiplication, such as *three hundred
books*. As illustrated in (8), we propose a fully recursive structure, in which
the lexical NP is the sister of the lowest cardinal, and the complex cardinal
three hundred does not form a semantic unit (in later chapters, we will provide
evidence that it also does not form a syntactic unit).[4]

(8)

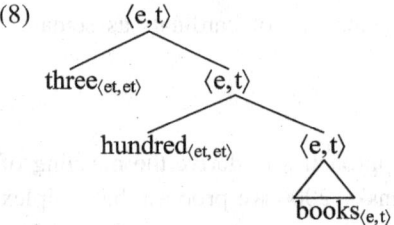

As shown by (7a), *hundred books* has the semantic type ⟨e,t⟩, which means
that it can be composed with a cardinal such as *three*, which has the denotation
in (4) and the type ⟨⟨e,t⟩,⟨e,t⟩⟩. In this way, we have achieved full compositionality, with the resulting denotation for *three hundred books* given in (9a)
and stated informally in (9b).

(9) a. ⟦three hundred books⟧
= $\lambda x \in D_e$. $\exists S$ [$\Pi(S)(x) \wedge |S| = 3$
$\wedge \forall s \in S \, \exists S'$ [$\Pi(S')(s) \wedge |S'| = 100$
$\wedge \forall s' \in S'$ ⟦book⟧(s')]]

   b. $\lambda x \in D_e$. x is a plural individual divisible into three nonoverlapping
   individuals p_i such that their sum is x and each p_i is divisible into one
   hundred nonoverlapping individuals p_k such that their sum is p_i and
   each p_k is a book

We note that on our proposal, the lexical NP that combines with a cardinal
is semantically singular: in (4), a member of the set denoted by *three books*
is a plural individual consisting of three *atomic* books. Atomicity is required:
without it, a member of the set denoted by *three books* could be, for example,
an individual divisible into three sets of books each of which consists of two
books—that is, six books total. The implementation of this atomicity requirement is left until chapter 3.

The atomicity requirement appears puzzling in light of the morphology on the lexical NP: if *three books* really means *three book*, then where is the plural *-s* coming from? We will lay this issue aside for now and come back to it in chapter 4, where we will (a) show that in many languages (e.g., Finnish, Hungarian, and Turkish), the lexical NP that combines with a cardinal is indeed morphologically singular; and (b) propose that in languages such as English, where the lexical NP occurring with a cardinal is morphologically plural, the plural marking should be analyzed as an instance of agreement.

2.1.3 Overgeneration, and the Role of Extralinguistic Constraints

An apparent problem for our proposal is that it overgenerates. Since on our proposal all simplex cardinals have the semantic type $\langle\langle e,t\rangle,\langle e,t\rangle\rangle$, nothing in the semantics prevents combinations such as *three-twenty* (meaning 'sixty') or *five-seven* (meaning 'thirty-five'). However, in fact languages place constraints on which cardinals may serve as complements for other cardinals. We rely on Hurford 2003 for our data and generalizations.

Each language defines a set of higher cardinals that may be used as multiplicands. In general, only 'ten' and some of its powers can be the head of the complement of a cardinal (Hurford 2003). The class of multiplicands usually contains most powers of the base (which is usually 'ten', but 'twenty', 'fifteen', and 'five' are also attested), and a small set of others ('twelve', 'twenty', 'sixty' ...), which may be subject to further constraints. For example, 'twenty' serves as a multiplicand productively in Celtic languages (see (10a)), Mixtec, and Yoruba, and sporadically in Danish and French (Hurford 1975, 2003).

(10) a. de-ugain   Biblical Welsh
         two-twenty
         'forty'
         (Hurford 1975, 149)
      b. kolme-kymmenta   Finnish
         three.NOM-ten.PART
         'thirty'

The constraint on what can serve as a multiplicand yields a number of less general constraints on possible complements of a cardinal. For example, a cardinal may not take a nonbase such as *six* in English as its complement (*two-six* is ungrammatical and cannot be used to mean 'twelve'). Note that the reverse is not true: not all bases may serve as complements of cardinals. For instance, while *ten* is a mathematical base, *three-ten* is not a possible complex cardinal of English, but is perfectly fine in Finnish, as seen in (10b).

We believe that the constraint on which cardinals may serve as multiplicands in a language is entirely extralinguistic, and the whole issue is not specific to our approach but arises for any theory addressing the existence and productivity of complex cardinals. For the sake of simplicity we assume the existence of a diacritic feature on some cardinals (multiplicands) that the multiplier selects for.[5] This can also explain why a cardinal cannot take a coordinated cardinal as its complement: for example, *two thousand and hundred books* cannot be used to mean 'two times one thousand one hundred (i.e., 2,200) books', with the bracketing in (11) (see chapter 5 for the analysis of NP deletion in (11)). The complex cardinal in (11) is impossible for the same reason that *two-thirty books* is impossible: *thousand and hundred* is not a base any more than *thirty* is.

(11) *[two [[thousand ~~books~~] and [hundred books]]]

The constraint on which cardinals can be multiplicands is where our arguments against the extralinguistic nature of cardinals do not apply: whereas syntactic phenomena inside complex cardinals (such as case assignment and number marking) show that they are combined in syntax, the distinction between multiplicands and nonmultiplicands does not to the best of our knowledge systematically correlate with any syntactic or morphological property relevant to the computational system.[6]

We will come back to the issue of overgeneration in section 2.5, when we discuss the proposal of Rothstein (2013, 2016, 2017), which, in an attempt to prevent overgeneration, assigns different semantic types to lexical powers and other cardinals. We will show that this proposal runs into a number of problems, and we will argue against a binary distinction between lexical powers and other cardinals.

2.2   Existential and Predicative Interpretations of Cardinal-Containing Indefinites

As discussed above, on our analysis, NPs containing cardinals have the type of predicates (type $\langle e,t \rangle$), as shown in (9). Such NPs should be able to combine with determiners and quantifiers, as indeed they do, for example in (12). The ability of cardinal-containing NPs to combine with determiners provides further evidence that cardinals are not themselves determiners. On our analysis, the cases in (12) are straightforwardly derived through the composition of a determiner (type $\langle\langle e,t \rangle,\langle\langle e,t \rangle,t \rangle\rangle$) with an NP such as *three books* (predicate type $\langle e,t \rangle$), resulting in a generalized quantifier (type $\langle\langle e,t \rangle,t \rangle$). (Alternatively, the definite article may map the predicate to a type e R-expression.)

(12) the three books/those six children/every five miles/all eight cats

We next consider what happens when there is no determiner in front of a cardinal-containing NP. In a sentence like (13a), the NP *three students* is associated with existential force, but there is no overt element that could be judged responsible for this, unlike in (13b).

(13) a. Three students came.
     b. A student came.

In the rest of this section, we discuss how existential readings of cardinal-containing indefinites can be derived, and we show that our analysis is compatible with existing mechanisms for deriving existential readings.

2.2.1 Cardinal-Containing Indefinites as Existential Quantifiers

As discussed in Ionin and Matushansky 2006, any standard theory of indefinites can account for the behavior of indefinites containing cardinals. One traditional view is that predicate NPs become generalized quantifiers (type $\langle\langle e,t\rangle,t\rangle$) as a result of a type-shifting operation (see Partee 1986 and Landman 2003). As a generalized quantifier, the cardinal-containing NP can then scope freely, resulting in the availability of both wide-scope readings, as in (14a), and narrow-scope readings, as in (14b).

(14) a. I am looking for *two secretaries*—they left the office for coffee break this morning and haven't been seen since.
     b. I am looking for *two secretaries*—one to do the typing and one to do the filing.

Alternatively, we can adopt the analysis of Krifka (1999). Krifka's semantic analysis of cardinals, based on that of Link (1987), is different from ours, but like us, he treats cardinals as being inside the NP, and he proposes that the empty D position is interpreted as an existential quantifier or that, alternatively, the existential quantifier is introduced via global existential closure, per Heim 1982. (Yet another proposal, Krifka 1990b, addresses the fact that cardinal-containing NPs can also be interpreted as degrees, further supporting the hypothesis that in the absence of a D layer a cardinal-containing NP is semantically a predicate.)

While the above analysis successfully accounts for the existential force of cardinal-headed indefinites, it faces the same problem as the generalized-quantifier analysis of *a* indefinites and *some* indefinites—namely, the availability of long-distance-scope readings (cf. Fodor and Sag 1982 and much subsequent literature). We turn to this next.

2.2.2 Cardinal-Containing Indefinites as Choice Functions

There is much literature on exceptional scope readings of indefinites, in which indefinites containing unmodified cardinals (*three students*, *five books*, etc.) have been shown to behave much like indefinites headed by *a* or *some* (see, among many others, Farkas 1981; Ludlow and Neale 1991; Ruys 1992; Winter 1997, 2001a, 2005; Kratzer 1998b).

We have already seen that, like other indefinites, indefinites containing cardinals can take either wide scope (14a) or narrow scope (14b). Furthermore, indefinites containing cardinals show the exceptional scope-taking properties of *a* and *some* indefinites, being able to scope out of islands: see (15a), where *two people I know* can scope above the *if* clause. The possibility of intermediate-scope readings, well established for *a* and *some* indefinites (see Farkas 1981; Ludlow and Neale 1991; Ruys 1992; Reinhart 1997; Winter 1997, among many others), is present for indefinites containing cardinals as well. Thus (15b) has a reading on which the doctors vary with the patients who died: the cardinal-containing NP scopes below the universal quantifier but above the *because* clause.

(15) a. If *two* people I know are John's parents then he is lucky.
         (Winter 1997)
      b. Every one of them died because *one* doctor from the hospital was
         on vacation (at the time).
         (Kratzer 1998b, 183, (40))

On the standard quantificational analysis of indefinites, wide-scope and intermediate-scope readings out of islands are not expected, since other quantifiers (such as *every*) cannot scope out of islands. The availability of long-distance readings of indefinites has thus led to proposals which treat all but bare-plural indefinites as choice-function indefinites rather than generalized quantifiers (or at least posit an ambiguity between choice-function and generalized-quantifier readings of indefinites): see Reinhart 1997; Winter 1997, 2001a; and Kratzer 1998b, among many others. A choice function is of type $\langle\langle e,t\rangle,e\rangle$: it applies to a set and returns an element of this set. On Reinhart's and Winter's analyses, a choice function must be existentially closed: existential closure at the topmost level results in the widest-scope reading of indefinites (15a), while closure at an intermediate level results in the intermediate-scope reading (15b). On Kratzer's analysis, the choice-function variable is free, contextually determined by the speaker's state of mind, and intermediate-scope readings arise from (hidden or overt) bound pronouns (see also Schwarzschild 2002b for a similar analysis that relies on implicit domain restriction rather than choice functions).

In the case of *a* indefinites and *some* indefinites, it has often been argued that *a* or *some* is in fact an expression of the choice function. However, an overt determiner is not necessary for a choice function to be present: on the analysis of Winter (2001a, 2005), cardinals, as well as *a*, are predicate modifiers located in Spec,NP, while the D position is occupied by a phonologically null choice-function operator. (In contrast, *some* is itself a choice-function operator in D.) While we disagree with Winter's syntactic analysis of cardinals (since we treat them as heads, not specifiers), we can adopt Winter's proposal of a phonologically null choice-function operator located in D as a possible implementation of the existential readings of cardinals. The view that lexical items which are not themselves in D can be associated with choice-function readings is supported by the existence of such adjectives as *certain, specific,* and *particular* (for analyses of *certain* in terms of choice functions, see Schwarz 2001 and Winter 2001a, among others).

Winter's proposal, coupled with our structure for cardinals, results in the DP structure in (16) for indefinites containing cardinals.[7]

(16)

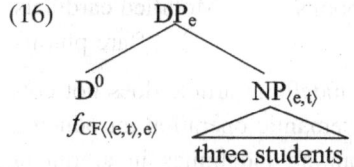

On this proposal, a choice function f applies to the set of all plural individuals x, such that each x is divisible into three nonoverlapping individuals, each of which is a student, and returns a single such x. A DP such as *three students* thus has type e: it is a plural individual (consisting of three nonoverlapping individuals, each of which is a student), which is picked out by the choice function f from the set of such plural individuals.

2.2.3 Indefinite-Article Insertion

As discussed above, any standard theory of indefinites combined with the semantics in (4) yields the existential force of indefinite cardinal-containing NPs. However, the question arises why cardinal-containing indefinite NPs behave like *a/some* indefinites rather than bare plurals: why is existential closure, or a null D^0, available to *two students* but not to *students*?

It is well-established that bare plurals such as *students* generally have only narrow-scope readings relative to a higher operator (Carlson 1977a,b; Chierchia 1998; etc.), which means that they cannot combine with either choice functions or existential quantification. The puzzle for us is why bare plurals, unlike cardinal-containing NPs, are not able to combine with a null determiner.

We do not have a straightforward answer to this question, but we note that the status of the D^0 position is not the same in bare NPs as in cardinal-containing indefinite NPs: bare plurals never combine with the indefinite article in English, whereas cardinal-containing NPs can (indeed, must) combine with it under certain circumstances. First, the indefinite article appears obligatorily when the leftmost cardinal is one of the so-called semilexical cardinals (*hundred*, *dozen*, etc.), as shown in (17).

(17) a. *(a) hundred/thousand/million/dozen books   Semilexical cardinals
       b. (*a) twenty/thirty/five/twelve/one thousand books

Second, indefinite-article insertion is obligatory in the *modified-cardinal construction*, as in (18a) (for more discussion of this construction, see Jackendoff 1977; Babby 1985; Gawron 2002; Ionin and Matushansky 2004; and chapter 8). Nothing comparable happens with English bare plurals (18b) (but see Bennis, Corver, and Den Dikken 1998 for indefinite-article insertion in Dutch plurals).

(18) a. a stunning one thousand/twenty-five books   Modified cardinals
       b. (*a) (stunning) books   Bare plurals

We hypothesize that the appearance of the indefinite article does not correspond to any semantic operation. Rather, a semantic operation introducing existential force applies to all cardinal-containing indefinites in argument position (in contrast to bare plurals), but is marked by *a* only under certain conditions: that is, the indefinite article marks rather than introduces existential force. In chapter 8, we will come back to the question of why the indefinite article is inserted in (17a) and (18a).

2.2.4   Exactly or at Least?

So far in this chapter, we have been treating simplex cardinals such as *two/ three* as having the meaning 'exactly two'/'exactly three' (see the lexical entry in (4)). However, the traditional view of cardinals (e.g., Barwise and Cooper 1981) treats *three* as having the meaning 'at least three', because a sentence like (19) is also true in a situation where more than three birds sang.

(19) Three birds sang.

The fact that (19) is usually interpreted as saying that *only* three birds sang is explained as a scalar implicature generated under the Maxim of Quantity proposed by Grice (1975; cf. Horn 1972; Levinson 1983; and Krifka 1999, among others). As discussed in chapter 1, other analyses have treated cardinals as underspecified for 'exactly' versus 'at least' readings (e.g., Carston 1998), or have treated the 'exactly' reading as primary and the 'at least' reading as derived (e.g., Geurts 2006; Breheny 2008).

Given that it clearly does exist, we need to be able to capture the 'at least three' reading of *three* under our analysis. One possibility is to put 'at least' into our lexical entry for simplex cardinals, as shown in (20).

(20) $[\![three]\!] = \lambda P \in D_{\langle e, t \rangle} . \lambda x \in D_e . \exists S \; \Pi(S)(x) \; [|S| \geq 3 \wedge \forall s \in S \; P(s)]$

This change to the semantics of simplex cardinals has rather interesting consequences for the semantics of complex cardinals, as shown in (21).

(21) $[\![three \; hundred \; books]\!] \approx \lambda x \in D_e . x$ is a plural individual divisible into *at least three* nonintersecting nonempty subindividuals p_i such that their union is x and each p_i is divisible into *at least one hundred* nonintersecting nonempty subindividuals p_k such that their union is p_i and each p_k is a book

Intuitively, *three hundred books* does not mean, as stated in (21), 'three (or more) of hundred (or more) of books'. However, while (21) appears to be rather odd, it is truth-conditionally undistinguishable from the traditional 'at least' meaning of *three hundred books*, so there seems to be no way to rule it out.

The alternative, which we adopt here, is to keep the semantics in (4) (i.e., *three* means 'exactly three') and to treat the 'at least' reading as arising from the logic of existential quantifiers. If we adopt the structure in (8), (19) has the (informally stated) meaning in (22) once the cardinal-containing NP is existentially quantified over.[8]

(22) $[\![three \; birds \; sang]\!] = \exists x \; [[x$ is a plural individual divisible into three nonintersecting nonempty subindividuals p_i such that their union is x and each p_i is a bird] and [x sang]]

Crucially, thanks to the pragmatics of existential quantification, (22) is compatible with the existence of a plural individual divisible into four or more birds who sang (or for that matter, three birds plus a diva, all of whom sang): the existence of a plural individual divisible into four birds necessarily entails the existence of a plural individual divisible into three birds. Thus, if we treat *three birds* in (19) as an existential, we can derive the 'at least' reading without positing the intuitively odd semantics in (21) (see Krifka 1999 for an approach deriving the 'at least' reading of cardinal-containing NPs from the pragmatics of alternative sets).

The same logic applies to the choice-function analysis of indefinites. Adopting the syntax in (16), and assuming existential closure over choice functions (per Winter/Reinhart), we end up with the (informally stated) semantics in (23).

(23) [[three birds sang]] = [there exists a choice function f that (a) applies to the set of plural individuals such that every x in this set is a plural individual divisible into three nonintersecting nonempty subindividuals p_i such that their union is x and each p_i is a bird and (b) returns a plural individual y] and [y sang]

Suppose there are more than three birds singing (e.g., the four birds a,b,c,d), which provides us with four plural individuals corresponding to the description *three birds*. These four plural individuals form the set denoted by *three birds* in our analysis:

(24) [[three birds]] = {a $\oplus$ b $\oplus$ c, a $\oplus$ c $\oplus$ d, a $\oplus$ b $\oplus$ d, b $\oplus$ c $\oplus$ d}

This means that there exists a choice function f that can apply to this set and return one of these individuals (e.g., a $\oplus$ b $\oplus$ c), and therefore *Three birds sang* is true even if more than three birds did. To sum up, in order to derive the 'at least' readings of cardinals, it is not necessary to write the 'at least' meaning into the semantics of cardinals.

In this section, we have shown that, assuming the mechanism of either existential closure or choice functions, the 'at least' reading is the basic reading of cardinal-containing NPs. The 'exactly' reading then arises through scalar implicature (cf. Horn 1972; Levinson 1983; and Krifka 1999, among others). However, as discussed in chapter 1, some researchers argue that the 'exactly' reading is in fact basic: for example, for Geurts (2006), the operation that applies to a cardinal-containing NP is not existential closure (which derives the 'at least' reading) but instead a type-shifting operation that derives the 'exactly' reading, which is primary. The 'at least' reading is secondary, requiring another type-shifting operation. Our analysis of cardinal semantics is just as compatible with this view as it is with the classic view, on which the 'at least' reading is primary. As our main concern here is with the derivation of complex cardinals, not with 'at least'/'exactly' readings, we have no stake in this matter.

2.2.5 Cardinal-Containing NPs as Predicates

The hypothesis that cardinals are modifiers (semantic type $\langle\langle e,t\rangle,\langle e,t\rangle\rangle$) necessarily entails, as discussed above, that the combination of a cardinal and a lexical NP is interpreted as a predicate (semantic type $\langle e,t\rangle$). This means that we expect to find cardinal-containing NPs in predicate positions (25) and do not expect cardinals themselves to function as predicates. (Examples (25a–b) are attested examples from a Google search.) The second prediction would seem to be contrary to fact, as the grammaticality of (26a–b) shows.

(25) a. how/why Pakistan and India became *two different countries*
        (Found via Google)
     b. Signs and normalcy seem *two different and contradictory phenomena.*
        (Found via Google)
     c. The animals in the shipment were *fifty ferocious lions.*
        (Landman 2003)

(26) a.   The apostles *are twelve.*
          (Higginbotham 1987)
     b.   They *are three.*
          (Winter 1997, fn. 7)
     c.   ??The chairs in this room are twelve.
     d.   ??These cats are four.

We note that both examples in (26a–b) involve copular clauses with human subjects, and that using nonhuman subjects with a bare cardinal predicate is rather infelicitous (26c–d). It is possible that the putative bare cardinals in the predicate position actually represent cases of NP ellipsis, which cardinals license without a linguistic antecedent (27). If this hypothesis is on the right track, cases like (26a–b) do not contradict our hypothesis that cardinals are modifiers rather than predicates.

(27) a. There were ten in the bed and the little one said …
     b. We had a suite since there were four in our party.

We return to this issue in section 2.5.3 in greater detail.

2.2.6 Summary

We have seen that the existential force of NPs containing cardinals can be accounted for under any standard theory of indefinites. In argument positions, indefinites containing cardinals behave like other indefinites: they are either generalized quantifiers or choice functions, depending on one's theory. Nothing in our analysis of cardinals hinges on a particular view of indefinites. As neither choice functions nor existential quantification are compatible with bare plurals, which are known to generally have narrow-scope readings only (Carlson 1977a,b; Chierchia 1998; etc.), indefinites containing cardinals behave like *a/some* indefinites rather than bare plurals.

Similarly, there are multiple ways of accounting for the 'at least' readings of indefinites containing cardinals—either by incorporating 'at least' into the lexical entry for simplex cardinals or by deriving the 'at least' readings from the logic of existential quantifiers. We choose the latter alternative, as it avoids the intuitively odd semantics in (21).

Thus, nothing special needs to be said about the existential force of cardinal-containing indefinites. For the remainder of this work, we will concentrate on predicate-denoting cardinal-containing NPs.

2.3   Cardinals inside Measure Phrases

One of the arguments in favor of treating cardinals as number-denoting comes from measure phrases, as many people have suggested that the cardinal is an argument of the measure noun (see Krifka 1990b; Schwarzschild 2005; Rothstein 2011a,b; and Scontras 2014, among others; for some details of these proposals, see section 2.5). In this section, we will present the proposal of Matushansky and Zwarts (2017), on which cardinals in measure phrases are treated exactly as cardinals in regular NPs.

The starting point for Matushansky and Zwarts (2017) is that measure nouns should be treated as having the usual nominal predicative denotation (type $\langle e,t \rangle$). However, measure nouns differ from other count nouns in that they are not individuated in the same way. To capture this fact, Matushansky and Zwarts propose that measure nouns do not denote in the same spatial domain as regular nouns: while the denotata of regular nouns are located in the regular three-dimensional space, the denotata of measure nouns are located in various one-dimensional spaces. Evidence for this view comes from prepositional measures, as in (28).

(28) a. Don't touch the steering wheel if you have drunk *over five glasses of wine*.
   b. I ate *around a pound of jam*.
   c. The mass of the meteorite was estimated at *under sixty-six tons*.
   d. I was swimming *between a kilometer and a mile* four days a week.

In order to explain why locative prepositions can combine with measure phrases, and how this combination is interpreted, Matushansky and Zwarts propose that measure nouns denote containers in one-dimensional space and the prepositions retain their regular locative interpretation, as illustrated in (29) for the transitive use of measure nouns.

(29) a. The picture is *over* the mantel.
      Over expresses a vertical relation between two material objects in three-dimensional space.
   b. I ate *over* a pound of jam.
      Over expresses a vertical relation between two abstract containers in one-dimensional space.

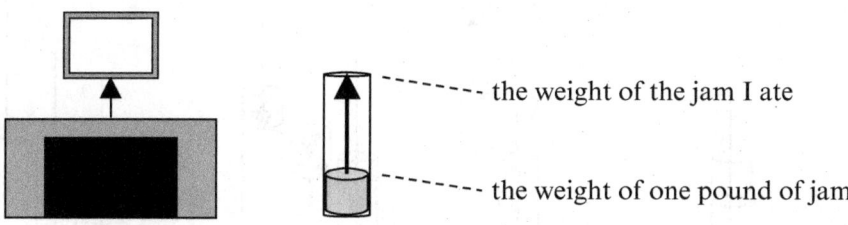

the weight of the jam I ate

the weight of one pound of jam

As a result, the interpretation of prepositional measures is derived compositionally using standard machinery, for example, vector-space semantics (Zwarts and Winter 2000). Likewise, cardinals in measure phrases can have the semantics that we propose. Assuming the lexical entry for *liter* in (30a) and our nonintersective semantics for cardinals, repeated in (30b), *three liters* in (31) can be treated exactly as *three books* except in one-dimensional space. (For transitive uses of measure nouns, as in *three liters of vodka*, see Matushansky and Zwarts 2017.)

(30) a. $[\![\text{liter}]\!] = \lambda x \in D_e^{1D}$. x is a liter entity in the one-dimensional volume space

b. $[\![\text{three}]\!] = \lambda P \in D_{\langle e, t \rangle}$. $\lambda x \in D_e$. $\exists S \in D_{\langle e, t \rangle}$ $[\Pi(S)(x) \wedge |S| = 3$
$\wedge \; \forall s \in S \; P(s)]$

(31) a. $[\![\text{three liters}]\!] = \lambda x \in D_e$. $\exists S \; [\Pi(S)(x) \wedge |S| = 3$
$\wedge \; \forall s \in S \; [\![\text{liter}]\!](s)]$

b. $\lambda x \in D_e$. x is a plural individual divisible into three nonoverlapping individuals p_i such that their sum is x and each p_i is a liter

The next step for Matushansky and Zwarts (2017) is to derive prepositional measures like *under three liters*. Using the vector-space semantics from Zwarts and Winter 2000, Matushansky and Zwarts appeal to the function LOC, which gives for each entity its eigenspace (Wunderlich 1991), which they redefine as the boundary of the region that the entity occupies (rather than the region itself). In a one-dimensional space, this gives us the top region of the container. Then the preposition applies, mapping this region to the region of spatial points below it: $[\![\text{under}]\!](\text{LOC}([\![\text{three}]\!]([\![\text{liter}]\!])))$. We now need to return from regions to entities, and the function LOC⁻ applies, mapping this region to the set of containers that have their top in that region, that is, that are lower than three liters. This is illustrated in the picture below.

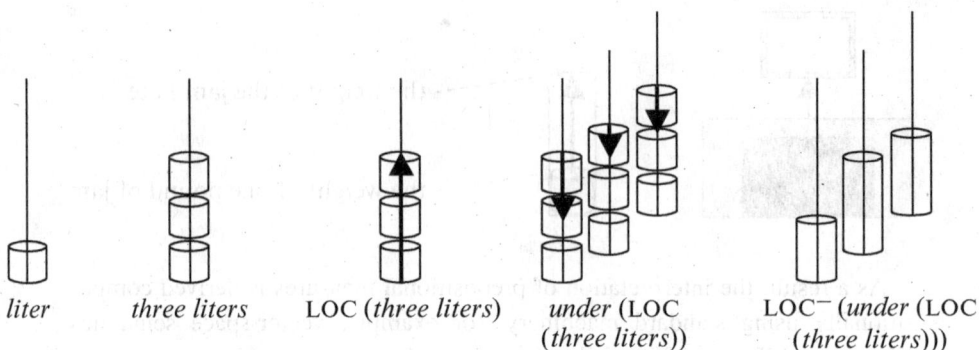

liter   *three liters*   LOC (*three liters*)   *under* (LOC   LOC⁻ (*under* (LOC
                                                     (*three liters*))   (*three liters*)))

We conclude that it is not only desirable but also possible to have the same semantics for cardinals in measure phrases as for cardinals elsewhere. Furthermore, this proposal makes it possible to derive the meaning of cardinal-containing NPs that are used as measure phrases, even when the lexical noun is not a measure noun, as in *twenty students fewer than last year*, without changing our central proposal concerning cardinal semantics.

2.4 The Relationship between NP-Internal and Arithmetical Uses of Cardinals

So far in this work we have focused on NP-internal uses of cardinals; the question naturally arises how they relate to arithmetical uses of cardinals, such as the ones in (32).[9]

(32) a. Two and two is/are four.
        (Rothstein 2013)
     b. Two is the only even prime number.

In our prior work, Ionin and Matushansky 2006, we claimed that cardinals inside arithmetical expressions are nominalizations, without specifying any further details. This seems similar to the view expressed by Hofweber (2005), who proposes that there is an elided lexical NP inside arithmetically used cardinals, and to that of Moltmann (2012a,b; but not Moltmann 2017), who suggests the presence of a covert sortal in arithmetically used cardinals. Rothstein (2013) argues against this proposal on the basis of examples like (33).

(33) a.   I counted (up) to thirteen (*things).
     b.   #Two things are even/a prime number.

Making our earlier proposal more specific, we propose instead that names of numbers can be derived from the corresponding cardinal predicate (type $\langle e,t \rangle$) or modifier (type $\langle \langle e,t \rangle, \langle e,t \rangle \rangle$)[10] by the nominalizing suffix NOMNUM, which returns for any cardinal the cardinality corresponding to it, as shown in (34).[11]

(34) a. NOMNUM($\text{card}_{\langle e,t \rangle}$) = ιx . $\forall y$ [$|y| = x \rightarrow \text{card}(y)$]   Predicate

b. NOMNUM($\text{card}_{\langle \langle e,t \rangle, \langle e,t \rangle \rangle}$) = ιx . $\forall g_{\langle e,t \rangle}$. $\forall y$ [$*g(y) \wedge |y| = x$ Modifier
$\rightarrow \text{card}(g)(y)$]

On the semantic side, it would seem to be counterintuitive to derive the arithmetical use from the NP-internal use, as the interpretation of cardinals appears to rely not only upon the notion of cardinality but also on a specific number for each cardinal. We should not, however, be misled by the notation. It is not at all clear that we have any direct intuitive access to the notion of cardinality, which means that the lexical meaning of *seven*, as an abstract number or as an NP-internal modifier, can be achieved by whichever circuitous way necessary (perhaps even making reference to the counting sequence), just as we need not be concerned by the exact interpretation of the adjective *blue*. Furthermore, there is independent evidence for the direction of derivation that we propose.

The first piece of evidence comes from languages where the arithmetical use is derived morphologically, such as Tamil (Caldwell 1913). For example, the Tamil word for the number *two*, used in arithmetical expressions, is *iraṇḍu*, yet the NP-internal word for *two* is *iru*, also used in derivation, as in *iruvar* 'two persons' or *irubadu* 'twenty'. The same is true for other lower cardinals up to *eight* (216–240). At the same time, to the best of our knowledge, no language morphologically derives the NP-internal use from the arithmetical use. Any analysis that derives the NP-internal use from the arithmetical use has to account for why the morphological derivation goes in the opposite direction in at least some languages.

Another piece of evidence for deriving the arithmetical use from the NP-internal use is the fact, mentioned in Ionin and Matushansky 2006, that in their arithmetical use cardinals show full syntactic uniformity: unique form for the adjectival lower cardinals (no variation in gender or number), lack of gender and number features capable of triggering agreement, and a distribution identical to that of DPs or proper names. As the translations in (35) indicate, no determiner is necessary to use cardinals arithmetically in argument positions.

(35) a. Dva/*dve pljus tri   budet/bylo   četyre.   Russian
         two.M/N/F plus three be.FUT.3SG/be.PAST.N.SG four
         'Two plus three will be/was four.'
      b. My pribavili tysjaču   k   trëm.
         we added   thousand.ACC towards three.DAT
         'We added one thousand to three.'

On our proposal, the NP-internal use is basic; that is, lexical entries for
cardinals can be specified as being more or less nominal or adjectival, and they
can bear specific (values of) φ-features: number, gender, and so on (see chapters 3 and 6 for more discussion). As a result, the different morphosyntactic
properties of different cardinals are not surprising. If, on the other hand, the
NP-internal use were derived, its morphosyntactic variability would be unexpected. We will come back to the arithmetical use of cardinals on other proposals in section 2.5.3.

2.5   Against Different Semantic Types for Different Cardinals

So far in this chapter, we have shown that the analysis of simplex cardinals as
semantic modifiers can account for the composition of complex cardinals more
successfully than more traditional analyses of cardinals as predicates or determiners. We now consider yet another approach to the semantics of simplex
cardinals, on which they have the semantic type of degrees (d) or numbers (n).
This view has been advanced in several recent papers (Scontras 2013, 2014;
Kennedy 2013, 2015; Rothstein 2013, 2016, 2017; Ouwayda 2014; cf. also
Scha 1981; Hackl 2000; and Nouwen 2010; see also Zabbal 2002, 2005).
While Kennedy (2013, 2015) and Scontras (2013, 2014) assume one basic
semantic type for cardinals, Rothstein (2013, 2016, 2017) allows them to have
the types $\langle e,t \rangle$ and $\langle n,\langle e,t \rangle \rangle$ as well. Rothstein's proposal is the only one that
deals with the composition of complex cardinals, and we therefore focus on
her work here.

For Rothstein, there are two sources for variation in the semantics of cardinals. One is the availability of the predicative use (type $\langle e,t \rangle$) in addition to
the arithmetical use and the NP-internal use. The other is the distinction
between lower and higher cardinals. Specifically, Rothstein distinguishes
lower numerals (up to 'ninety-nine'), which have the semantic type n, from
multiplicands (*lexical powers* in her terminology), which for her have the more
complex type $\langle n,\langle e,t \rangle \rangle$. One apparent advantage of this proposal is that it does
not overgenerate: since lexical powers and other cardinals are assigned differ-
ent semantic types, this proposal can readily explain the nonexistence of
complex cardinals such as *five-seven* (with the meaning of 'thirty-five'; see

section 2.1.3). However, as shown below, this apparent advantage is out-weighed by a number of both theoretical and empirical problems.

2.5.1 Theoretical Problems for Rothstein's Approach

In the following sections, we will point out a number of empirical problems with the divisions proposed by Rothstein (2013, 2016, 2017). Additionally, a major theoretical problem for Rothstein arises from the need (fueled and required by the assumption that cardinals denote numbers) to place arithmetical operations, such as addition, in the grammar. Two predictions are made as a result. On the one hand, in order to use complex cardinals, children must have mastered arithmetical addition (which developmentally does not happen until fairly late if at all—see, e.g., Gilmore, McCarthy, and Spelke 2007). On the other hand, addition being distinct from the semantic sum operation, it is not expected that the same morpheme ('and') should be used for both; hence, the crosslinguistic tendency to use conjunction markers or prepositions in complex cardinals involving addition (see chapter 5 for the details) is totally unexpected.

Another problem for Rothstein's proposal is the following. As first noted by Hurford (1975), the syntax of cardinals crosslinguistically forms a cline from adjectives to nouns: whereas lower cardinals are generally more adjectival, higher cardinals are more nominal (see chapter 3 for more discussion). Strangely enough, from the semantic standpoint, Rothstein's proposal looks like the exact opposite: while the lower cardinals denote a sort of individuals (type n), the higher cardinals have a complex semantic type making them very different from nouns (type $\langle e,t \rangle$ or perhaps k for kinds) and more similar to adjectives (type $\langle e,t \rangle$ or $\langle d,\langle e,t \rangle \rangle$). While we do not expect a tight correspondence between the semantic type and the lexical category, a total mismatch is unexpected.

2.5.2 Against the Special Status of Lexical Powers

One of the two basic hypotheses advanced by Rothstein (2013, 2016, 2017) is that cardinals that can serve as bases for multiplication (lexical powers) should be distinguished from those that cannot. As we will now show, this distinction does not appear to translate into a robust crosslinguistic distinction in syntactic behavior. While, on the one hand, the numeral cline may contain as many as seven potential distinctions (see chapter 6 on Russian), on the other hand, the specific diagnostics for the division advanced by Rothstein do not yield crosslinguistically uniform results. Indeed, the distinction between lower cardinals and lexical powers is supposed by Rothstein to manifest itself in three different ways.

First, while the lower cardinals disallow a multiplier (*two-twenty*), lexical powers require it (*one/a/*∅ hundred*). This is accounted for by the appeal to their semantic types. Lexical powers are given the semantic type $\langle n, \langle e, t \rangle \rangle$ and as a result, they require a multiplier for both NP-internal and predicative use.

Second, only lexical powers are asserted to permit the derivation of the approximative use (*hundreds of books*).[12] To account for this fact, Rothstein proposes a particular semantics for the operation APPROX used to derive approximative readings, which existentially quantifies over and pluralizes the n argument slot.

Third, lexical powers have exactly the same semantic type as measure nouns (*liter*, *meter*, etc.), predicting the similarities attested between the two (cf. Ionin and Matushansky 2006).

While we do not deny the fact that lexical powers and the lower cardinals frequently do not exhibit the same syntactic behavior, we disagree that the division is binary and as simple and straightforward as Rothstein claims. The three properties discussed above (the requirement to have a multiplier, the derivation of approximative use, and the similarity with measure nouns) need not go together crosslinguistically. This is clear even when we examine the more familiar languages.

For example, in French, just as in English, *cent* 'hundred' and *mille* 'thousand' are lexical powers and *dix* 'ten' is not (36a). However, *cent* and *mille* do not require a multiplier, and indeed cannot appear with *un* 'one/a' (36b); furthermore, the approximative use is possible for nonpowers with the proper suffix (37).[13]

(36) a. deux cents/mille/*dix   personnes   French
        two hundred.PL/thousand/ten person.PL
        'two hundred/thousand people'
     b. (*un) cent/dix   personnes
        one hundred/ten person.PL
        'one hundred/ten people'

(37) a. une centaine/dizaine   de personnes
        a   hundred.APPROX/ten.APPROX of people
        'about a hundred/ten people'
     b. des   centaines/dizaines   de personnes
        INDEF.PL hundred.APPROX.PL/ten.APPROX.PL of people
        'hundreds/tens of people'

It could be claimed that it is the overtness of the suffix that makes a difference with respect to the approximative use. However, the same pattern recurs in Dutch, where *honderd* 'hundred' and *duizend* 'thousand' appear bare in the singular; they differ in this respect from Dutch measure nouns, as shown in (38).

(38) a. honderd/duizend boeken   Dutch
          hundred/thousand books
          'a hundred/thousand books'
     b. een liter jenever
          a   liter Dutch.gin
          'a liter of Dutch gin'

To be interpreted approximatively, lexical powers are preceded by the indefinite article (39), as are other cardinals denoting more or less round numbers, including complex cardinals formed by addition (40). Given the absence of overt morphology deriving this approximative use, the difference between English and Dutch is unexpected on the view that lexical powers, and only lexical powers, require a multiplier.[14]

(39) a. een honderd/duizend boeken
          a   hundred/thousand books
          'about a hundred/thousand books', *'a hundred/thousand books'
     b. Hij heeft een tachtig boeken.
          he has   an   eighty books
          'He has about eighty books.'
     (Broekhuis and Den Dikken 2012, 874, 741)

(40) een honderd twintig personen
     an   hundred twenty people
     'some one hundred and twenty people'

A division of cardinals into multipliers and lexical powers would seem therefore to give rise to wrong predictions in languages where the ability of a cardinal to function as a multiplicand does not correlate with the need for a determiner or with the ability to form approximative numerals. Even for English, however, this division turns out to be problematic, as we will see when we confront addition (see section 2.5.5).

2.5.3 The Arithmetical Use of Cardinals on Rothstein's Proposal

To argue that lower cardinals (as opposed to lexical powers, discussed above) are systematically ambiguous between type n and the predicative type $\langle e,t \rangle$, Rothstein (2013, 2016, 2017) observes that all cardinals can appear in arithmetical statements (41) and (in some languages but not in others) as predicates (42).

(41) a. Two and two is/are four.

     b. Two is the only even prime number.

     (Rothstein 2013)

(42) a. My reasons are two.

     b. The children were two.

The arithmetical use of cardinals gives Rothstein an independent reason for postulating that cardinals, in addition to type $\langle e,t \rangle$ (motivated by their predicative use, more on this below), can also have a dedicated semantic type n. While for a number of authors, including Kennedy (2013, 2015) and Scontras (2013, 2014), n is a subtype of degrees, Rothstein treats it as a sort of individuals. Adopting the hypothesis (Chierchia 1985) that every property (type $\langle e,t \rangle$) has an entity correlate (type π) derived by the nominalization type shift ($\cap$), Rothstein proposes that while the predicate-type cardinals denote the property of having a given cardinality (43a), type n cardinals denote their entity correlates along the lines of (43b).

(43) a. $[\![\text{four}_{\langle e,t \rangle}]\!] = \lambda x \,.\, |\{y : y \sqsubseteq_{\text{ATOMIC}} x\}| = 4$

     b. $[\![\text{four}_n]\!] = {}^{\cap}\lambda x \,.\, |x| = 4$

Conversely, the $\cup$ operator applied to a number yields the set of plural individuals having that cardinality, as in (44).

(44) ${}^{\cup}[\![\text{four}_n]\!] \asymp {}^{\cup\cap}[\lambda x \,.\, |x| = 4] \asymp [\lambda x \,.\, |x| = 4] \asymp [\![\text{four}_{\langle e,t \rangle}]\!]$

The notation here gives rise to some confusion. Chierchia's nominalization operator should yield for the predicate $three_{\langle e,t \rangle}$ the property of having the cardinality 3, which does not seem like a good candidate for the interpretation of *three* in (45): it is not the property of being three in number that is an argument in arithmetical statements, it is the number 3 itself. In other words, the type shifts between n and $\langle e,t \rangle$ cannot be equated with Chierchia's $\cap$ and $\cup$, and are not, therefore, independently motivated.

(45) a. Just Molly and me, and baby makes three.

     b. Three is a prime number.

Furthermore, the same predicates that can be used with number-denoting subjects can also have regular DP subjects, as illustrated in (46), from Tennant 2015, a philosophy text. This necessarily entails that in their arithmetical use cardinals should have the type e (contra Kennedy 2013, 2015 and Scontras 2013, 2014), which means that n, if distinguished as a separate category at all, should be viewed as a sort of the type e, and we do not see so far any reason to postulate such a sort.

(46) a. Hegel believed that eight is greater than seven.
     b. Hegel believed that the number of planets is greater than seven.
     (Tennant 2015, 102)

Summarizing, while an entity denotation is necessary to account for the use of cardinals in arithmetical statements, it does not seem to be connected to their predicate denotation by the regular type-shifting rules linking predicates to their entity correlates (Chierchia 1985). The arithmetical use of cardinals, furthermore, does not offer any insight into their NP-internal use because in arithmetical use cardinals exhibit syntactic and morphological uniformity not found NP-internally. For our analysis of arithmetical uses of cardinals, please see section 2.4.

2.5.4 Cardinal Predicates and the Predicative Denotation of Cardinals

While we argue in Ionin and Matushansky 2006 that cardinals should be treated as modifiers (type $\langle\langle e,t\rangle,\langle e,t\rangle\rangle$), a much more frequent view, also shared by Rothstein (2013, 2016, 2017), is that they are predicates. While for some languages this is clearly the case (see chapter 3 for a discussion of Seri, where cardinals are verbs), such crosslinguistic semantic uniformity of cardinals is far from evident in the more familiar languages. As the examples in (47a–d) show, in Russian and in Dutch cardinals cannot function as predicates, suggesting a different denotation. This contrasts with English and French, where cardinals can function as predicates (47e–f).

(47) a. *Deti   byli dva/dvoe.   Russian
        children were two/two.COLL
     b. Detej   bylo dvoe/*dva.
        children.GEN   were two.COLL/two
        'There were two children.'
     c. Man en vrouw zijn één/*twee.   Dutch
        man and woman are one/two
        'Man and woman are one.'
        (Barbiers 2005)
     d. De kinderen zijn twee.
        the children are two
        'The children are two years old.'
        (Not 'The children are two in number.')
     e. The children were two.   English
     f. Les   saisons   sont quatre: le printemps, l'été,   French
        the.PL season.PL are four   the spring   the.summer
        l'automne et   l'hiver.
        the.fall   and the.winter
        'The seasons are four: spring, summer, fall, and winter.'

It would seem to be the case that crosslinguistically, the NP-internal use is always available alongside the arithmetical one, yet the predicative use is heavily restrained. In the absence of an independent explanation, Rothstein's theory incorrectly predicts that (47a–d) should be grammatical. We believe that it is far easier to appeal to the well-known peculiarities of the copula *be* to explain how the predicative use of cardinals in English is derived from their NP-internal $\langle\langle e,t\rangle,\langle e,t\rangle\rangle$ type than to assume that the predicative denotation is basic and then try to rule out the predicative use in Russian and in Dutch. We conclude therefore that the predicative use of cardinals, like the arithmetical use, should not be used as an argument for treating them as type $\langle e,t\rangle$ NP-internally.

2.5.5 Complex Cardinals Involving Addition: Against Different Semantic Types

While complex cardinals involving multiplication (with the exception of the multiples of ten, which are lexically constructed) are constructed in syntax for Rothstein (2013, 2016, 2017), she derives complex cardinals involving addition by two different routes. As complex cardinals below 'one hundred' exhibit certain individual quirks—for example, the reverse order and the obligatorily overt 'and' in Dutch, as in (48)—Rothstein assumes they are constructed in the lexicon (49a). Conversely, complex cardinals above 'one hundred' are constructed in the syntax (49b).

(48) twee-*(en)-twintig   Dutch
        two-and-twenty
        'twenty-two'

(49) a. nine-ty-nine   Lexical composition
        b. one thousand one hundred and two   Syntactic composition

Addition is, however, semantically identical in both modules, requiring that the two elements that it conjoins have the semantic type n, as in (50a).

(50) a. $[\![and_{\langle n,\langle n,n\rangle\rangle}]\!] = \lambda m\lambda n \, . \, n + m$
        b. $[\![two\ hundred\ and\ four]\!] = [\![and]\!]\,(\cap [\![two\ hundred_{\langle e,t\rangle}]\!])([\![four_n]\!])$
        c. $[\![two\ hundred]\!] = [\![hundred_{\langle n,\langle e,t\rangle\rangle}]\!]([\![two_n]\!]) = \lambda x \, . \, |x| = 200$

As shown in (50b), in order to compute the denotation of *two hundred and four* it is necessary to shift *two hundred* from type $\langle e,t\rangle$ (arising after Functional Application of *hundred*$_{\langle n,\langle e,t\rangle\rangle}$ to *two*$_n$, as in (50c)) to its entity correlate (type n). In order to then combine the resulting entity-denoting constituent with an NP, it is necessary to type shift it to $\langle e,t\rangle$, as in (51).

(51) $[\![two\ hundred\ and\ four\ books]\!] = {}^{\cup}[\![two\ hundred\ and\ four]\!]([\![books]\!])$

The repeated use of type shifting is not a problem here, but the division of labor between the two modules is. On the one hand, the supposedly purely syntactic addition above 'one hundred' can be demonstrated to also have lexical quirks (e.g., *one hundred *(and) two* vs. *one hundred (and) twenty*). Given that *one hundred* is constructed syntactically, nontrivial assumptions are required if such quirks are to be resolved only in the lexicon. On the other hand, exactly the same meaning (that of the arithmetical operation of addition) turns out to have three distinct lexicalizations: the phonological zero morpheme, a covert *and*, and a freestanding *and*. Although Rothstein notes that the last two also realize nominal coordination, she provides no explanation for this fact. Thirdly, Rothstein assumes that the nontrivial procedure of addition is part of the functional lexicon and leaves unexplained the systematic coincidence of the addition marker in cardinals with the coordination marker, the comitative preposition, or a locative preposition (see chapter 5). In other words, the appeal to a separate semantic type n is counterproductive when it comes to the need to account for addition: the same result can be achieved with a more conventional semantics for coordination (see chapter 5).

Summarizing, the particular mechanics (lexical vs. syntactic derivation, the use of type n) that Rothstein assumes in order to construct and interpret complex cardinals involving addition seem to be problematic, although not in a fatal way.

2.5.6   Numerals in Measure Phrases

The hypothesis that the lower cardinals denote their corresponding numbers (or sets) whereas lexical powers are functions from numbers to sets yields correct compositional semantics for complex cardinals involving multiplication, but does not appear to make correct predictions otherwise, as shown above. While the arithmetical use of cardinals does not show the same morphosyntactic properties as their NP-internal use, their ability to be used as predicates is restricted both to and within particular languages. Neither use, therefore, would seem to provide any insight into their NP-internal semantics.

It could be claimed, however, that a strong argument in favor of Rothstein's approach is that it permits a unified treatment of lexical powers and measure nouns, while accounting at the same time for the composition of measure phrases in pseudopartitives. However, in order to make use of this advantage, problematic assumptions about the syntax of pseudopartitives must be made.

Krifka (1990b), Rothstein (2013, 2016, 2017), and Scontras (2013, 2014) treat the cardinal in measure phrases as an argument of the measure noun.

Suggesting that the semantics of a measure noun involves a measuring function and a projection to the scale of natural numbers, they use the lexical entries in (52) to construct the interpretation of pseudopartitives. While Rothstein (52a) and Krifka (52b) presuppose the syntactic-selector structure in (53), for Scontras, as is evident from the order of arguments, the substance NP is the first argument of the measure noun (52c) and this is the view to which we also subscribe.

(52) a. $\llbracket kilo \rrbracket = \lambda n \lambda x$. measure$_{WEIGHT,KILO}(x) = n$
    (Rothstein 2016)

   b. $\llbracket liter \rrbracket = \lambda n \lambda P \lambda x$. $P(x) \wedge liter'(x) = n$
    (Krifka 1990b)

   c. $\llbracket kilo \rrbracket = \lambda k \lambda n \lambda x$. $^\cup k(x) \wedge \mu_{kg}(x) = n$
    (Scontras 2014)

(53)

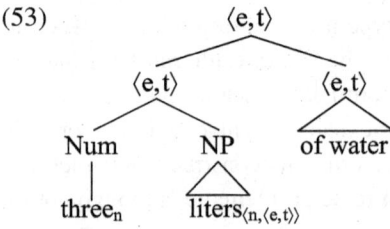

For Rothstein, therefore, lexical powers have the same type as measure nouns, which allows her to explain why the morphosyntax of measure nouns is often similar to the morphosyntax of lexical powers (Ionin and Matushansky 2006). However, the hypothesis that the measure noun and the cardinal form a constituent to the exclusion of the substance NP does not find clear support in crosslinguistic data.

The most straightforward evidence against the structure in (53) comes from Hebrew, where pseudopartitives, unlike true partitives, require construct-state morphology (Rothstein 2011a,b). Whereas the construct state in (54b) exhibits the hallmark of pseudopartitive constructions in being ambiguous between the container interpretation and the measure interpretation, the free-genitive construction in (54a) can only denote containers; only the construct state can have the free-portion reading (55).

(54) a. *True partitive*   Hebrew
      šloša bakbukim šel yayin
      three bottle.PL of wine
      'three bottles full of wine'

b. *Pseudopartitive*
šloša bakbukey   yayin
three bottle.PL_BOUND wine
'three bottles of wine'

(55) Yeš   od   šaloš ka'arot   marak / #šaloš ka'arot šel marak
there   more three bowl.PL_BOUND soup   three bowl.PL of soup
ba   sir.
in.DEF pot
'There are three more bowls of soup in the pot.'

As known since Ritter 1987, 1988 (see also the discussion of Hebrew in chapter 7), the construct-state morphology is a clear sign of the head–complement relation, which entails that Hebrew pseudopartitives, at least, do not have the structure in (53) and the structure in (56) should be assumed instead.[15] Further crosslinguistic evidence in favor of the structure in (56) comes from case assignment, agreement, prepositional measures, and word order; see Matushansky, Ruys, and Zwarts 2017 for details.

(56) *Cascade constituency for pseudopartitives*

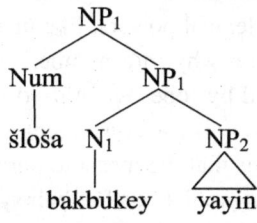

'three bottles of wine'

Conversely, Ruys (2017) demonstrates that the cardinal cannot be an argument of the measure noun. For one thing, it can be absent in the singular (*a liter of vodka*), and the Dutch examples in (57) show that the indefinite article cannot be taken as a phonologically weak substitute for the cardinal. Likewise, as (58a) shows, the Dutch equivalent of *a liter of vodka* cannot plausibly be analyzed via head movement of the cardinal to the indefinite-article position (which would predict that the full variant *één* should also be grammatical), and (58b) rules out the analysis of (57a) as modification of *liter* prior to its combination with the cardinal.

(57) a. een goeie (*één/*een) liter vodka   Dutch
a   good   one/a   liter vodka
'a good liter of vodka'

    b. een goeie tien liter vodka
       a   good ten liter vodka
       'a good ten liters of vodka'

(58) a. #één goeie liter vodka
       a   good liter vodka
       'one "slightly-over-a-liter container" of vodka'
    b. *een tien goeie liter vodka
       a   ten good liter vodka

Given the syntactic evidence against the hypothesis that the cardinal and the measure noun can form a constituent to the exclusion of the substance NP, Rothstein's proposed assimilation of lexical powers, even if successful, to measure nouns cannot be taken as an argument for treating complex cardinals as constituents to the exclusion of the lexical NP.

2.5.7 Summary: A Single Semantic Type for All Cardinals

While Rothstein is one of the very few people working on a compositional semantics for complex cardinals, her concrete proposal assuming the semantic types n and $\langle n, \langle e, t \rangle \rangle$ for simplex cardinals does not make the right crosslinguistic predictions. More precisely, assuming that lexical powers, like measure nouns, require a type n argument does not explain why in a number of languages lexical powers do not have to be preceded by 'one' when used in the singular. The hypothesis that complex cardinals form a constituent to the exclusion of the lexical NP, just like the hypothesis that the measure phrase in a pseudopartitive is a constituent, leads to incorrect predictions for their syntax.

Importantly, Rothstein's approach does not resolve the two major problems arising from the syntactic approach to complex-cardinal formation: the lexical idiosyncrasies internal to complex cardinals and the asymmetric properties of addition (it is often the rightmost cardinal that assigns case/number; see chapter 5). As will be shown in chapters 3 and 5, while these problems seem to be resolvable by hypothesizing that complex cardinals are heads, such a "solution" does not explain case assignment and agreement inside complex cardinals.

As discussed at the beginning of section 2.5, the major advantage that Rothstein's approach appears to have over ours is that it does not generate nonexistent complex cardinals such as *six-ten* meaning 'sixty'; it avoids this overgeneration by positing different semantic types for those cardinals that can and cannot be multiplicands. However, as discussed above, there is no principled motivation for such a binary division. Furthermore, Rothstein's approach does not solve the overgeneration problem for cardinals involving addition, only for cardinals involving multiplication: why is *twenty-two* possible, but not *twenty-seventeen* ('twenty plus seventeen', i.e., 'thirty-seven')? This over-generation is as much a problem for Rothstein's proposal as for ours; as discussed in section 2.5.5, Rothstein has additive cardinals below 'one hundred' compose in the lexicon, but this simply moves the problem from one domain (syntax) to another (lexicon) and, as discussed above, gives rise to other problems. We argue, instead, that constraints on which cardinals can combine via multiplication or addition are extralinguistic (see section 2.1.3 on cardinals involving multiplication, and chapter 5 on cardinals involving addition). In light of these issues, we conclude that Rothstein's approach does not offer any advantages over ours.

Other proposals treating cardinals as numbers are far less detailed than Rothstein's and do not discuss the full range of her data. The problem of the internal structure of complex cardinals is not touched upon by most other semantic work that assumes type n for cardinals. The same is true for the difference between measure phrases and numeral NPs, which gives rise to some problems. While treating cardinals as type n seems natural for measure phrases, where the cardinal is viewed as an argument of the measure noun, extending this analysis to regular numeral NPs, as Kennedy (2015) does, requires some adjustments. Kennedy, relying upon Hackl 2000, hypothesizes a covert *many* that combines with a plural NP. Two types of arguments can be advanced against this proposal. On the one hand, there are good reasons against hypothesizing true plurals in numeral NPs (see chapter 4). On the other hand, there is absolutely no evidence from overt morphology in other languages for the presence of such a covert *many*, either as an independent syntactic constituent or as a suffix on the cardinal. While Rothstein resolves this dilemma by assuming that different types of cardinals are used with measure nouns versus other nouns (see section 2.5.6), it is also possible to construct a semantic treatment of cardinal-containing NPs that are used as measures, e.g., *twenty degrees centigrade* or *five guests fewer than last time*, that is compatible with our semantics of cardinals (see Matushansky and Zwarts 2017).

To conclude, we see no advantages to accounts of cardinals that treat them as number-denoting. Before concluding this chapter, we turn to a different topic, the ability of cardinals to combine with relational nouns, and propose a semantic analysis of such combinations.

2.6   Relational Nouns

Our analysis appears to run into a problem with plural relational NPs of the kind illustrated in (59) (see Dowty 1986; Eschenbach 1993; Hackl 2002;

Staroverov 2007). There exist multiple compositional treatments of relational nouns (see Partee 1999 and Le Bruyn, De Swart, and Zwarts 2013, among many others); however, they do not address what happens when relational nouns combine with cardinals.

The italicized plural NPs in (59) are interpreted reciprocally: (59a) is about houses that are different *from each other*, (59b) is about sisters *of each other*, (59c) is about outcomes which are equally possible *with respect to each other*, and (59d) is about solutions which mutually exclude *each other*.

(59) a. My friends live in *different houses*.
  b. I saw *sisters* walking down the street.
  c. These are *equally possible outcomes*.
  d. These are *mutually exclusive solutions*.

The reciprocal interpretation is maintained when the NP is preceded by a cardinal, as in (60). On the other hand, the singular counterparts of these plurals either have a nonrelational (sortal) interpretation, as in (61a–c), or are infelicitous, as in (61d).[16]

(60) a. My friends live in *five different houses*.
  b. I saw *two sisters* walking down the street.
  c. These are *seven equally possible outcomes*.
  d. These are *four mutually exclusive solutions*.

(61) a. Sue lives in *a different house* (from the one Mary lives in).
  b. I saw *a sister* (of John).
  c. This is *an equally possible outcome* (to the one you mentioned).
  d. *This is *a mutually exclusive solution*.

Our proposal that the lexical NP combining with a cardinal is semantically singular appears to face a problem with relational plural NPs that combine with cardinals, as in (60): how can the essential plurality of the lexical NP predicate in (60) be reconciled with the requirement that the lexical NP sister of a cardinal denote an atomic set? In what follows, we will show that relational plural NPs are not a problem for us.

2.6.1 Relational NPs: The Proposal
The starting point of our proposal is the fact that in the singular, relational NPs require an internal argument (cf. Löbner 1985),[17] as shown in (62). (Our focus here is on relational readings only; the relational NPs in (62) can also have sortal readings, e.g., *brother* in (62b) can mean somebody who is a member of the set of male individuals that have siblings, or a monk.)

(62) a. Daniel's/*a sibling walked in.
     b. Let's invite Edgar's/*a brother.

The internal argument need not be explicit, and may be supplied by the context, as in (63), or by the verb *have*, as in (64). We further observe that relational NPs may have plural internal arguments, as in (65).

(63) a. A friend just called.
     b. Let's invite a colleague.

(64) a. Everyone has a friend.
     b. Thomas doesn't have a single colleague working on head movement.

(65) a. Betsy and Claudia's sibling
     b. their next-door neighbor

The subset of relational nouns that we are concerned with permits the distributive reading only: there are no relational nouns that, like the verbal predicate *combine*, would only be true of a plural internal argument. Furthermore, plural relational nouns only give rise to Strong Reciprocity, not Weak Reciprocity (Langendoen 1978): assuming that *these people* in (66) refers to Annie, Bob, Claudia, Donovan, Elizabeth, and Fred, (66) is not true if there exists a pair of individuals in this group (for instance, Annie and Elizabeth) that aren't friends with each other, while all others are each other's friends.[18]

(66) These people are very good friends.

Given Strong Reciprocity, we appeal to the distributive operator DIST for both argument slots, as in (67).

(67) a. $\text{DIST}_s = \lambda f \in D_{\langle e,t \rangle} \cdot \lambda X \in D_e \cdot \forall x \in X \,[\text{atom}(x) \to f(x)]$
     b. $\text{DIST}_o = \lambda f \in D_{\langle e,\langle e,t \rangle\rangle} \cdot \lambda X \in D_e \cdot \lambda y \in D_e \cdot \forall x \in X \,[\text{atom}(x)$
$$\to f(x)(y)]$$

We will presently show that the reflexive operator REFL, defined in (68), will suffice to derive the reciprocal interpretation of relational nouns, though the bulk of the motivation for its use will be presented in the next section. Crucially, we hypothesize that the reflexive operator can (and in our case, as we will show in the next section, must) undergo Quantifier Raising (QR).

(68) $[\![\text{REFL}]\!] = \lambda P_{\langle e,\langle e,t \rangle\rangle} \cdot \lambda x \cdot P(x,x)$

Putting together the DIST operator, the reflexive operator, and our analysis of cardinals, we end up with the structure in (69) for *four siblings*.[19] As a result, the compositional semantics of (69) is spelled out in (70) and paraphrased informally in (71).

(69)

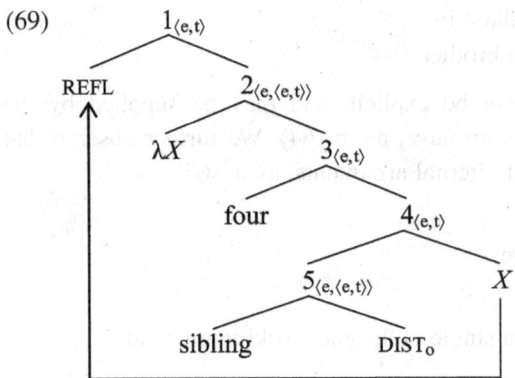

(70) a. $[\![sibling]\!] = \lambda x \in D_e . \lambda y \in D_e . sibling(x)(y)$

b. $[\![5]\!] = \lambda X \in D_e . \lambda y \in D_e . \forall x \in X\ sibling\ (x)(y)$

c. $[\![4]\!] = \lambda y \in D_e . \forall x \in X\ sibling(x)(y)$

d. $[\![3]\!] = \lambda Z \in D_e . \exists S \in D_{\langle e,t\rangle} [\Pi(S)(Z) \wedge |S| = 4 \wedge \forall s \in S\ \forall x \in X$
$$sibling(x)(s)]$$

e. $[\![2]\!] = \lambda X . \lambda Z \in D_e . \exists S \in D_{\langle e,t\rangle} [\Pi(S)(Z) \wedge |S| = 4 \wedge \forall s \in S\ \forall x \in X$
$$sibling(x)(s)]$$

f. $[\![1]\!] = \lambda X . \exists S \in D_{\langle e,t\rangle} [\Pi(S)(X) \wedge |S| = 4 \wedge \forall s \in S\ \forall x \in X$
$$sibling(x)(s)]$$

(71) a set of plural individuals X such that there exists a partition of X into
four nonintersecting nonempty parts s such that for every individual x
in X, s is the sibling of x

However, the effect of using a reflexive operator together with distribution
to atoms on both argument slots seems to give rise to truth conditions that are
slightly stronger than we might wish: we are predicting that every individual
in *four siblings* is also their own sibling. We will now show how this incorrect
prediction is to be avoided.

2.6.2 The Formal Link between Reflexives and Reciprocals

It is well known that crosslinguistically, reflexive and reciprocal verbs often
bear identical morphology. For instance, in Russian, the reflexive clitic *-sja*
may, depending on the verb it combines with, correlate with reciprocal or
reflexive interpretation, as illustrated in (72) (see Letuchiy 2009).[20]

(72) a. myt'-sja 'wash (oneself)', brit'-sja 'shave (oneself)'   Russian

b. obnimat'-sja 'hug (each other)', celovat'-sja 'kiss (each
other)'

(Letuchiy 2009)

Likewise, in French, a reciprocal or reflexive verb requires the presence of *se*, be the verb derived from a transitive stem, as in (73) and (74), or lexically/inherently reflexive or reciprocal, as in (75) and (76), respectively (Labelle 2008; see also Embick 1997a,b).

(73) a. *Le ministre copie   lui-même.   French
          the minister copy.3sg him-self
      b. Le ministre se copie.
          the minister *se* copy.3sg
          'The minister copies himself.'
      (Labelle 2008)

(74) a. *Les voisins   détestent les uns les autres.
          the neighbors hate.3pl the ones the others
      b. Les voisins   se détestent (les uns les autres).
          the neighbors *se* hate.3pl   the ones the others
          'The neighbors hate each other.'
      (Labelle 2008)

(75) a. *Jean autoanalyse.
          Jean self-analyze.3sg
      b. Jean s' autoanalyse.
          Jean *se* self-analyze.3sg
          'Jean analyzes himself.'
      (Labelle 2008)

(76) a. *Les participants entreregardèrent.
          the participants entre-look.at.past.3pl
      b. Les participants s' entreregardèrent.
          the participants se entre-look.at.past.3pl
          'The participants looked at one another.'
      (Labelle 2008)

Setting aside the question of what the role of the reflexive clitic might be (see Reinhart and Siloni 2004 for some discussion), a comparison of the interpretation of a plural reflexive verb with that of a plural reciprocal verb immediately reveals the formal link between the two. The strongly reciprocal interpretation ends up being universal distribution down to atoms to the exclusion of atomic reflexives: that is, the interpretation of (77a) presupposes that (77b) is untrue. In other words, plural reciprocal VPs, just like plural relational NPs discussed above, can almost be treated as plural reflexives.

(77) a. The kids embraced (each other).
      b. The kids embraced themselves.

Crucial here is the notion of *irreflexivity* introduced by Barker (1999). Both Barker and Hackl (2002) note that some relational nouns presuppose the non-identity relation between their arguments, viewed by both authors as part of their lexical entry: one can't be one's own sibling, child, or neighbor. While Barker explicitly constrains irreflexivity to some but not all relational nouns, we hypothesize that the presupposition of nonidentity extends to all relational NPs.[21] This means that upon plural reflexivization of a relational NP, the "atomic reflexive" ($R(x,x)$) is automatically excluded from consideration, and therefore the reciprocal reading is straightforwardly obtained as a result of reflexivization, as in (69) above (as well as in (78), below).[22]

Finally, the irreflexivity hypothesis can also explain why the reflexive operator in (69) (and (78)) can neither stay in situ nor move to a position lower than the $dist_s$ operator or the cardinal. Though neither option is excluded on purely syntactic grounds,[23] a low position for the reflexive operator would lead to a violation of irreflexivity, as long as we continue to assume that irreflexivity is an inherent property of relational NPs rather than a lexical property of relational predicates. While we are convinced that such must be the case (see also note 21), we cannot advance any hypothesis for why this should be so.

2.6.3 Relational Nouns: Summary

To sum up, we derive the interpretation of cardinal-containing relational NPs by appealing to (a) the treatment of cardinals that imposes the atomicity constraint on the lexical NP sister of a cardinal, (b) the QRing reflexive operator REFL, and (c) the irreflexivity hypothesis. Combining these pieces with a pluralizing operator, we are now able to derive the correct meaning as well for plural relational NPs like *siblings*, as shown in (78). The derivation is given in (79) and an informal paraphrase in (80).

(78)

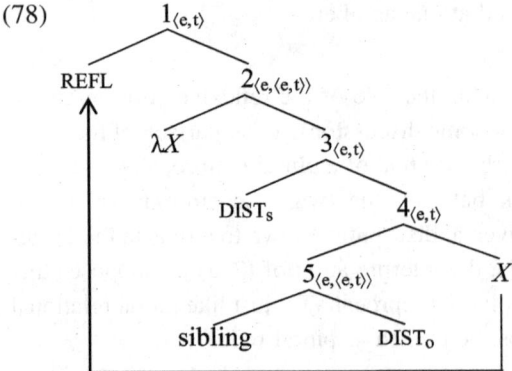

(79) a. $[\![5]\!] = \lambda X \in D_e \, . \, \lambda y \in D_e \, . \, \forall x \in X \, \text{sibling}(x)(y)$
      b. $[\![4]\!] = \lambda y \in D_e \, . \, \forall x \in X \, \text{sibling}(x)(y)$
      c. $[\![3]\!] = \lambda Y \in D_e \, . \, \forall y \in Y \, \forall x \in X \, \text{sibling}(x)(y)$
      d. $[\![2]\!] = \lambda X \in D_e \, . \, \lambda Y \in D_e \, . \, \forall y \in Y \, \forall x \in X \, \text{sibling}(x)(y)$
      e. $[\![1]\!] = \lambda Y \in D_e \, . \, \forall x,y \in Y \, \text{sibling}(x)(y)$

(80) a set of plural individuals X such that for any individuals x,y in X, x is a sibling of y

Importantly, every piece of this analysis is independently motivated: the dist$_s$ operator and its extensions to the internal-argument slot are needed for distributive readings elsewhere; the reflexive operator REFL is necessary in order to account for reflexive and reciprocal verbs; and the atomicity requirement is necessitated by the composition of complex cardinals. Finally, the reciprocal interpretation of the reflexive plural relational NP naturally follows from the independently motivated irreflexivity hypothesis.

2.7 Summary: The Compositional Semantics of Complex Cardinals

In this chapter, we showed that the compositional semantics of multiplicative complex cardinals necessitates that simplex cardinals have the semantic type of modifiers, rather than determiners or predicates. If cardinals were determiners or predicates, the construction of complex cardinals would require the semantics of simplex cardinals to vary as a function of their role (e.g., *hundred* would need to have different meanings in *a hundred* vs. in *four hundred* vs. in *a hundred thousand*). Importantly, though the hypothesis that cardinals are semantic modifiers requires a special mechanism introducing existential force for cardinal-containing indefinite NPs in argument positions, this mechanism is provided by any standard view of indefinites. We have also shown that our analysis of cardinals is compatible with measure phrases and with relational nouns, and we have provided evidence that number denotation (type n) is not the basic denotation of cardinals.

An alternative hypothesis is that complex cardinals are constructed extralinguistically and thus belong to a completely separate system (see, e.g., Wiese 2003; Zabbal 2005). Our objection to this view is twofold. First, if syntactic composition and semantic interpretation of complex cardinals can be seamlessly incorporated into standardly assumed syntax and semantics, such an appeal to an extralinguistic system is unnecessary.[24] Secondly, in chapter 3 we will show that complex cardinals are subject to such morphosyntactic operations as case assignment and number marking, which means that they are constructed by regular linguistic mechanisms. Even though we have to make an appeal to extralinguistic factors in order to explain certain ordering constraints in complex cardinals, we will argue in chapters 3 and 5 that similar factors come into play in complex measure phrases, such as *five feet five inches*, for which an extralinguistic analysis seems unwarranted.

Finally, we have considered an alternative approach, that of Rothstein (2013, 2016, 2017), on which cardinals come in two distinct semantic types. We have provided evidence against this approach and have argued that it holds no advantage over ours. While different types of cardinals exhibit different properties (as will be discussed in more detail in subsequent chapters), there is no binary distinction between those cardinals that can appear NP-internally and those that cannot, contra Rothstein's analysis. In chapter 3, we will show that different cardinals have different morphosyntactic properties. Our central proposal is that while the syntax of cardinals varies, their semantics stays the same. Given that there is independent crosslinguistic evidence (from facts of case assignment, gender and number agreement, etc.) for positing different syntax for different types of cardinals, we believe that it is more economical not to posit a different semantics as well in the absence of compelling evidence to the contrary.

3   Syntax–Semantics Mapping with Cardinals

In this chapter, we address the issue of how to map the semantics proposed for cardinals in chapter 2 into syntactic structure. We also provide evidence from the morphosyntactic behavior of cardinals in favor of our semantic proposal. The focus of this chapter is on multiplicative complex cardinals such as *two hundred* or *three hundred thousand*. We will consider additive complex cardinals such as *twenty-two* in chapter 5.

We adopt here the standard syntactic framework of Government and Binding and Minimalism, according to which case assignment and agreement occur only in syntax, with the former only done by heads in their c-command domain.

In section 3.1, we consider different approaches to the syntax of multiplicative complex cardinals and argue for a cascading structure where the complex cardinal does not form a constituent to the exclusion of the lexical NP. We show that analyses of simplex cardinals as either heads or adjuncts are in principle compatible with the semantics proposed in chapter 2, with the former reflecting the most direct semantics-to-syntax mapping. Section 3.2 considers some issues in the internal syntax of multiplicative complex cardinals, while section 3.3 takes a brief look at crosslinguistic and intralinguistic variation regarding which cardinals are complex versus simplex. In section 3.4, we review crosslinguistic evidence that cardinals can be nouns, adjectives, or (less commonly) verbs, rather than forming a separate functional class. Finally, section 3.5 takes up the fact that our semantics of cardinals, proposed in chapter 2, requires the lexical NP combining with a cardinal to denote an atomic set. In this section, we show how this requirement can be implemented; we leave the question of plural morphology on the lexical NP, in languages like English, until chapter 4.

The ultimate conclusion of this chapter is that cardinals are lexical rather than functional, and that in most languages, they exhibit (some of) the properties of nouns or adjectives. We argue that nominal cardinals should be analyzed as heads and adjectival cardinals as adjuncts, and that as a result, complex cardinals are built using existing syntactic mechanisms.

3.1 Syntax of Cardinal-Containing NPs

The simplest syntax–semantics mapping is one on which the semantic representation in (1) translates directly into a syntactic tree.

(1)

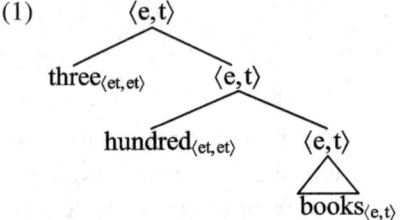

The tree in (1) can correspond to various syntactic configurations: each cardinal may be a head of its own projection, a specifier of some functional projection, or an adjunct to the lexical NP (*books*). In this section, we discuss each of these possibilities in turn and then provide evidence showing that the complement and adjunct analyses are both attested crosslinguistically and compatible with our semantic proposal, whereas the specifier analysis is not compatible with the proposed semantics.

3.1.1 Cardinals as Heads

The view that a cardinal is a head taking the lexical NP as its complement has been advanced in several proposals (e.g., Ritter 1991; Giusti 1991, 1997; and Zamparelli 1995, 2002a) that analyze cardinals as a functional head, Num^0 or Q^0, as schematized in (2).[1]

(2)

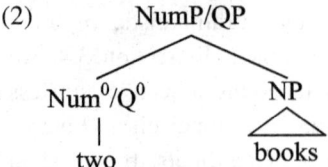

In order to accommodate complex cardinals on the structure in (2), it would be necessary to hypothesize either that more than one NumP (QP) can be present in the extended NP or that the entire complex numeral is composed in the morphology before being inserted into a single Num^0 (Q^0). The hypothesis that multiple NumPs (QPs) are stacked on top of each other in the syntactic tree is not easily compatible with standard assumptions about functional structure.

The hypothesis that the entire complex cardinal is inserted into a single functional head assumes that complex cardinals are built entirely in the morphology, that is, that they are basically morphological compounds (albeit very complex ones, such as *three hundred forty thousand and one hundred eighty-five*). On this approach, all cardinals within a complex cardinal must combine with each other before combining with the lexical NP. As discussed in chapter 2, this is not compatible with the semantics that we have proposed for cardinals.

We argue that (2) is almost the right structure for complex cardinals in the more familiar languages, excepting the labels: we will argue that cardinals project lexical rather than functional projections, as in (3), for instance. In this structure, each cardinal takes an NP as its complement. Evidence for this structure comes from case assignment, a commonly accepted diagnostic for complementation.[2]

(3)

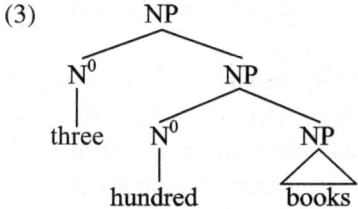

In languages with overt morphological case marking, a lexical NP combining with a cardinal may surface with a particular, language-specific case that is the same case assigned within the NP in other configurations. This is illustrated in (4)–(6) for Finnish, Latin, and Russian.

(4) a. aski   tupakkaa   Finnish
       packet tobacco.PART
       'a packet of tobacco'
   b. neljä   kynää
       four.NOM pencil.PART
       'four pencils'

(5) a. pars   Galliae   Latin
       part.NOM Gaul.GEN
       'a part of Gaul'
   b. cum tribus milibus   clavum
       with three thousand.PL.ABL key.PL.GEN
       'with three thousand keys'

(6) a. syn   Maši   Russian
       son.NOM Mary.GEN
       'Mary's son'
    b. korobka karandašej
       box.NOM pencil.PL.GEN
       'a box of pencils'
    c. šest'   šagov
       six.NOM step.PL.GEN
       'six steps'

However, it is often assumed that nouns cannot assign structural case (Chomsky 1986), which is why the insertion of the dummy preposition *of* is necessary in English. If so, examples (4)–(6) involve additional structure, potentially containing a null equivalent of the English preposition *of*. Does this mean that case assignment is not a reliable diagnostic for the structural con-figuration in which a cardinal and the lexical NP it combines with find themselves?

Fortunately, in addition to crosslinguistic variation in the apparent ability of nouns to assign case, there also exists language-internal variation. As illus-trated in (7) for Dutch, the choice of the head noun determines whether an overt preposition *van* 'of' is necessary: with a measure head noun (i.e., in so-called *pseudopartitives*) it is, in fact, prohibited.[3]

(7) a. drie liter (*van) water   Dutch
        three liter   of   water
        'three liters of water'
    b. drie eigenschappen *(van) water
        three properties   of   water
        'three properties of water'

Under any assumptions, *water* 'water' in (7a) requires case, and the simplest possible view is that the head noun of the NP, *liter* 'liter', assigns genitive to it. Conversely, the head noun *eigenschappen* 'properties' in (7b), even though semantically transitive as well, cannot assign case and the preposition *van* 'of' is necessary to do so.[4]

It is, of course, possible that (7a) contains the null version of the overt preposition *van* 'of' in (7b). For our purposes this possibility makes no differ-ence, since in order to enforce the choice between the overt and covert preposi-tions it would be necessary in any case to assume that the PP ~~van~~ *water* 'of water' appears as the complement of the head noun in order to create the right configuration for c-selection. We opt for the simpler option here, assuming that it is the head noun itself that assigns case to the NP it combines with.[5]

To summarize, whatever assumptions we adopt regarding the ability of a noun to assign case, in examples (4)–(6) the lexical NP is either itself the complement of the cardinal or is contained inside the PP complement of the cardinal. In other words, case assignment within cardinal-containing NPs is a reliable diagnostic of the complementation structure.

Does complementation also permit us to diagnose the lexical category of the cardinal? On the one hand, it is well known that adjectives can take internal arguments, just like nouns, and these internal arguments can be introduced by the preposition 'of', as in (8a), or by genitive case, as in (8b).

(8) a. fier   de son   travail   French
       proud of 3SG.POSS.M.SG work
       'proud of her/his work'
    b. polnyj vody   Russian
       full   water.GEN
       'full of water'

On the other hand, however, the combination of a cardinal and an NP is always itself an NP, which means that, if the cardinal behaves like a head in that it assigns case to the lexical NP, it must be nominal. Therefore, the ability of a cardinal to assign case seems to be a clear diagnostic for the structure in (3), repeated below.

(3)

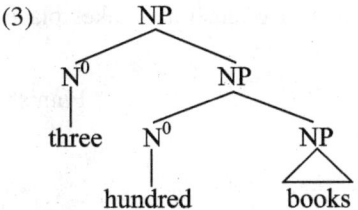

Support for our conclusion that it is the cardinal itself that assigns case comes from two sets of data, from Russian and from Inari Sami, where the choice of the cardinal affects the case on the lexical NP. In Russian, the lower cardinals 'two', 'three', and 'four' assign the so-called paucal case (9) (see Franks 1994; Halle 1994; and Garde 1998, among others), whereas all other cardinals assign genitive (10a), like "regular" nouns (10b).

(9) a. dva   stola   Russian
       two.NOM table.PAUC
       'two tables'
    b. četyre   šagá
       four.NOM step.PAUC
       'four steps'

(10) a. šest'   šagov
        six.NOM step.PL.GEN
        'six steps'
    b. syn   Maši
        son.NOM Mary.GEN
        'Mary's son'

In Inari Sami, lower cardinals from 'two' to 'six' assign accusative while
higher cardinals assign partitive, as shown in (11) (Nelson and Toivonen
2000).[6] (For more information on case assignment by cardinals crosslinguistically, see Hurford 2003.)

(11) a. kyehti/kulmâ/nelji/vittâ/kuttâ päärni   Inari Sami
        two/three/four/five/six   child.ACC
        'two/three/four/five/six children'
    b. čiččâm/kávci/ovce/love/ohtnubáloh/kyehtnubáloh/čyeti/... pärnid
        seven/eight/nine/ten/eleven/twelve/hundred   child.PART
        'seven/eight/nine/ten/eleven/twelve/one hundred/ ... children'

   Further support for the cascading structure in (3) comes from evidence of
case assignment within complex cardinals. In both Russian and Finnish, case
assignment within complex cardinals behaves just like case assignment from
cardinals to lexical NPs. Finnish cardinals 'two' and above assign partitive
case to the lexical NP; partitive-case assignment in Finnish also takes place
within complex cardinals, as shown in (12).

(12) a. kolme-kymmenta   Finnish
        three.NOM-ten.PART
        'thirty'
    b. viisi-sataa
        five.NOM-hundred.PART
        'five hundred'

   In Russian the cardinals 'two', 'three', and 'four' assign paucal case to the
lexical NP, while 'five' and higher assign genitive case. In a complex cardinal
like 'four thousand' or 'five thousand' in Russian, the case on 'thousand'
depends on the preceding cardinal. While 'four' assigns paucal case to 'thousand' (13a), 'five' assigns genitive (13b).

(13) a. četyre tysjači   šagov   Russian
        four   thousand.PAUC step.PL.GEN
        'four thousand steps'
    b. pjat' tysjač   šagov
        five thousand.PL.GEN step.PL.GEN
        'five thousand steps'

Case assignment within complex cardinals provides evidence against the view that the entire complex cardinal is contained within a single functional head (as proposed in Ritter 1991; Giusti 1991, 1997; Zamparelli 1995, 2002a; see above), and argues for the structure in (3), in which each cardinal is a nominal that assigns case to its sister, be it the lexical NP or another cardinal.

3.1.2 Cardinals as Adjuncts

In the previous section, we have shown that the complementation structure in (3) is compatible with the semantics that we have proposed, in that it allows for iteration. However, complementation is not the only structure that allows for iteration: another structure that has this property is adjunction, since adjuncts can be infinitely iterated. Under this view, a cardinal-containing NP looks like (14).

(14)

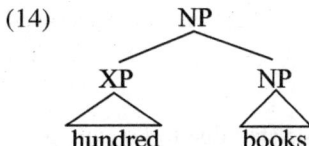

This view can deal with iteration easily, as it is built into the syntax of adjunction, and does not rely on additional functional projections. The structure in (14) does not work for case-assigning (nominal) cardinals, which require the structure in (3), on the assumption that case assignment is a diagnostic for complementation; however, the structure in (14) does appear to successfully capture the behavior of some lower cardinals found crosslinguistically that behave like adjectives rather than nouns, exhibiting agreement for gender, number, and/or case. Some examples are given in (15) and (16), from Latin and Russian respectively. (In section 3.4, we will demonstrate that such cardinals also behave like adjectives morphologically.)

(15) a. cum uno   misero   servo   Latin
         with one.M.SG.ABL poor.M.SG.ABL slave.M.SG.ABL
         'with one poor slave'
      b. duarum miserarum   puellarum
         two.GEN poor.F.PL.GEN girl.F.PL.GEN
         'of (the) two poor girls'

(16) a. pro   odnu   stranu   Russian
         about one.F.ACC country.ACC
         'about one country'
      b. iz   odnix   mest
         from one.PL.GEN place.PL.GEN
         'from one and the same area'

The cardinals in (15) and (16) behave like adjectives, and adjectives are generally analyzed as adjuncts. Does this mean that number, gender, and case agreement should be taken as a reliable diagnostic of the adjunction structure in (14)?

Theoretically, the answer is negative. Agreement can in principle arise in all of the three configurations available within a maximal projection: a head can agree with (something in) its complement, for example, *la* agreeing with *dame* in (17a); an adjunct can agree with the XP it adjoins to, for example, *belle* agreeing with *dame* in (17a); and a specifier can agree with the head, as in (17b).[7]

(17) a. la   belle   dame   French
        DET.F.SG beautiful.F.SG lady.F
        'the beautiful lady'
     b. trouver [sc la   dame belle]   French: Spec-head
        find.INF   DET.F.SG lady.F beautiful.F.SG
        'find the lady beautiful'

However, in fact our options are a lot more restricted due to lexical specifications: it is generally assumed that nouns enter the derivation with their own set of ϕ-features and as a result cannot show agreement for gender. This means that, for instance, in examples like the Russian (16), 'one' cannot be analyzed as the head of the cardinal-containing NP because had it been the head, it should have been a noun (to permit the maximal projection it appears in to function as an argument), and had it been a noun, it wouldn't have shown agreement in gender. Likewise, agreement for number, as in (16b), would have been unexpected had 'one' been a noun.

It is possible, of course, that a cardinal can be a functional head and project a functional projection within the extended NP (e.g., Ritter 1991; Giusti 1991, 1997; Zamparelli 1995, 2002a) as schematized in (2), repeated below. However, as discussed in section 3.1.1, this would require the appearance of more than one functional head of the same type within the extended NP for complex cardinals, which is undesirable. Therefore, we rule out the possibility that a cardinal can be a head that agrees with (the head of) its complement.

(2)   NumP/QP

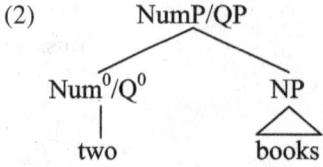

Thus, we have seen that cardinals can in principle be either (case-assigning) heads, as in (3), or (agreeing) adjuncts, as in (14). We now ask whether they can ever function as specifiers.

3.1.3   Cardinals as Specifiers

A common approach to the syntax of cardinals treats them as specifiers; this is proposed by Selkirk (1977), Jackendoff (1977), Li (1999), Haegeman and Guéron (1999), Gawron (2002), Gärtner (2004), and Zweig (2006), among others. On these proposals, cardinals occupy Spec,NumP/QP, while Num^0/Q^0 is suggested to contain the singular/plural features of the NP. This is schematized in (18). Most of these proposals do not address complex cardinals. One that does is Zweig 2006: for Zweig, complex cardinals are formed by dedicated processes, which suggests that numerals are combined with the rest of syntactic structure by means of a functional head, Ord or Card.[8]

(18)

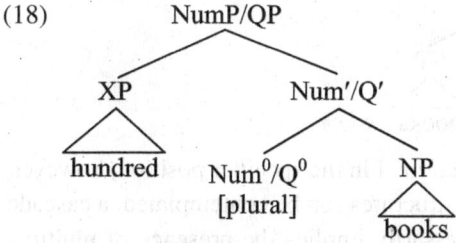

We now consider what a language would look like if it placed its cardinals in the specifier position more or less uniformly. Given that the only way of detecting the complementation structure is case assignment and that agreement is equally compatible with adjunction and specification, morphosyntactic evidence can only tell us so much. For languages where cardinals behave morphologically like adjectives or nouns, the adjunction or complementation option, respectively, seems preferred. However, in languages where no evidence can be drawn from the morphology of cardinals, as in English, or where not only some (lower) cardinals but also demonstratives and quantifiers surface with adjectival inflection, as in Russian, it is impossible to draw upon morphological evidence to decide whether cardinals can occur in specifier position or be adjoined.

Nonetheless, once we turn to complex cardinals like (19), it turns out that there are good reasons to decisively exclude the structure in (18) for the languages considered so far.

(19) nine hundred thousand books

Both the complementation and adjunction structures can straightforwardly deal with (19), as illustrated in (20).

(20) a.   Complementation structure

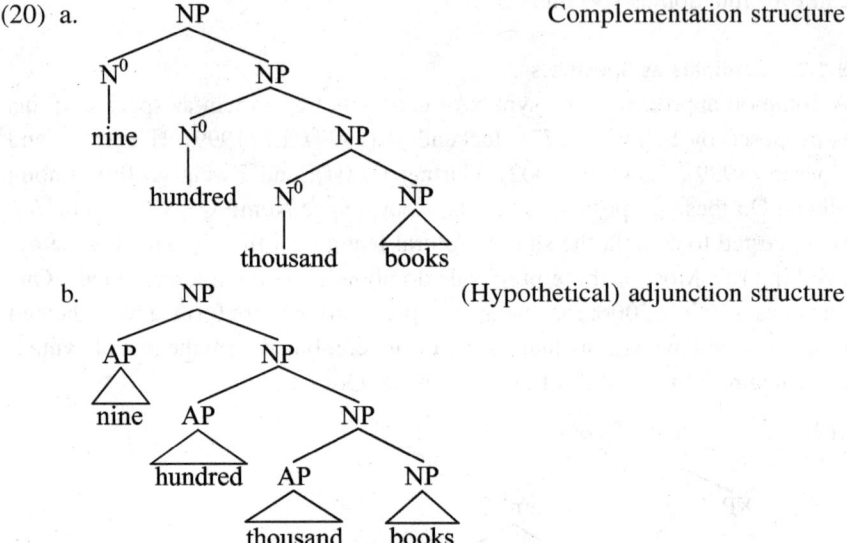

   b.   (Hypothetical) adjunction structure

Attempting to place the complex cardinal in the specifier position, however, gives rise to problems. Two possible structures can be contemplated: a cascade structure similar to (20), which necessarily implies the presence of multiple projections to host the specifier cardinals, as in (21a), or a single-specifier configuration like (18), adjusted for our example in (21b).

(21) a.   (Hypothetical) cascading-specifier structure

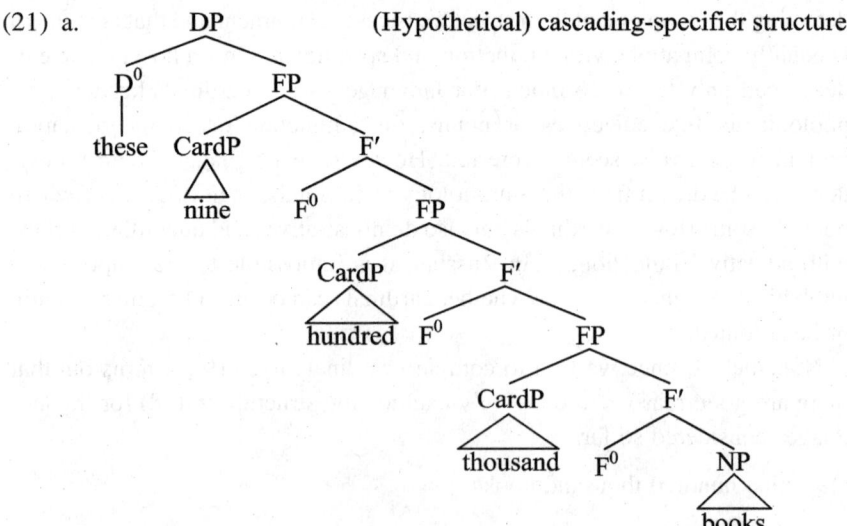

b.   (Hypothetical) single-specifier structure

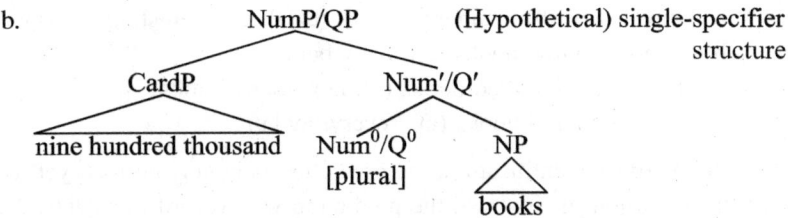

The configuration in (21a), while compatible with our semantic proposal, lacks any independent motivation and has the additional disadvantage of postulating a functional head F with no phonological content and no discernible semantics. As for (21b), it is incompatible with our semantics: *thousand*, being of the semantic type $\langle\langle e,t\rangle,\langle e,t\rangle\rangle$, requires a sister of the semantic type $\langle e,t\rangle$, which the structure in (21b) does not provide. Thus, (21b) fails on semantic grounds as long as we assume the simplest possible mapping between syntax and semantics. In section 3.5.1 we will consider possible ways of salvaging the specifier analysis by including additional syntactic structure inside the specifier and will show that they too run into problems.

3.1.4 Interim Summary

We have shown that both the complementation and adjunction analyses of complex cardinals, unlike the specifier analysis, allow for direct mapping between the proposed semantics and the syntax. We have furthermore suggested that cardinals that have the complementation structure are nominal while cardinals that have the adjunction structure are adjectival. This goes against the standard view that cardinals are a special lexical class. In section 3.4, we will present evidence from the crosslinguistic morphological behavior of cardinals that they are not a separate lexical class but are (most commonly) either nouns or adjectives. First, however, we will consider the internal syntax of multiplicative complex cardinals, as well as cardinal derivation.

3.2 The Internal Syntax of Complex Cardinals

The structure we propose in (3) predicts that within *three hundred books*, the NP *hundred books* should behave just like a regular lexical NP. In other words, we incorrectly predict the possibility of adjectival modification internal to the complex cardinal (22a), and we expect the combination of *hundred books* with all types of quantifiers (22b–d).

(22) a. *three interesting hundred books (cf. three interesting books)
     b. ?many hundred books (cf. many books)
     c. ?how many hundred books (cf. how many books)
     d. every hundred books (cf. *every books)

With regard to quantification, the prediction is largely correct, yet requires further discussion. In contrast, the prediction with regard to adjectival modification is clearly wrong. We discuss the two in turn.

3.2.1 Cardinal-Internal Modification

The fact that an NP-internal intermediate cardinal cannot be modified, as in (22a), is not expected under our proposal, given our cascading structure. While we do not have a ready explanation for this fact, we observe that the same constraint against modification of intermediate cardinals holds for vague-cardinal constructions (23a) and even for numerical nouns (23b–c).[9]

(23) a. several (*interesting) hundred books   (cf. *several interesting books*)
     b. many (*excited) thousands of people   (cf. *many excited people*)
     c. several (*amazing) dozen(s of) books   (cf. *several amazing books*)

Thus, if one were to explain the ungrammaticality of (22a) by treating *three hundred* as a nonnominal constituent, one would have to say that *several hundred*, *many thousands*, and *several dozen(s)* are also nonnominal. Treating vague cardinals as nonnominal is potentially viable, but numerical nouns are generally treated as nouns.

In light of the above, we speculate that there are only two places for modification inside an extended NP: at the level of the lexical NP (*five good books*) and just below the determiner (*an amazing twenty thousand books*—see chapter 8 for more discussion of this construction).

3.2.2 Cardinals and Quantifiers

As shown in (22b–d), NP-internal cardinals like *hundred* are able to combine with a variety of quantifiers. The fact that *hundred* can combine with *every* (22d), which requires a singular complement (*every book*, not **every books*), speaks in favor of our analysis of *hundred books* as semantically singular. NP-internal cardinals also combine readily with quantifiers like *several* or *a few* (*several/a few hundred books*). Combination with *many* or *how many* is marginal for some speakers (22b–c) but nevertheless attested, as the following examples from the Corpus of Contemporary American English (COCA; Davies 2008–) and Google show:

(24) a. the surprising work of God in the conversion of many hundred souls in Northamton, of New England
(https://books.google.nl/books?id=7DdfAAAAcAAJ)

b. How many hundred Schmidts would there be in this area?
(COCA)

c. If you ever watch the Olympics, the weight lifters can lift so many hundred pounds.
(COCA)

d. I've received many hundred e-mails, mostly supportive.
(COCA)

Our proposal readily accounts for the existence of quantifier-cardinal combinations, without requiring vague cardinals such as *many* or *how many* to have different semantics when they combine with a regular lexical NP (*(how) many books*) versus a cardinal-containing NP (*(how) many hundred books*).

3.2.3 Ordering Constraints on Complex Cardinals

It is reasonable that languages should encode complex cardinals involving multiplication and addition differently: for example, no language should use exactly the same identical form to express both *three hundred* and *one hundred and three*. Some languages distinguish the two forms by case marking (e.g., Russian; see chapter 6) or by the use of prepositions and/or overt conjunctions for addition (see chapter 5). In addition to these means, the more commonly studied languages use word order to distinguish the two: for example, in English, *five thousand* cannot be interpreted, through addition, to mean 'five plus one thousand (i.e., 1,005)'; and conversely, *a thousand five* is not interpreted as 'a thousand times five (i.e., 5,000)'. The discussion below is largely based on the description in Hurford 2003 of crosslinguistic constraints on the order of cardinals. We note in passing that these constraints are only relevant in the absence of other marking: for instance, in Mano (a Mande language of Liberia and Guinea), higher cardinals precede lower ones in both multiplication and addition, but the latter is signaled by the presence of a preposition (Khachaturyan 2015, 58; see also chapter 5).

Turning to other languages, in the case of multiplication, the higher cardinal generally follows the lower cardinal: for example, the number 200 is pronounced as (the equivalent of) *two hundred*, not **hundred two*. The only exceptions discussed by Hurford (2003) are languages in which cardinals follow nouns, such as Shona and Yoruba. This is consistent with our analysis. In these languages, NPs are head-final and cardinal-containing NPs are no exception. A simplex cardinal like 'three' follows both lexical NP complements (the equivalent of *houses three* for 'three houses') and cardinal complements (*hundred three* for 'three hundred').

An illustration of the above generalizations is found in some lines from the Russian poem "Konjok-Gorbunok," by Yeršov, given in (25). As shown by the second sentence, the unconventional form *dva-pjat'*, literally 'two-five', is interpreted as 'ten' (multiplication) rather than 'seven' (addition), consistent with the low-high order being indicative of multiplication.

(25) Dva-pjat' šapok   serebra. To est', èto budet desjat'.   Russian
     two-five hat.PL.GEN silver   that is   that be.FUT ten
     'Two-five hats of silver. That is, that will be ten.'

This said, unconventional arithmetic is not easy to interpret: an informal survey suggests that quite a few native Russian speakers confronted with the first sentence in (25) are not certain as to what 'two-five' means and do consider both 'seven' and 'ten' as possibilities.

More puzzling for our proposal are languages like Sinhala and Maori, in which, according to Hurford, the cardinal follows the noun (26a) but the ordering of cardinals is nevertheless low-high (26b). (This results in the order shown in (26c) for NPs containing complex cardinals.) According to Hurford, the reverse scenario is unattested in any language.

(26) a. ge-val   tun-ak   Sinhala
        house-PL three-INDEF
        'three houses'
        (Hurford 2003)
     b. tun   si:yə
        three hundred
        'three hundred'
        (Hurford 2003)
     c. ge-val   tun   si:y-ak/siya-y-ak
        house-PL three hundred-INDEF
        'three hundred houses'
        (Wijeratne Wijesinghe, pers. comm.)

All nominal dependents including relative clauses precede the noun in Sinhala, making the NP head-final. Cardinals being postnominal (with the exception of 'one'), we expect the multiplicand to precede the multiplier, contrary to fact. Yet there might be reasons to believe that cardinals do not combine with the lexical NP in a complementation structure in Sinhala. Specifically, as noted by Chandralal (2010), while numerals follow inanimate nouns directly (27a), with animate nouns, numerals require a connective (27b), which Hurford (2003) refers to as a classifier. The indefinite article is suffixed to this connective (27c).

(27) a. gas   deka/tuna/dolaha
          tree.PL two/three/twelve
          'two/three/twelve trees'
          (Chandralal 2010, 52)
       b. lamai   dedena/tundena/dolosdena
          child.PL two.CONN/three.CONN/twelve.CONN
          'two/three/twelve children'
          (Chandralal 2010, 52)
       c. balallu tun.den-ek
          cat.PL   three.CONN-INDEF
          'three cats'
          (Childers 1876)

Alternatively, this connective can also be viewed as an agreement marker for animacy. We leave this question open.

3.3 Cardinal Derivation

There is clearly crosslinguistic as well as within-language variation as to which cardinals are linguistically simplex or complex. For instance, in Russian, *sorok* 'forty' is a linguistically simplex cardinal, derived from the Old Nordic *sekr* 'furs' (see Wiese 2003, chap. 3). On the other hand, the Russian *trinadcat'* 'thirteen' is a complex cardinal (literally 'three on ten'). In English, *fifty* is arguably derived from *five-ten* (the same holds for *forty*, *sixty*, *eighty*, etc.); however, in light of the phonological changes, it may be preferable to treat it as a simplex cardinal (the same holds for *forty*, *eighty*, etc.). Our analysis has nothing to say about why the same cardinal (e.g., 'forty') would be syntactically simplex in one language but complex in another. We recognize the need for internal morphological structures in cardinals like *forty* in English or *ducenti* 'two hundred' in Latin, just as we recognize the internal complexity of nouns like *bicarbonate* or *hexagon*, without attempting to provide their compositional semantics. The question arises, of course, where to draw the line between morphological and syntactic complexity. The status of each complex cardinal as either composed in the morphology or composed in the syntax needs to be decided individually on the basis of such criteria as internal case marking and agreement, productivity, and phonological integration.

Finally, Wiese (2003, chap. 3) points out that in Hebrew and Arabic, words for base-ten cardinals like 'sixty' are formed using plural morphology: for instance, in Arabic, 'sixty' is *sittuna*, literally the plural form of *sittun* 'six'. Wiese uses this idiosyncratic application of plural morphology as an argument for the treatment of complex cardinals as extragrammatical elements. We note, however, that the use of plural morphology in unconventional ways is attested outside the domain of cardinals. For instance, in Russian, the suffix *-y* is a regular plural suffix, as shown in (28); when *-y* is combined with the noun for 'hour', the result is *either* a regular plural *or* a brand new word, as shown in (29) (see also Veselinova 2004).

(28) a. stol   b. stoly   Russian
        table   table.PL
        'table'   'tables'

(29) a. čas   b. časy
        hour   hour.PL
        'hour'   'hours' or 'clock'

Thus, the use of plural morphology to derive new words does not speak in favor of an extragrammatical analysis. We treat *sittuna* and other base-ten cardinals in Arabic and Hebrew as simplex derived nouns.

3.4 The Lexical Category of Cardinals

Traditional as well as generative grammarians unite in assuming that cardinals (or numerals in general) form a lexical category of their own. However, when we consider languages with richer morphology than English, we immediately discover a lot of evidence that they actually belong to some of the core lexical categories: noun, adjective, and verb (cf. Andersen 2005). In this section, we discuss evidence from four typologically different languages (Finnish, Latin, Russian, and Seri) that indicates that cardinals are compatible with the core lexical categories. The example of Russian furthermore shows that different cardinals within the same language may belong to different lexical categories—a hypothesis supported by the discussion of other languages in chapter 7.

3.4.1 Cardinals as Nouns: Finnish
Hurford (1975) notes that lower cardinals in most languages behave differently from higher cardinals, and in particular, that lower cardinals are more likely than higher cardinals to be adjectival. In the case of Finnish, the cardinal *yksi* 'one' behaves like an adjective, agreeing in case with the NP it combines with

throughout the paradigm, as in (30a). All Finnish cardinals above 'one', in contrast, assign partitive case to the lexical NP, just like Finnish nouns expressing quantity or a measure: compare (30b) to (31).

(30) a. yksi   kuva   Finnish
        one.NOM   picture.NOM
        'one picture'
     b. neljä   kynää
        four.NOM pencil.PART
        'four pencils'

(31) a. aski   tupakkaa
        packet   tobacco.PART
        'a packet of tobacco'
     b. joukko   ihmisiä
        group   people.PART
        'a group of people'

The least complicated conclusion that we can draw from these facts is that cardinals in Finnish are a subset of measure nouns, assigning partitive case to their sister.[10] Setting aside potential differences between cardinals and measure nouns in the functional layers projected in the lower noun phrase, this idea can be schematized as follows:

(32)

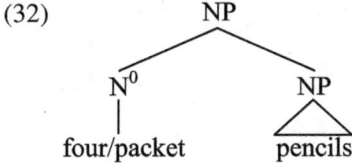

The immediate advantage of this view is that it allows for a straightforward mapping from the semantic structure in (1) to the syntactic representation. In the case of complex cardinals, the semantic structure in (1) translates into the syntactic structure in (3). In section 3.1.1, we showed that this analysis is further supported for Finnish by the facts of case assignment within complex cardinals.

However, we have seen that the syntactic structure in (3) is not the only syntactic structure that maps straightforwardly to our semantics: the other possibility is the adjunction structure in (14), in which cardinals are adjectives. An example of a language for which (14) appears to be the right structure is Latin, as discussed next.

3.4.2 Cardinals as Adjectives: Latin

Latin cardinals seem to be adjectival, although they do not all behave in the same way from the point of view of their morphology (or syntax), as shown in tables 3.1 through 3.3. As in many other languages (Hurford 1975), Latin lower cardinals are different from higher cardinals in that they decline like somewhat irregular adjectives. The three lowest Latin cardinals agree in gender, number, and case with the lexical NP, as already shown in (15), repeated below.[11]

(15) a. cum uno   misero   servo   Latin
        with one.M.SG.ABL poor.M.SG.ABL slave.M.SG.ABL
        'with one poor slave'

Table 3.1
Latin cardinals: unus 'one', singular forms

	M	F	N
NOM	unus	una	unum
VOC	une	una	unum
ACC	unum	unam	unum
GEN	unius	unius	unius
DAT	uni	uni	uni
ABL	uno	una	uno

Table 3.2
Latin cardinals: *duo* 'two'

	M	F	N
NOM, VOC	duo	duae	duo
ACC	duo *or* duos	duas	duo
GEN	duorum	duarum	duorum
DAT, ABL	duobus	duabus	duobus

Table 3.3
Latin cardinals: *tres* 'three'

	M/F	N
NOM, VOC	tres	tria
ACC	tres	tria
GEN	trium	trium
DAT, ABL	tribus	tribus

b. duarum miserarum puellarum
two.GEN poor.F.PL.GEN girl.F.PL.GEN
'of (the) two poor girls'

Cardinals from *quatuor* 'four' to *centum* 'hundred' behave like indeclinable adjectives: they show no gender or case agreement and the NP they combine with surfaces in the case that the entire DP receives, as shown in (33).

(33) Saturni stella triginti fere annis cursum suum
Saturn.GEN planet thirty almost years.ABL course 3SG.POSS
conficit.
complete.PRES.3SG
'The planet of Saturn completes its course in about thirty years.'

Cardinals from *ducenti* 'two hundred' to *nongenti* 'nine hundred' decline like regular first- and second-declension plural adjectives, as shown in table 3.4. They agree with the lexical NP in case, as shown in (34); gender agreement is undetectable here.

(34) Et vixit Malalehel [...] octingentis triginta annis.
and live.PERF.3SG Malalehel eight.hundred.ABL thirty years.ABL
'And Malaleel lived [after he begot Jared] eight hundred and thirty years.'

The cardinal *mille* 'thousand' behaves like an indeclinable adjective when not preceded by another cardinal (35)[12] and as a regular declinable noun when combining with cardinals higher than one (36); in the latter case, the lexical NP that it combines with is marked genitive plural.

(35) a. cum mille hominibus *Mille*: singular
with thousand man.M.PL.ABL
'with a thousand men'

Table 3.4
Latin cardinals: *ducenti* 'two hundred'

	M	F	N
NOM	ducenti	ducentae	ducenta
VOC	ducenti	ducentae	ducenta
ACC	ducentos	ducentas	ducenta
GEN	ducentorum	ducentarum	ducentorum
DAT	ducentis	ducentis	ducentis
ABL	ducentis	ducentis	ducentis

b. mille   claves
   thousand key.F.PL.NOM
   'a thousand keys'

(36) a. octo milia   equorum   *Mille*: plural
        eight thousand.PL.NOM horse.PL.GEN
        'eight thousand horses'
     b. cum tribus   milibus   clavum
        with three.ABL thousand.PL.ABL key.PL.GEN
        'with three thousand keys'

To sum up, Latin cardinals up to 'thousand' behave like more or less regular adjectives, which fits in perfectly with the structure in (14). Note, however, that once genuine multiplication becomes necessary (as it does with thousands), the multiplicand becomes nominal (36). In the following sections, we will see that this generalization holds for other languages (Russian and Seri) as well.

3.4.3 Mixed Pattern: Russian

Russian provides a more complicated case, as its cardinals exhibit heterogeneous behavior on the scale between adjectives (for lower cardinals) and nouns (for higher ones). Once again we begin with *odin* 'one' and work our way up.

As discussed earlier, *odin* 'one' behaves completely like an adjective, agreeing with the NP it combines with in gender, number, and morphological case, as in (16), repeated below.

(16) a. pro   odnu   stranu   Russian
        about one.F.ACC   country.ACC
        'about one country'
     b. iz   odnix   mest
        from   one.PL.GEN place.PL.GEN
        'from one and the same area'

Beyond a minor idiosyncrasy in the nominative masculine form, *odin* 'one' declines like a regular (functional) adjective, with the same paradigm as demonstratives, possessives, and the universal quantifier (see Halle and Matushansky 2006), as shown in table 3.5.

The so-called paucal cardinals *dva* 'two', *tri* 'three', and *četyre* 'four' are considerably less adjectival than *odin* 'one': only *dva* shows a gender distinction (feminine vs. nonfeminine) and then only in the morphological nominative, where it agrees in gender with the lexical NP. However, all three paucal cardinals follow the declension pattern of plural adjectives (table 3.6), which

Table 3.5
Russian cardinals: *odin* 'one'

	[–feminine]	[+feminine]	PL
NOM	odín/odn-ó	odn-á	odn-í
ACC		odn-ú	
GEN	odn-ogó	odn-ój	odn-íx
DAT	odn-omú	odn-ój	odn-ím
LOC	odn-óm	odn-ój	odn-íx
INS	odn-ím	odn-ój (u)	odn-ími

Table 3.6
Russian paucal cardinals

	'two'	'three'	'four'
NOM	dv-á/dv-é	tr-í	četyr-e
GEN	dv-ú-x	tr-é-x	četyr-é-x
DAT	dv-ú-m	tr-é-m	četyr-é-m
LOC	dv-ú-x	tr-é-x	četyr-é-x
INS	dv-u-mjá	tr-e-mjá	četyr'-mjá

Note: Accusative case is not included in the tables, because it is syncretic with the nominative when the lexical NP is inanimate and syncretic with the genitive when the lexical NP is animate.

suggests that they also agree with the lexical NP in number (see chapter 6 for more discussion).

On the other hand, these cardinals assign paucal case to the NP they combine with (see Franks 1994; Halle 1994; and Garde 1998, among others), as shown in (9), repeated below.[13] The paucal cardinals therefore combine some properties of adjectives (declension pattern, agreement) with some properties of nouns (case assignment, since Russian nouns assign genitive case to their complements: see (10b), repeated below).

(9) a. dva   stola
       two.NOM table.PAUC
       'two tables'
   b. četyre   šagá
       four.NOM step.PAUC
       'four steps'

Cardinals from 'five' to 'hundred' are still less adjectival and still more nominal. Morphologically they behave like nouns of the third declension (with additional complications in complex cardinals like *p'atdes'át* 'fifty'; see chapter 6) and do not agree with the lexical NP. Furthermore, they assign genitive case to the lexical NP, just like regular nouns, as shown in (10), repeated below.

(10) a. šest'   šagov
         six.NOM step.PL.GEN
         'six steps'
     b. syn   Maši
         son.NOM Mary.GEN
         'Mary's son'

However, unlike with regular nouns, the genitive case assigned by cardinals disappears when the entire NP receives an oblique case: everything within an NP containing a cardinal is assigned the same oblique case (see Babby 1987 and Franks 1994). This holds for both higher and lower cardinals, as shown in (37).

(37) a. šest'ju šagami
         six.INS step.PL.INS
         'with six steps'
     b. dvumja šagami
         two.INS step.PL.INS
         'with two steps'
     c. s   synom Maši
         with son.INS Mary.GEN
         'with Mary's son'

With higher cardinals this effect disappears: although *tys'ača* 'thousand' shows mixed behavior (with some speaker variation attested as well), higher cardinals ('million', etc.) behave completely like nouns, assigning regular genitive case.[14] We will come back to this in chapter 6.

Russian is certainly not the only language to show such a lexical split in cardinals. For instance, in Polish (Rutkowski and Maliszewska 2007), the cardinals from 'one' to 'four' are adjectival, cardinals higher than 'thousand' are purely nominal, and cardinals in between behave in more or less the same fashion as the Russian cardinals from 'five' to 'hundred' (see Neidle 1988; Franks 1994; and Veselovská 2001, among others; see chapter 6 for more discussion). In Lithuanian (Andersen 2005), on the other hand, the cardinals from 'one' to 'nine' behave like adjectives. The same patterns obtain, for example, in Bantu languages (see Hyman 2003 on Basaa, where cardinals 'one' to 'seven' are adjectival, whereas the rest are nominal; Corbett 1978 on Chinyanja; and Zweig 2006 on Luvale), and in Biblical and Modern Hebrew (Halle 1994; Zweig 2006). See Corbett 1978 for further discussion.

The natural question arises how this lexical split is to be related to the semantics of cardinals. Two major strategies can be envisaged for dealing with this issue. One possibility is to maintain the type $\langle\langle e,t\rangle,\langle e,t\rangle\rangle$ and the semantics above for nominal cardinals while treating the lower adjectival cardinals as intersective predicates, as usually suggested. Given, as we will show below, that a strict division of cardinals into adjectival and nominal is not possible, we will not seriously entertain this hypothesis. The alternative, which we will adopt here, is to relegate the issue to syntax by adopting the same semantics for all cardinals and dealing with their different syntactic behavior in syntax.

As proposed in section 3.1.2, adjectival cardinals naturally merge as adjuncts to the NP that they combine with semantically, yielding a straightforward syntax–semantics mapping, as in (14), repeated below. Conversely, the same sisterhood relation is obtained for nominal cardinals by using standard complementation, as in (3), repeated below.

(14)

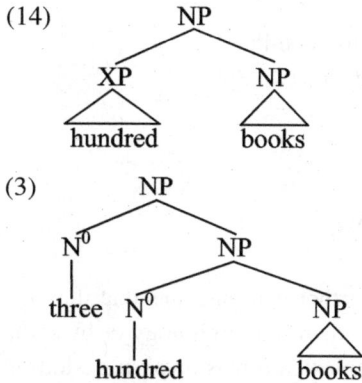

(3)

Note that the only difference between (3) and (14) is syntactic: the sisterhood and thus c-command relations obtaining between cardinals and the lexical NP they combine with is unaffected. Though additional syntactic issues arise with cardinals that show mixed behavior (see chapter 6), in essence the syntax–semantics mapping here is totally transparent if we assume that cardinals do not form a category of their own but rather are nouns, adjectives, or verbs, with some language-internal as well as language-external variation.

3.4.4   Cardinals as Verbs: Seri

Having seen adjectival and nominal cardinals, we should also expect verbal cardinals in some languages. We are aware of two languages where simplex cardinals behave as verbs: Yurok (Blevins 2004) and Seri (Moser and Marlett 1994).

In Seri as in many other languages, the first cardinal, *tazo* 'one', is morphologically an adjective when appearing with a noun, as shown in (38a); it is historically derived from the verb *azoj* 'to be one, alone' (38b).[15]

(38) a. Haxz tazo hyooho.   Seri
       dog   one saw.1SG
       'I saw one dog.'
     b. Ihpoozojo ...
        be.one.GRD.1SG
        'If I am alone ...'
     (Moser and Marlett 1994)

For cardinals higher than 'one' the verbal connection is much stronger: they are verbs underlyingly and as such are primarily used as predicates. This is illustrated for the cardinal *oocj* 'to be two', functioning as a predicate (39a) and as a modifier (39b); for the latter case a construction resembling a relative clause is used.

(39) a. Hoyacalcam   ctamcö quih coocjiha.
        those.that.are.my.siblings men   det be.two.3pl
        'I have two brothers.'
     b. Coftj   quih coocj quih hyooho.
        coralillo det be.two det saw.1sg
        'I saw two coralillos (a kind of snake).'
     (Moser and Marlett 1994)

The syntax of complex cardinals in Seri further demonstrates that they are formed by essentially syntactic means (rather than in morphology or by some unspecified extralinguistic mechanism). Complex cardinals involving addition rely on what amounts to a comitative construction, as in (40a), whereas complex cardinals involving multiplication very clearly use embedding in what looks like a possessive construction, with the multiplicand apparently turning into a noun, as in (40b).

(40) a. thanl   tapxa   cöquiih
        being.ten being.three is.with (= have)
        '(be) thirteen'
     b. ihanl   ihanl   coitom
        3.POSS.tens 3.POSS.tens be.five
        'five hundred'
     (Moser and Marlett 1994)

In other words, even though Seri cardinals are underlyingly verbal, when they appear NP-internally, they show the behavior of participles or reduced relative clauses. This is as expected, since true verbs cannot be modifiers, and follows from our hypothesis that NP-internal cardinals are semantic modifiers.

3.4.5 Vague Cardinals

Crosslinguistically, there are a number of similarities between true cardinals and quantifiers like *several/a few/many* (also known as *vague numerals*). First, a number of languages treat QPs headed by these quantifiers in the same distinctive manner as NPs containing numerals: the lexical NP is morphologically singular. This is the case, for instance, for *many* in Turkish (Gulsat Aygen, pers. comm.) and for *many* and *several* in Hungarian (Szabolcsi 1997; Kiss 2002; and Farkas and De Swart 2003, among others). Furthermore, in Russian, the lack of suppletion in the *čelovek* ('person') complements of cardinals (see section 6.6.2) is paralleled by the behavior of this noun with *neskol'ko* 'several', but not *mnogo* 'many', as shown in (41).

(41) a. pjat' čelovek/*ljudej   Russian, nonsuppletive
      five person.PL.GEN/people.GEN
      'five people'

    b. neskol'ko čelovek/*ljudej   Russian, nonsuppletive
      several   person.PL.GEN/people.GEN
      'several people'

    c. mnogo čelovek/ljudej   Russian, nonsuppletive/suppletive
      many   person.PL.GEN/people.GEN
      'many people'

Finally, *several* and *a few* behave just like cardinals in that they can only combine with cardinals that are multiplicands: *several/a few hundred books* versus **several/a few forty books*. In this, *several* and *a few* differ from quantifiers like *every*, which can combine with any cardinal: *every forty books*. (*Many* is different in that it does not take cardinal-containing NPs as complements: ??*many hundred/forty books*). The similarity in the behavior of cardinals and vague cardinals provides further evidence that cardinals do not form a separate lexical category.

At the same time, vague cardinals differ from true cardinals in that their distribution inside additive complex cardinals is more constrained. We can easily exclude **several hundred and twenty books* on the reading 'several sets each consisting of a hundred and twenty books': *hundred and twenty* is not a base. A more puzzling question is why **several hundred and twenty books* cannot be used to mean 'several hundred books and twenty books'. We rule this out on pragmatic grounds: it is pragmatically odd to combine a large approximate amount (denoted by *several hundred books*) with an exact small amount (denoted by *twenty books*). The same explanation accounts for the impossibility of *about two hundred and twenty books* on the reading 'about two hundred books and exactly twenty books'. We observe that the same pragmatic effect is found outside of the domain of coordinated cardinals: note the impossibility of *She grew about six feet and six inches* on the reading 'She grew about six feet and exactly six inches'. The reverse (combining a large exact amount and a small approximate amount) appears to be unproblematic, as shown by (42).

(42) a. Two thousand and several hundred years ago people passed along proverbs, sayings, funny stories, tearjerkers, and scraps of wisdom. (http://www.firstpaloalto.com/entry.php?which=000110)
    b. She grew six feet and about six inches.

If the restricted distribution of *several* inside complex cardinals is due to pragmatics rather than syntax, it is reasonable to conclude that vague cardinals are no different from true cardinals. Vague cardinals therefore provide additional evidence for treating complex cardinals as formed in the syntax.

3.4.6 Interim Summary

As shown above, verbal cardinals are quite rare crosslinguistically. On the other hand, nominal and adjectival cardinals are quite common crosslinguistically and may coexist within a single language. We furthermore have provided convergent evidence from morphological form and case assignment in favor of the complementation and adjunction analyses of complex cardinals. Some cardinals crosslinguistically are morphologically nominal and assign case to their sisters, while other cardinals are morphologically adjectival and agree with the lexical NP in gender and/or number. In chapters 6 and 7, we will examine crosslinguistic evidence in more detail and discuss cardinals that show mixed behavior, such as the "paucal" cardinals in Russian (section 6.3.2).

3.5 The Atomicity Requirement

In the literature on the distinction between mass and count nouns, it is commonly assumed that only atoms can be counted (Kratzer 1989; Chierchia 1998). On most standard approaches to cardinals (Link 1987; Verkuyl 1993; Carpenter 1998; and Landman 2003, among others), a cardinal combines, as in (43), with an NP denoting a set of entities and returns a subset consisting of those members of the set that have the relevant cardinality, that is, contain the relevant number of atoms. This explains why we can say *two chairs* but not *two furnitures*: cardinality is not defined for mass nouns because mass nouns have an undefined number of atoms.

(43) a. $[\![\text{two}]\!] = \lambda x \in D_e \,.\, |x| = 2$ Standard view

   b. $[\![\text{two books}]\!] = \lambda x \in D_e \,.\, [\![\text{books}]\!](x) \land |x| = 2$

As we showed in chapter 2, the semantics in (43) does not allow for the composition of complex cardinals. In order to derive complex cardinals, we need the more complex lexical entry in (44), on which a member of the set denoted by *two books* is a plural individual consisting of two atomic books.

(44) $[\![\text{two}]\!] = \lambda P \in D_{\langle e, t \rangle} \,.\, \lambda x \in D_e \,.\, \exists S \in D_{\langle e, t \rangle}\ [\Pi(S)(x) \land |S| = 2$
$\land\ \forall s \in S\ P(s)]$

In order for the semantics in (44) to work, the lexical complement of a cardinal has to be atomic—otherwise, a member of the set denoted by *two books* would be able to denote a plural individual divisible into two sets of books (and thus a plural individual of unknown cardinality), which is not what we want. Likewise, *two hundred books* has the extension in (45), where, crucially, each p_k needs to be a single book rather than a set of books.

(45) $\lambda x \in D_e \,.\, x$ is a plural individual divisible into two nonoverlapping individuals p_i such that their sum is x and each p_i is divisible into one hundred nonoverlapping individuals p_k such that their sum is p_i and each p_k is <u>a book</u>

Thus, our compositional analysis of complex cardinals requires the sister of a cardinal to be a *singular* count noun (which denotes an atomic set, i.e., a set of entities of the cardinality 1), rather than just a count noun (plural or singular).[16] This is a central part of our proposal.

We see two ways of ensuring this requirement: (1) a (null) classifier; or (2) a special constraint that would exclude plural lexical NPs as complements of cardinals. Depending on which of these two implementations of the atomicity requirement is assumed, different syntactic structures must be adopted. Here, we discuss both alternatives and argue that the second one (a constraint against a plural lexical NP with a cardinal) is preferred.

3.5.1 Implementing Atomicity, Take 1: Classifiers

One way of implementing the atomicity requirement is via classifiers. In languages without a grammaticized singular–plural distinction, atomizing conversion is usually associated with overt classifiers (see Downing 1984, 1996; Chierchia 1998; Cheng and Sybesma 1999; Kobuchi-Philip 2003; and references cited therein). The atomicity requirement imposed by cardinals on their lexical-NP sister can thus also be explained by positing a null classifier in languages like English, which exhibit plural marking inside cardinal-containing NPs (see Borer 2005 for a similar proposal). Thus if *books* in *three books* is a semantically plural predicate, then we need an operation that converts it into a semantically singular predicate. We define such an operation in (46).

(46) $\text{CL} = \lambda f \in D_{\langle e,t \rangle} . \lambda z \in D_e . z : \exists x \ [f(x) \wedge z \leq_i x \wedge \neg \exists x_0 \leq_i z]$

CL applies to a predicate f (a set of individuals, singular or plural) and returns another predicate denoting the set of all individuals z such that z is an individual part of some individual in f and z is atomic (i.e., there is no individual that is an individual part of z). When applied to a set consisting of atoms only, CL returns that set; when applied to a set containing plural individuals, CL returns atoms that form those plural individuals. In other words, CL is a classical default classifier and may be additionally specified for what kind of atoms to extract (e.g., people, flat objects, cylindrical objects, etc.) when it is overt.

In Ionin and Matushansky 2006, we discuss two possible implementations of the classifier analysis, and ultimately reject them both for languages without overt classifiers. The two possible structures correspond to (47a), where CL has the semantic type of a modifier ($\langle \langle e,t \rangle, \langle e,t \rangle \rangle$) and appears between the lexical NP and the lowest (simplex) cardinal (cf. Cheng and Sybesma 1999); and to (47b), in which CL is a predicate (type $\langle e,t \rangle$) and is a complement of the innermost simplex cardinal, with the lexical NP merged as the sister of the entire complex cardinal.[17]

(47) a.

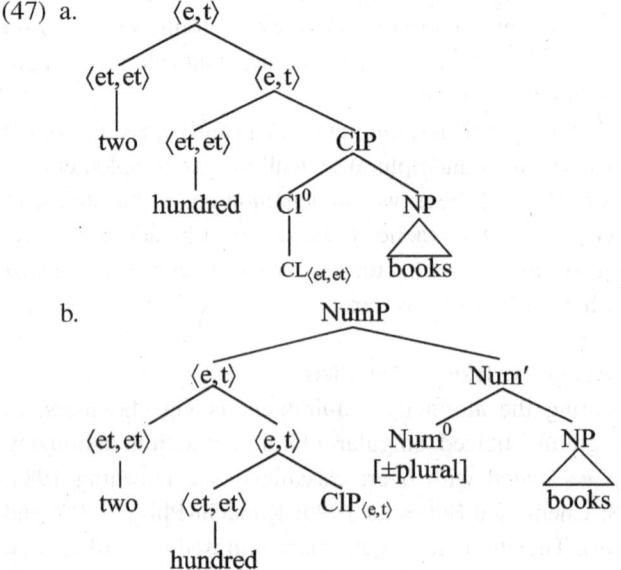

Syntactically, (47a) can map into either of the two cascading structures in (20), repeated below, modulo one additional functional projection, ClP, between the innermost cardinal and the lexical NP.

(20) a.   Complementation structure

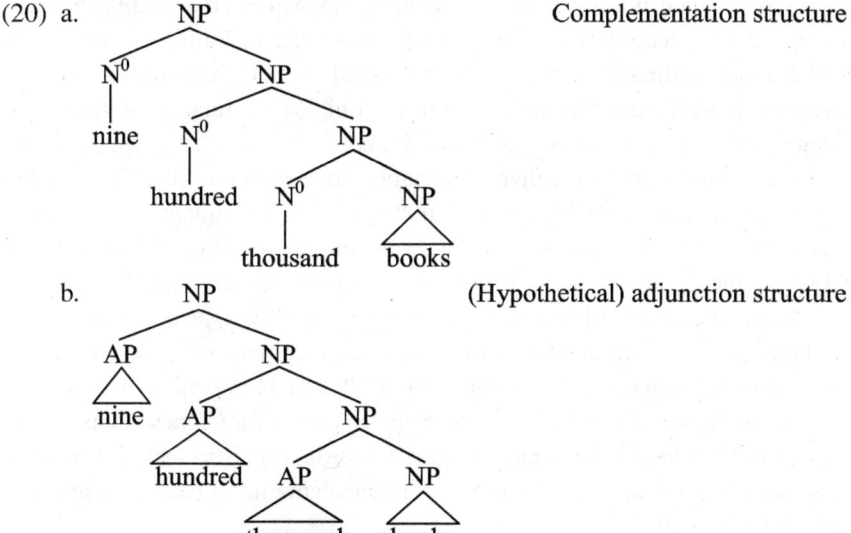

b.   (Hypothetical) adjunction structure

For languages such as Russian and Latin, this would make it possible to maintain roughly the same configurations and still assume that the plural morphology inside the lexical NP has its usual semantic value.[18] If ClP is not projected, the resulting structure is not interpretable due to the constraint introduced in the next section.

The specifier structure in (47b), on the other hand, is incompatible with the cascading structures in (20). This means that for languages where cardinals are clearly nouns (the complementation structure in (20a)) it cannot be used, as case-assignment effects cannot be handled by it (see section 3.1.1).

On the other hand, for languages where cardinals are clearly adjectives (the adjunction structure in (20b)), adjunction is theoretically possible as an option inside the ClP specifier, as in (48).

(48)

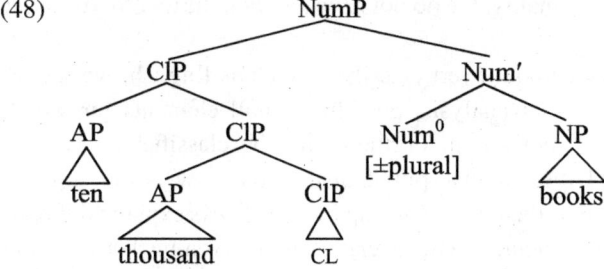

However, in practice the ClP then starts looking so similar to an NP (in the sense of having a lexical head, permitting modification by an adjective, etc.) that the question arises why ClP cannot be replaced with a lexical NP. Furthermore, we are not aware of any language in which all multiplicands are adjectival (see section 3.4.2 for evidence that even in Latin, a language in which most cardinals are adjectival, 'thousand' is not). This means that the structure in (48) cannot be adopted for all complex cardinals in a given language, and an alternative structure is required.

To sum up, both the iterative complementation structure in (47a) and the specifier structure in (47b) can potentially be used to implement the atomicity requirement. For languages with overt classifiers, such as Japanese or Chinese, the choice between these two structures must be made on a language-specific basis. Although the classifier generally appears linearly next to the cardinal, the cardinal-classifier sequence combines with the lexical NP in a variety of ways (see Downing 1984, 1996 and Muromatsu 1998, among others, for Japanese facts; Cheng and Sybesma 1999 for Chinese facts; Jenks 2011 for Thai facts; and Simpson 2005 for crosslinguistic facts). This suggests that more than one structure may be available in a given language and crosslinguistically.

As discussed in Ionin and Matushansky 2006, the semantics of the structures in (47a) and (47b) maps straightforwardly into classifier languages. We follow Chierchia (1998, 2004) in analyzing all lexical NPs in classifier languages as mass-denoting. For (47a), this means that a classifier maps the denotation of a mass-denoting NP onto a set of atoms. In Japanese and Chinese, overt classifiers are subject to additional constraints on which atoms to consider (e.g., humans, groups of humans, long cylindrical objects, etc.; see Downing 1984). Cardinals in classifier languages cannot combine with an NP directly because cardinals require atomic complements, and mass-denoting NPs are not atomic (see Chierchia 2004). Once a classifier has converted the mass-NP denotation into an atomic one, combination with a cardinal is possible. For (47b), the combination of the lexical NP and the ClP (via Predicate Modification) yields the intersection of the denotation of the lexical NP with the set of plural individuals of a particular cardinality. We do not have anything further to say about classifier languages.

Turning to languages without overt classifiers, such as English, we see no reason to apply the classifier analysis, postulating null elements for which there is no independent motivation. Furthermore, the classifier analysis is problematic in view of the singular–plural distinction: how is the singular marking in *one book* versus the plural marking in *three books* explained if both phrases have the same structure, with a classifier intervening between the cardinal and the lexical NP? Below, we propose an alternative implementation of the atomicity requirement for languages without overt classifiers, one that does not posit null classifiers.

3.5.2 Implementing Atomicity, Take 2: A Constraint against Plural Complements

If null classifiers are not an option, then how else might we implement the atomicity requirement while maintaining the cascading structure in (20)? At first blush, it would seem to be enough to propose that cardinals impose a constraint on their internal argument: they can only combine with atomic sets. This alternative is in fact proposed by Chierchia (2004) and it works straight-forwardly for cases like *three books*, as indicated by the informal denotation in (50), which uses the lexical entry for *three* in (49).

(49) $[\![\text{three}]\!] = \lambda P \in D_{\langle e,t \rangle} \cdot \lambda x \in D_e \cdot \exists S \in D_{\langle e,t \rangle} [\Pi(S)(x) \wedge |S| = 3 \wedge$
$$\forall s \in S [\underline{At(s)} \to P(s)]]$$

(50) $[\![\text{three books}]\!] \approx \lambda x \in D_e \cdot x$ is a plural individual divisible into three nonintersecting *atomic* subindividuals s such that their union is x *and each s is a member of the set of books (i.e., s is a single book).*

However, this approach fails in the case of an NP containing a complex cardinal, for example, *three hundred books*, as shown in (51): here, each sub-individual p_i must at the same time be an atom and be divisible into a hundred atoms. Since atoms are by definition indivisible, this is a contradiction.

(51) $[\![\text{three hundred books}]\!] \approx \lambda x \in D_e \cdot x$ is a plural individual divisible into three nonintersecting *atomic* subindividuals p_i such that their union is x and each p_i is divisible into one hundred nonintersecting *atomic* subindividuals p_j such that their union is p_i and each p_j is a member of the set of books (i.e., a single book).

Chierchia's solution to this problem is to relativize the definition of an atom to include such entities as a hundred books. However, the property of being a "nonplural atom" (necessary, for example, for the quantifier *each* or for plural/singular marking in contexts not involving cardinals) then has to be redefined.

A different approach, one that does not run into such problems, would be not to modify the definition of an atom but to introduce a constraint on what can be counted; this is the approach we adopt in Ionin and Matushansky 2006. Informally speaking, only individuals of the same (known) cardinality can be counted. Formally, as given in (52), this means that the sister of a cardinal can only denote a set of (singular or plural) individuals x such that there exists a number n such that for every x, $|x| = n$. In other words, a cardinal can only combine with a quantized predicate (Krifka 1990b).

(52) $[\![\text{two}]\!] = \lambda P \in D_{\langle e,t \rangle} . \lambda x \in D_e . \exists S [\Pi(S)(x) \wedge |S| = 2 \wedge \forall s \in S \, P(s)]$
$[\![\text{two}]\!](P)(x)$ is defined only if $\exists n \forall z [P(z) \rightarrow |z| = n]$

The constraint in (52) ensures that true plurals (e.g., *students*, which denotes a set of pluralities of students) cannot combine with cardinals, because different pluralities do not necessarily have the same cardinality. Instead in *two students*, the lexical NP is semantically singular, denoting a set of atomic entities, all of which by definition have the same cardinality. In Ionin and Matushansky 2006, we suggest that the constraint in (52) is probably due to pragmatics, since counting pluralities of an unknown size is pointless.

On this proposal, cardinals are able to combine with NPs that denote atomic sets, as well as with NPs headed by another cardinal, such as *hundred students* or *thousand students*: all members of the set denoted by *hundred students* have the same cardinality, since they are plural individuals that are divisible into one hundred students.

3.5.3 Supporting Evidence for the Atomicity of the Lexical NP
Supporting evidence for the proposal that the atomic NP appearing with a cardinal is semantically singular comes from the existence of languages such as Hungarian (Farkas and De Swart 2003), Welsh (53), Turkish (54), Inari Sami (Nelson and Toivonen 2000), and Finnish (55), where the lexical NP that the cardinal combines with is morphologically singular.[19]

(53) y   tair   cath   ddu   hynny   Welsh
     the.PL three.F cat.F.SG black.SG that.PL
     'those three black cats'
     (Mittendorf and Sadler 2005)

(54) Yüz   kedi   gel-di-∅.   Turkish
     hundred cat.SG come-PAST-3SG
     'A hundred cats came/arrived.'
     (Gulsat Aygen, pers. comm.)

(55) Yhdeksän omena-a   puto-si   maa-han.   Finnish
     nine.NOM apple-PART.SG fall-PAST.3SG earth-ILL
     'Nine apples fell to earth.'
     (Nelson and Toivonen 2000)

Let us see what these data mean for the two proposals discussed in the two preceding subsections. Suppose first that in a language like English, the atomicity requirement is implemented via null classifiers (section 3.5.1):

in *two students*, the lexical NP is semantically as well as morphologically plural, and a null classifier is required for atomization. Since, in a language like Finnish, the lexical NP combining with a cardinal is morphologically singular, it is reasonable to assume that it is semantically singular as well, hence no null classifier is required in order to ensure atomicity. This posits that languages with a fully grammaticized mass–count distinction and singular–plural distinction fall into two different types with regard to the existence of null classifiers: English and Russian, for example, have null classifiers, but Finnish and Welsh do not. While this is a logical possibility, it is extremely counterintuitive.

On the other hand, consider the proposal on which there are no null classifiers in languages such as English, and instead the constraint in (52) ensures that cardinals can combine only with NPs whose denotations contain entities of the same cardinality, that is, atomic entities. On this view, English and Finnish have exactly the same semantics (the cardinal takes a semantically singular NP as its sister), and the only difference is in the morphology: the semantically singular NP is marked plural in English but not in Finnish. This is the analysis that we adopt. In chapter 4, we will address the question of *why* a semantically singular NP is marked plural in languages like English and Russian: we will argue that this is an instance of agreement.

3.5.4 Duals and Trials

In light of the proposal in section 3.5.2, two issues arise: the incompatibility of higher cardinals with the morphological dual or trial in languages that have it (56) and the availability of what looks like a plural under cardinal in Finnish (57).

(56) a. dvá   gospóda/*gospódje   Slovenian
        two.DU gentleman.DU/gentleman.PL
        'two gentlemen'
     b. tríje gospódje/*gospóda
        three gentleman.PL/gentleman.DU
        'three gentlemen'
        (Derbyshire 1993, 57)

(57) a. kolme   saapasta   Finnish
        three.NOM boot.PART
        'three individual boots'
     b. kolmet   saappaat
        three.PL.NOM boot.PL.NOM
        'three groups (pairs) of boots'
        (Hurford 2003, 586)

On the proposal in section 3.5.2, duals and trials are expected to be compatible with cardinals, since they denote sets whose members all have the same cardinality (two or three atoms, respectively). Yet as (56) shows, the morphological dual or trial is not always compatible with cardinals. To explain this, we propose to rely upon the hypothesis (Dvořák and Sauerland 2005) that dual is introduced as a presupposition rather than as part of the assertion. Irrespective of whether we decompose it into two features ([singular] and [plural] or [minimal] and [group]: cf. Noyer 1992; Harbour 2008) or not, if the asserted semantic value of the dual is the same as that of the plural (i.e., the resulting predicate is not quantized), it will be incompatible with a higher cardinal.

Turning to (57), it may appear surprising that a language like Finnish, which normally requires the lexical-NP sister of a cardinal to be morphologically singular (57a), also allows what looks like a plural (57b). However, it can be immediately seen from the examples in (57) that the syntax of such plurals differs from that of normal cardinal-containing NPs: not only is the cardinal itself marked plural, the plural lexical NP that it combines with does not bear the partitive case normally assigned by the cardinal. The same pattern arises with pluralia tantum nouns, as shown in (58) (*weddings* and *faces* are pluralia tantum nouns in Finnish, having no singular form; see Hurford 2003, 586 and the references cited therein).

(58) a. kahdet   häät
        two.PL.NOM wedding.PL.NOM
        'two weddings'
     b. kahdet   sakset
        two.PL.NOM scissors.PL.NOM
        'two pairs of scissors'
     c. kahdet   kasvot
        two.PL.NOM face.PL.NOM
        'two faces'
     (Hurford 2003, 586)

We propose therefore that the plural marker on the noun here reflects the productive process of forming a pluralia tantum noun. The number marking on the cardinal then reflects agreement of the cardinal with the lexical NP: the cardinal is singular when combining with a morphological singular and plural when combining with a morphological plural.

The question now arises why singular cardinals assign partitive case whereas plural cardinals fail to do so. We see two possible explanations. The first is that the presence of plural morphology on the cardinal somehow removes its ability to assign case. The second explanation is that all cardinals assign adnumerative case to their complement, which is syncretic with the partitive case in the singular and with the nominative case in the plural. The underlying adnumerative case is different from the true partitive, as evidenced by the fact that, just like the genitive case assigned by cardinals in Russian (see chapter 6), it is overridden by the oblique case in nondirect-case positions, as shown in (59) (Karlsson 2002; Karttunen 2006).

(59) a. kolme-ssa maa-ssa
       three-INE   country-INE
       'in three countries'
   b. kahde-lla-tuhanne-lla   mark-lla
       two-ADE-thousand-ADE mark-ADE
       'for two thousand marks'
   c. kaks-i-lla   saks-i-lla
       two-PL-ADE scissors-PL-ADE
       'with two pairs of scissors'
       (Karlsson 2002, 117–118)

Oblique-case assignment overrides both the partitive case in the singular (59a–b) and nominative case in the plural (59c), supporting the hypothesis that both are instances of adnumerative.

3.5.5 Interim Summary

To sum up, we have proposed two ways of ensuring that the lexical NP inside a cardinal-containing expression is atomic. For languages with overt classifiers, atomicity is ensured by the classifier, via one of the structures in (47). For languages without overt classifiers, however, the classifier structures are not motivated, and the variant with the cardinal and classifier inside the specifier (47b) runs into additional problems. Instead, we argue that languages without overt classifiers use the cascading structures in (20), and that the pragmatic constraint in (52) ensures that the lexical NP denotes a set of atoms. The question that we have not yet addressed is why, if the lexical NP is semantically singular, it appears with plural marking in languages like English and Russian. We turn to this in the next chapter.

3.6 Summary

This chapter represents a refinement of Ionin and Matushansky 2006, showing that the semantics of cardinals that we propose maps straightforwardly to two types of syntactic structures: cardinals as lexical heads and cardinals as phrasal adjuncts. The former structure is available to nominal cardinals and the latter to adjectival cardinals. We have shown that both nominal and adjectival cardinals are attested crosslinguistically; we have also shown that the much rarer case of cardinals as verbs is also compatible with our proposal. To the best of our knowledge ours is the only proposal attempting to derive the syntax of complex cardinals from independently available syntactic structures, lexical categories, and processes.

We have also proposed two ways of implementing the requirement that the lexical sister of a cardinal is atomic. In chapter 4, we will come back to the question of why, if the NP sister of a cardinal denotes an atomic set, it often occurs with plural marking.

4   Concord and Agreement with Cardinals

As discussed in chapters 2 and 3, we analyze cardinals as requiring a semantically singular NP as their complement. Yet in languages like English, the lexical NP that appears with a cardinal is morphologically plural (1a), and cardinal-containing NPs generally require plural marking both with the predicate (1b) and with the determiner (1c).

(1) a. two books / *two book
    b. Two books are / ??/*is lying on the table.
    c. I like these / ??/*this five books.

At the same time, as will be discussed below, in many other languages (e.g., Turkish, Hungarian, Finnish, Western Armenian), the lexical NP that combines with the cardinal is morphologically singular, and both the predicate and the determiner can or even must occur in singular form. In this chapter, we provide evidence for analyzing the plural marking on the lexical NP (1a) and on the predicate and the determiner (1b–c) as cases of agreement.

The goals of this chapter are threefold. The first goal is to provide evidence that the lexical NP appearing with a cardinal is semantically singular crosslinguistically, despite the appearance of plural marking in languages like English. The second goal is to demonstrate that plural marking on the lexical NP, as well as on the predicate and the determiner, is a case of semantic agreement reflecting the semantic plurality of the cardinal-containing NP: even though *books* in *three books* is analyzed as semantically singular, the entire NP *three books* is semantically plural, since it denotes a plural individual divisible into three nonintersecting parts. The third goal of this chapter is to sketch out a possible approach to conditioned semantic number agreement. Developing a full formal theory of semantic number agreement is well beyond the scope of this work, so our account necessarily leaves some open questions for future research.

This chapter is organized as follows. First, in section 4.1, we discuss languages like English, in which the lexical-NP sister of a cardinal is morphologically plural, and provide evidence against treating it as semantically plural. In section 4.2, we move on to languages in which the lexical-NP sister of a cardinal is morphologically singular and provide evidence against the hypothesis that this lexical NP is number-neutral (i.e., it cannot be the case that cardinals combine with a set consisting of both singular and plural entities). Evidence for our argument comes from the fact that in several languages where cardinals obligatorily combine with singular (morphologically unmarked) nouns, there are no other environments where such nouns fail to have the singular denotation. In section 4.3, we provide evidence that both the plural marking on the lexical NP within cardinal-containing NPs and verbal agreement with such NPs can be conditioned by the same factors; we go through a number of crosslinguistic examples to show that plural marking on the lexical NP can be conditioned by such factors as individuation, animacy, definiteness/ specificity, and grammatical gender. The data all point to the conclusion that plural marking inside cardinal-containing NPs does not reflect true semantic plurality of the lexical NP but instead corresponds to agreement. Finally, in section 4.4, we argue in favor of analyzing plural marking with cardinals as conditioned semantic agreement, analogous to what happens with group nouns. However, a formal theory of semantic agreement lies beyond the scope of this work.

4.1 Evidence That the Plural-NP Sister of a Cardinal Is Not Semantically Plural

We start with languages like English, Russian, and Dutch, in which the lexical NP that appears with a cardinal higher than 'one' is (in most cases) morphologically plural. We provide evidence, first from morphologically derived cardinal-containing structures and then from measure NPs, that the lexical-NP sister of a cardinal is in fact *not* semantically plural.

4.1.1 Word-Internal Lack of Number Marking
In many languages, including English, Dutch, and Russian, morphological derivation from cardinal-containing NPs (including compounding) requires the morphologically unmarked form of the lexical noun; in English this constraint extends to measure phrases in attributive APs. This is illustrated in (2) through (6).[1]

(2) *Attributive adjectives*
    a. een drie   maanden lange voettocht   Dutch
       a   three month.PL long   trek
       'a three-month-long trek'
    b. a three-year/*years-old girl   English

(3) *Compounding*
    a. een drie-duim-(*en)-s-plank   Dutch
       a   three-inch-PL-LKR-board
       'a three-inch board'
    b. five-inch/*inches nails   English

Morphological derivation

(4) a. a four-legged/*legsed animal   English
    b. a five-person/*people vehicle

(5) a. twee-jar-(*en)-ig   ⇐ *twee jaren*   Dutch
      two-year-PL-ADJ   'two years'
      'of two years'
    b. drie-arm-(*en)-ig   ⇐ *drie armen* 'three arms'
      three-arm-PL-ADJ
      'three-armed'
    c. de *één*/zes-daag-s-e   oorlog ⇐ *één dag* 'one day'/
      the one/six-day-ADJ-AGR war   *zes dagen* 'six days'
      'the one-/six-day war'

(6) a. trëx-sekund-n-yj   ⇐ *tri sekundy*   Russian
      three-second-ADJ-AGR   'three seconds'
      'three-second'
    b. trëx-čas-ov-oj   ⇐ *tri časa* 'three hours'
      three.GEN-hour-ADJ-AGR
      'three-hour'
    c. soroka-nog-ij   ⇐ *sorok nog* 'forty legs'
      forty.GEN-leg-AGR
      'forty-legged'

If cardinals combine with semantic plurals, the question arises why the plural marker fails to be realized in these environments. Regarding morphological derivation, the absence of plural morphology appears to be a fairly general phenomenon, which has been attributed to level ordering (cf. Kiparsky 1982), but in compounding the empirical picture is more complicated. As noted by Selkirk (1982; see also Sneed 2002), while, in general, regular plural morphology is strongly dispreferred on the first member of the compound (7), exceptions can be easily found (8).

(7) a. trouser/*trousers leg
    b. rat/*rats catcher
    c. mouse/mice catcher

(8) Human Services Administration, parks commissioner, admissions department ...

To account for these exceptions, Alegre and Gordon (1996) propose that the first member of a compound can be either a word or a phrase (see also Berent and Pinker 2007). If it is only the latter that can contain plural morphology, the contrast between modified and unmodified first members of a compound, as in (9), can also be explained.

(9) a. new-books shelf, American-cars exposition
    b. *books shelf, *cars exposition

The cardinal-based compounds in (3) do not seem to be word-based, since their first member consists of a cardinal and a noun, making it most likely to be a maximal projection. Still, irregular plurals are likewise impossible there, as shown in (10).[2]

(10) a. four-footed/*feeted
     b. a three-foot/*feet stick
     c. a three-foot/*feet-long stick

Since the first member of cardinal-based compounds, as in (3) and in (10b–c), is an NP rather than a single word, we do not expect agreement processes to fail inside cardinal-based compounds, even if these processes are hypothesized to fail word-internally. The lack of plural marking in cardinal-based compounds is therefore an indication of the lack of semantic plurality. We therefore conclude that NP-internal plural marking with cardinals results from agreement.

4.1.2 Plural Marking with Measure NPs

Another argument for treating the lexical NP that combines with a cardinal as semantically singular comes from the behavior of measure nouns.

Take a language like English, where the lexical NP appearing with a cardinal is morphologically plural. For morphologically plural nonmeasure NPs such as *chairs*, the meaning would, on the surface, appear to be the same with a cardinal (*I bought three chairs*) and without a cardinal (*I bought chairs*): in both cases, we are talking about a group of chairs. However, with measure

NPs, the two meanings are quite distinct. As noted by Ruys (2017), the plural forms of measure nouns such as *liter* or *pound* only have the "plural of abundance" interpretation. While Ruys notes this for Dutch (11a), the same holds for English (11b).

(11) a. Jan dronk liters   wine.   Dutch
         Jan drank liter.PL wine
         'Jan drank an excessive number of liters of wine.'
         (Ruys 2017)
     b. The kids ate pounds of cake during the birthday party!
     c. The kids ate two pounds of cake.

In (11b), *pounds* can only mean 'lots of pounds': this sentence would not be felicitous if the kids ate only two or three pounds of cake. In contrast, when measure NPs combine with cardinals higher than 'one', plural marking does not give rise to a "plural of abundance" interpretation, as evidenced by the fact that measure NPs can combine with cardinals as low as 'two': (11c) clearly lacks the "plural of abundance" interpretation.

The fact that (11b) has only a "plural of abundance" interpretation indicates that measure nouns do not have a normal plural. We propose that the lack of a regular plural for measure nouns results from the inability of measure phrases to undergo semantic pluralization, that is, to combine with Num. Thus, the presence of plural marking on the measure NP next to a cardinal (as in (11c)) cannot be the result of semantic pluralization. This is consistent with our proposal, on which plural marking on any lexical NP (measure or nonmeasure) next to a cardinal is not due to semantic pluralization but is instead a result of agreement. In section 4.3, we will provide further evidence that plural marking on the lexical NP appearing with a cardinal is a result of agreement. First, however, we consider languages where the lexical NP appearing with a cardinal does *not* bear plural marking.

4.2 Singular Marking with Cardinals Is Not General Number

As discussed in chapter 3, the standard view in the literature is that cardinals combine with a semantically plural NP. Link (1987), Verkuyl (1993), and Landman (2003), among others, augment this assumption with the hypothesis that cardinals are intersective modifiers (rather than determiners). A problem for this view is the existence of languages where cardinals obligatorily combine with morphologically singular lexical NPs. Such languages include Finnish, Welsh, Turkish, Hungarian, and Tagalog, as (12) illustrates.

(12) a. Yhdeksän omena-a   puto-si   maa-han.   Finnish
        nine.NOM apple-SG.PART fall-3SG.PAST earth-ILL
        'Nine apples fell to earth.'
        (Nelson and Toivonen 2000)
     b. y   tair   cath   ddu   hynny   Welsh
        the.PL three.F cat.F.SG black.SG that.PL
        'those three black cats'
        (Mittendorf and Sadler 2005)
     c. iki/beş/on adam   Turkish
        two/five/ten man.SG
        'two/five/ten men'
        (Görgülü 2012)
     d. két/néhány konyv   Hungarian
        two/some book.SG
        'two/some books'
        (Kenesei, Vágó, and Fenyvesi 1998, 255)
     e. tatlumpúʔ-'ng aklát   Tagalog
        thirty-CONN   book
        'thirty books'
        (Potet 1992)

    A potential solution for this problem, proposed by Bale, Gagnon, and Khan-
jian (2011), is to relate singular marking on NPs that combine with cardinals
to a phenomenon known as general number. In this section, we discuss evidence in favor of and against this proposal and conclude that the morphologically singular NP that combines with a cardinal should be analyzed as
semantically singular, that is, as *not* corresponding to general number. This is
consistent with our proposal in chapters 2 and 3 that the lexical-NP sister of
a cardinal denotes an atomic set.

4.2.1 Languages with General Number
The term *general number* (Corbett 2000; see also Schroeder 1999; Wiese
2003; and Acquaviva 2005, which use the term *transnumeral*) refers to the
possibility of using an NP without reference to number, enabling it to denote
both singular and plural entities. For example, in Bayso, a Cushitic language
of southern Ethiopia, a morphologically unmarked noun bears no number
information (13a), with all other numbers (singular, paucal, and plural)
expressed by affixes (13b–d).

(13) a. Lúban foofe.   Bayso
        lion   watch.1SG.PAST
        'I lion-watched.'

b. Lubán-titi foofe.
   lion-SG   watch.1SG.PAST
   'I watched a lion.'
c. Luban-jaa foofe.
   lion-PAUC watch.1SG.PAST
   'I watched a few lions.'
d. Luban-jool foofe.
   lion-PL   watch.1SG.PAST
   'I watched (a lot of) lions.'
(Corbett 2000, 11)

While in Bayso (as well as in some dialects of Fula and Arabic) general number is expressed by a separate morphological form (which, judging by the examples given by Corbett, seems to always be the bare stem), in other languages general number is morphologically identical to the unmarked singular. This is the case in Turkish (Schroeder 1999), where the unmarked form can be used for singulars and plurals ((14a), (15a)) while the plural form can only denote plural entities ((14b), (15b)), as well as in Western Armenian (Sigler 1992, 1996; Donabédian 1993; Bale, Gagnon, and Khanjian 2011).

(14) a. Çanta-da kitap var.   Turkish
        bag-LOC book exist
        'There is a book/there are books in the bag.'
     b. Çanta-da kitap-lar var.
        bag-LOC book.PL exist
        'There are books in the bag.'
     (Görgülü 2010)

(15) a. Ahmet kitap oku-du.
        Ahmet book read-PAST
        'Ahmet read a book/books.'
     b. Ahmet kitap-lar oku-du.
        Ahmet book-PL read.PAST
        'Ahmet read books.'
     (Görgülü 2010)

The situation is broadly the same in Hungarian, where the morphologically unmarked form can be number-neutral in two environments besides cardinal-containing NPs. Farkas and De Swart (2010) note that while Hungarian cardinals cannot combine with a plural lexical NP (16a), they share this property with other quantifiers and vague numerals (16b); moreover, incorporated singular NPs are likewise number-neutral (17).

(16) a. három gyerek-(*ek)   Hungarian
         three   child-PL
         'three children'
      b. sok   gyerek-(*ek)
         many child-PL
         'many children'
      (Farkas and De Swart 2010)

(17) Mari verset   olvasott   ma   délután.
      Mary poem.ACC read.PAST today afternoon
      'Mary read a poem/poems/poetry this afternoon.'
      (Farkas and De Swart 2003, 101)

Bale, Gagnon, and Khanjian (2011) appeal to general number in their analysis of Turkish and Western Armenian cardinal-containing NPs. While in Turkish the cardinal obligatorily combines with a singular lexical NP (12c), in Western Armenian both the singular and the plural lexical NP can be used (see section 4.3.3 for more discussion of Western Armenian). Since both Turkish and Western Armenian can be independently shown to use the singular form in contexts where number is unspecified, Bale, Gagnon, and Khanjian propose that the unmarked lexical NP in cardinal-containing NPs is an instance of general number (rather than singular), comprising in its denotation both singular and plural entities; an intersective semantics can then be used for cardinals.

4.2.2 Languages without General Number

Bale, Gagnon, and Khanjian's (2011) analysis cannot, however, be applied to languages that lack general number. As discussed in the previous section, languages with general number (Hungarian, Turkish, Western Armenian ...) allow bare singular NPs to have plural as well as singular interpretations in a variety of contexts, not only with cardinals. By this diagnostic, Finnish and Welsh are languages that demonstrably *lack* general number. In these languages, singular NPs in argument positions are not number-neutral: (18a) and (19a) are only felicitous in a situation where one book was read or one dragon was seen, respectively. In order to express the event of the speaker reading multiple books or seeing multiple dragons, it is necessary to use the plural, as in (18b). Furthermore, predicate NPs in these languages must agree for number, as in (18c) and (19b).

(18) a. Luin   kirjan/kirjaa.   Finnish
         read.1SG book.ACC/PART
         'I read a book/the book.' (≠ 'I read (the) books.')

b. Luin   kirjat/kirjoja.
read.1SG book.PL.ACC/PART
'I read the books/books.' (≠ 'I read a/the book.')

c. Olemme   suomalaisia.
be.1PL.PRES Finnish.N.PL.PART
'We're Finnish.'

(19) a. Gwelodd   Rhiannon ddraig.   Welsh
see.3SG.PAST Rhiannon dragon.SG
'Rhiannon saw a dragon/*dragons.'

b. Roeddwn i   ac Emyr yn   ysgrifenwyr rhagorol.
was.1SG 1SG and Emyr PRED writer.PL   excellent
'Emyr and I were excellent writers.'
(Sadler 2003)

This shows that in Finnish and Welsh general number is not an independently attested option for a number-unmarked NP.[3] Yet, as was shown in (12a–b) above, these languages require the lexical NP that appears with a cardinal to be morphologically singular.

It can be argued, however, that it is enough to demonstrate the existence of general number in environments other than cardinal-containing NPs (as shown for Hungarian in Farkas and De Swart 2010). Thus in Welsh the vague cardinals 'many', 'few', and so on pattern like cardinals (Sadler 2000; Hurford 2003; Mittendorf and Sadler 2005) in that two options are in principle possible (King 2003, 125–130): some vague cardinals combine with singular NPs, as in (21a) (compare to (20a)), while others (the majority) combine with plural NPs and require the preposition *o* 'of', as in (21b) (compare to (20b)).[4]

(20) a. naw ddyn   Welsh
nine man.SG
'nine man'

b. naw o ddynion
nine of man.PL
'nine men'
(Hurford 2003)

(21) a. Sawl   llythyren sy   yn yr wyddor Gymraeg?
how.many letter.SG be.3SG in the alphabet Welsh
'How many letters are there in the Welsh alphabet?'

b. Mae   llawer o bobol   yn dal heb   gofrestru.
be.3SG many of people.PL in still without register.VN
'Many people still haven't registered.'
(King 2003, 126, 128)

The distribution of the two variants casts some doubts on the hypothesis that the singular or unmarked form corresponds to general number, but we will postpone the analysis of Welsh cardinals until chapter 7 and turn to the simpler case of Finnish. In Finnish, the quantifiers 'many' and 'few' obligatorily combine with plural NPs, as shown in (22a–b), irrespective of their ability to assign case, while cardinals obligatorily combine with singular NPs (22c).

(22) a. Ville kutsui   paljon mukavia   kavereita   Finnish
        Ville invite.3SG.PAST a.lot nice.PL.PART friend.PL.PART
        juhliin.
        party.ILL
        'Ville invited a lot of nice friends to the party.'
     b. Monet   isot   kaupungit   ovat   pääkaupunkeja.
        many.PL.NOM big.PL.NOM cities.PL.NOM be.3PL capital.PL
        'Many big cities are capitals.'
     c. Kolme poikaa   odottaa alakerrassa.
        three boy.PART wait.3SG downstairs.INE
        'Three boys are waiting downstairs.'
     (Dal Pozzo 2007, 115–116)

Descriptively, vague numerals in Finnish fall into two categories. *Paljon* 'a lot', *vähän* 'little/few', *hiukan* 'a bit', and *jonkin verran* 'some/somewhat' assign partitive case (22a), while *moni/monet* 'many', *harva/harvat* 'few', *usea/useat* 'many/several', and *muutama/muutamat* 'some/a few' pattern with the universal quantifiers *jokainen* 'each', *molemmat* 'both', and *kumpikin* 'each' and the cardinal *yksi* 'one' in that they agree with their lexical NP in both case and number (22b). In other words, vague numerals in Finnish can be either "nominal" (nonagreeing and assigning partitive case to their sister) or "adjectival," precisely like cardinals (cf. Hurford 2003), but both types require a plural lexical NP.[5]

The lack of general number in Finnish in argument and predicate positions (18) as well as with vague numerals (22a–b) strongly suggests that the singular marking on lexical NPs combining with cardinals higher than 'one' reflects their semantics and cannot therefore be analyzed in Finnish as a special case of general number. We propose to analyze the Welsh and Hungarian vague numerals *many*, *few*, and so on on a par with cardinals, as nonintersective modifiers combining with semantically singular NPs and returning a plural predicate. In Finnish, on the other hand, vague numerals can have the more usually assumed intersective interpretation that requires them to combine therefore with semantically plural NPs. On the assumption that vague numerals are scalar (Hackl 2000) and their existential force comes from elsewhere, the two options can be represented as follows. The lexical entry in (23a) represents an instance of *many* that takes a degree argument *d* and combines with plural lexical NPs (plurality indicated by the use of capital letters) and returns a set of plural individuals such that their cardinality is equal to this degree *d*. The lexical entry in (23b) returns exactly the same set, but combines with a singular lexical NP.

(23) a. $[[\text{many}]] = \lambda d \in D_{\text{Card}} \,.\, F \in D_{\langle e,t \rangle} \,.\, \lambda X \in D_e \,.$ Plural
$[F(X) \wedge |X| = d]$

   b. $[[\text{many}]] = \lambda d \in D_{\text{Card}} \,.\, \lambda f \in D_{\langle e,t \rangle} \,.\, \lambda X \in D_e \,.$ Singular
$\exists S \,[\Pi(S)(X) \wedge |S| = d \wedge \forall s \in S\, f(s)]$

To sum up, we have argued that the singular (unmarked) lexical NPs in cardinal-containing NPs cannot be uniformly treated as an instance of general number, whose denotation includes both plural and singular entities. Our argumentation against the general-number hypothesis is based on the fact that at least some of the languages where cardinals combine with the morphologically unmarked singular form, in particular Finnish, do not allow this form to be used to make reference to both singular and plural entities elsewhere. On our analysis, the lexical NP that combines with a cardinal is always semantically singular, regardless of whether it occurs with plural marking, as in English, in a singular form in a language that has general number, as in Turkish, or in a singular form in a language that does not have general number, as in Finnish. The Finnish data provide strong support for this analysis, since they show that the ability of a cardinal to combine with a morphologically singular NP does not depend on the availability of general number in a language.

Our proposal goes against the analysis proposed by Bale, Gagnon, and Khanjian (2011) on the basis of Turkish and Western Armenian; in section 4.3.3, we come back to optional plural marking in Western Armenian and show that it correlates with specificity (Sigler 1992; Donabédian 1993; Sigler 1996), providing further evidence in favor of our analysis of plural marking as agreement.

4.3 Plural Marking with Cardinals Is Agreement

Our proposal that the lexical NP that combines with a cardinal is semantically singular seems to erroneously rule out more familiar languages, including English, Russian, and Latin, where cardinals higher than 'one' must combine with morphologically plural NPs, as (24) shows.

(24) a. five books/*book   English
     b. šest'ju zvezdami/*zvezdoj   Russian
        six.INS star.PL.INS/SG.INS
        'with six stars'
     c. du-abus   miser-is/*ae   puell-is/*ae   Latin
        two-F.DAT poor-F.PL.DAT/SG.DAT girl-PL.DAT/SG.DAT
        'for/to (the) two poor girls'

With cardinals higher than 'one', the entire cardinal-containing NP is semantically plural because it contains plural individuals in its extension. Following Ionin and Matushansky 2006, we suggest that the plural marking on the lexical NP in examples like (24) results from semantic agreement with the interpretable number of the entire cardinal-containing NP, and is therefore not interpreted. While it is not our goal to propose a formal theory of number agreement, in this section we will provide evidence for this view from languages where plural agreement on predicates and plural marking inside numeral NPs are conditioned by the same factors. In the Chadic language Miya (Schuh 1989, 1998), both plural marking in cardinal-containing NPs and plural concord and agreement are conditioned by animacy, while in Dutch, numeral NPs headed by a measure noun fail to trigger plural agreement on the verb and do not show plural marking on the measure noun itself. The fact that cardinal-containing NPs in both languages behave differently from regular plurals provides further evidence that cardinals do not combine with plurals. The treatment of cardinals as nonintersective modifiers combining with an atomic predicate to return a set of plural individuals will thus be shown to constitute the optimal option. After presenting the facts of Miya and Dutch, we will discuss the case of Western Armenian and show that plural marking in this language is conditioned by specificity, which indicates that it is another case of agreement. Finally, we will consider heterogeneous number marking in several languages (i.e., cases where the lexical NP is singular with some cardinals and plural with others) and argue that this is also best analyzed as agreement.

4.3.1 Miya: Plural Agreement Conditioned by Animacy

In the Chadic language Miya, plural formation is fully productive not only for count nouns but also for mass nouns and even plurals (Schuh 1998, 199). While Schuh's grammar does not say anything special about the realization of number, it provides plural glosses without further comment, so it would seem that bare indefinite NPs in argument positions can be interpreted as plural, as in (25a) and (26a), even though plural marking is available in this language for indefinite NPs, as shown in (25b) and (26b). This suggests that in Miya, bare indefinite NPs with no plural marking are number-neutral. (All examples in this section are taken from Schuh 1998; the glosses have been adjusted where possible to provide the maximal amount of grammatical information, crossverified with Schuh and Ciroma Tilde Miya 2010.)

(25) a. Bàzaniy   d-aa   dəhə́nà *tsə́tsalìy* j-áabə́dà-zà.   Miya
young.woman *Dá̃*-IMPFV tie   caurie.SG it's-front-her
'A young woman would wear (a string of) cauries on her front.'

b. Də̀ ɗúwrà tál-ay   ɗoo *ndùwul-álàw.*
SBJV pour   beer-TOT in   pot-PL
'They pour the beer into pots.'
(Schuh 1998, 283, 306)

(26) a. Náka sə́n n-aaMangila tə́ d-àa-táa   ndəm(a) ée dòona *tlíwìy*
that man of-Mangila he *Dá̃*-PROG stroll   to seek animal.SG
n-aatsákən.
of-bush
'The man of Mangila was going about looking for wild animals.'

b. Kwáa ts(a) a   yùw ká, suw pà-pə́ràka   *cúw-ay.*
when appear PSM dawn PRM TOT REDUP-slaughter goat.PL-TOT
'When the dawn comes, then they slaughter goats.'
(Schuh 1998, 166, 390)

Schuh (1998) does not discuss this lack of number marking, but in most if not all examples in the grammar, morphologically unmarked NPs translated as plurals are nonspecific, like the examples in (27).[6]

(27) a. ee doona *tliwiy n-aatsakən*
'to hunt *wild animals* [animal.M of.M-bush]'[7]
b. də təkəna-ya *gaangan*
'they beat *drums* [drum] for him'
(Schuh 1998, 231)

While it is perhaps unwise to draw far-reaching conclusions from negative evidence, it seems uncontroversial that even if Miya has general number with kind-denoting NPs, NP plurality in general does not interact with animacy. In contrast, plural marking that arises as a result of agreement *is* sensitive to whether the noun is animate. As examples (28)–(29) show, while with plural animate nouns the demonstrative takes the plural form and therefore bears no indication of gender (i.e., nouns of both genders appear with the same plural forms), with plural inanimate nouns the demonstrative does not agree in number (though it does agree in gender). Attributive APs, relative pronouns, linkers, and so on behave the same,[8] showing that [−animate] nouns do not trigger number agreement.

(28) a. níykin dzáfə   Animate: number agreement
          this.PL man.PL
          'these men'
      b. níykin təmakwìy
          this.PL sheep.PL
          'these sheep'
          (Schuh 1998, 197)

(29) a. nákən   víyayúw-awàw   Inanimate: gender agreement only
          this.M.SG fireplace.M.PL
          'these fireplaces'
      b. tákən   tlərkáy-ayàw
          this.F.SG calabash.F.PL
          'these calabashes'
          (Schuh 1998, 197)

Strikingly, animacy also affects number marking in NPs containing cardinals higher than 'one', as well as vague numerals, like *càsə* 'much/many' and *ndàßə* 'a little, a few': while [+animate] lexical NPs are obligatorily marked plural in this environment (30), [−animate] ones can stay singular (31).

(30) a. təvam   tsə́r   Animate: obligatory number agreement
          woman.PL two
          'two women'
      b. *'ám   tsər
          woman.SG two
      c. 'ám   wutə̀
          woman.SG one
          'one woman'
          (Schuh 1998, 198)

(31) a. zə̀kij-áyàw vaatlə   Inanimate: optional number agreement
          stone-PL   five
          'five stones'
      b. zə̀kij   vaatlə
          stone.SG five
          'five stones'
          (Schuh 1998, 198)

Once again the grammar does not mention whether the optionality in (31) has any semantic effect, beyond observing that the unmarked form is preferred in inanimate cardinal-containing NPs (Schuh 1998, 258). However, it should be noted that both options seem to be available in definite NPs as well, as in (32) and (33), suggesting that it is not referentiality that is at stake.[9]

(32) a. táka   anguw   màatsəree-ya ká
         that.F.SG quarter.SG seven-F.SG   PRM
         'those seven quarters'
     b. takən   də́m-amaw gyaruw-ya kidi
         this.F.SG tree-PL   big.PL-F.SG three
         'these three big trees'
         (Schuh 1998, 259, 277)

(33) kàb-abáw tuwsə àkwaatiy dərɓitim
     gown-PL his   chest.SG twenty
     'the twenty chests of his gowns'
     (Schuh 1998, 259)

It seems clear from these facts that Miya cardinals do not combine with plural NPs: had this been the case, animacy would not have affected the availability of plural marking on lexical NPs embedded under a cardinal. If the [plural] feature is located on Num^0, we must conclude that NumP is projected higher than the cardinal.

Further analysis is precluded by the paucity of the data. The three examples above appear to exhaust the set of definite inanimate cardinal-containing NPs in the grammar, and none of them is free from problems. Both examples in (32) contain additional agreement morphology that Schuh notes but leaves unexplained: the cardinal in (32a) agrees with the lexical NP (an option apparently available only in anaphoric/familiar definites, with animates optionally triggering agreement in number and inanimates optionally triggering agreement in gender), while the adjective in (32b) bears a feminine marker on top of the plural form. Despite the translation it is unclear whether (33) is obligatorily definite.

Finally, none of these examples is provided in context, making it impossible to figure out the semantic or pragmatic conditions on the optionality of plural marking. The question therefore arises how to describe the syntactic effect of a cardinal: does it force plural marking on all animate NPs (regardless of their referentiality) or, if cardinal-containing NPs cannot function as semantically incorporated predicates, as in (27), does it sometimes block plural marking for inanimate NPs? In the absence of further data we conclude this subsection by reiterating that the distribution of plural marking in cardinal-containing NPs is conditioned by animacy, like plural agreement and unlike number neutrality in bare NPs. The hypothesis that cardinals combine with semantically singular NPs that become plural as a result of language-specific agreement processes is more compatible with these data than the more traditional view where cardinals combine with plural NPs, even though the question remains what semantic or pragmatic factor regulates plural marking in inanimate cardinal-containing NPs.

4.3.2 Dutch: Plural Agreement Conditioned by Individuation[10]

As noted by Klooster (1972) and Doetjes (1997), in Dutch many nouns referring to measuring units obligatorily occur in the singular in cardinal-containing NPs (34a), while nouns outside this lexical-semantic class require plural marking with cardinals higher than 'one' (34b).[11]

(34) a. drie/vijf/dertig kilo/*kilo's   Dutch
        three/five/thirty kilo.SG/PL
        'three/five/thirty kilos'
     b. drie/vijf/dertig dozen/*doos
        three/five/thirty box.PL/SG
        'three/five/thirty boxes'
    (Matushansky and Ruys 2014)

At first blush, this seems to be a lexical property of the measure noun: as noted by Klooster (1972, 9–10), while some measure nouns (the ones in (35a)) occur in their singular form with cardinals, others (the ones in (35b)) occur in their plural form.[12]

(35) a. *dollar* 'dollar'   "Integer-independent"
        gulden 'guilder'   (singular with cardinals)
        cent 'cent'
        ton 'ton'
        ons (metric 'ounce': one hundred grams)
        pond (metric 'pound': five hundred grams)
        kilogram 'kilo(gram)'
        gram 'gram'
        (kilo)meter '(kilo)meter', *decimeter* 'decimeter', etc.
        mijl 'mile'
        voet 'foot'
        vadem 'fathom'
        mud 'hectolitre'

(centi)liter '(centi)litre', etc.
jaar 'year'
uur 'hour'
kwartier 'quarter of an hour'
man 'man'
keer 'time'
maal 'time'
decibel 'decibel'
volt 'volt', *watt* 'watt', *farad* 'farad', *ohm* 'ohm', etc.
bunder 'hectare' (2.471 acres)

b. *dubbeltje* 'ten-cent piece'   "Integer-dependent"
   stuiver 'five-cent piece'   (plural with cardinals
   kwartje 'quarter' (of a guilder)   higher than 'one')
   seconde 'second'
   minuut 'minute'
   dag 'day'
   week 'week'
   maand 'month'
   decade 'decade'
   eeuw 'century'
   millennium 'millennium'
   vrouw 'woman'
   graad 'degree'
   bit 'bit' (information theory)[13]
   schepel 'bushel', 'deciliter'
   lichtjaar ('light-year')

At the same time, it is easy to demonstrate that the nouns in (35a) do not lack a morphological plural form. As the examples in (36) and (37) show, at least some measure nouns can appear in the plural in the absence of a cardinal while retaining their semantics as measure nouns.

(36) vele   kilo's/*kilo suiker
     many kilo.PL/SG   sugar
     'many kilos of sugar'
     (Doetjes 1997, 190)

(37) a. kilo's   en   kilo's   zand
       kilo.PL and kilo.PL sand
       'kilos and kilos of sand'

   b. Die kilo's   die ik ben aangekomen zitten voornamelijk op mijn
      the kilo.PL that I   am gained   sit.PL mostly   on my
      heupen.
      hips
      'The kilos that I have gained are mostly on my hips.'
   c. Kilo's zijn zwaarder dan ponden.
      kilo.PL are heavier   than pound.PL
      'Kilos are heavier than pounds.'
  (Matushansky and Ruys 2014)

Thus, we cannot simply say that *kilo* (and other nouns in (35a)) have only a singular form available to them. Rather, they have both singular and plural forms, but only the former is allowed when the measure noun occurs with a cardinal.

In a cardinal-containing NP, plural marking on a measure noun of the (35a) type, if at all possible, obligatorily gives rise to a special individuating or "packaging" interpretation, as shown in (38) (Doetjes 1997; Rothstein 2011a).

(38) Ik heb twintig liters   frisdrank bezorgd voor het feestje.
    I   have twenty liter.PL soft drink delivered for the party
    'I have delivered twenty liter bottles of soft drink for the party.'
    (Rothstein 2011a)

Intersective modifiers such as *crisp* or *new* are compatible only with the individuating interpretation (only individual dollar bills can be crisp and new, not a monetary amount). As shown in (39), in the presence of such modifiers, plural marking on measure nouns appearing with a cardinal higher than 'one' is obligatory. Syncategorematic adjectives, on the other hand, generally preserve the special properties of measure nouns, as shown in (40).

(39) Dat gaat je   alleen maar tien spiksplinternieuwe dollar-*(s) kosten.
    that goes you only   PRT ten crisp.new   dollar-PL cost.INF
    'That will only cost you ten crisp new dollars.'
    (Klooster 1972, 14)

(40) drie kubieke   centimeter
    three cubic.AGR centimeter
    'three cubic centimeters'
    (Klooster 1972, 15)

Strikingly, cardinal-containing singular measure NPs also give rise to singular agreement on the verb, as (41) shows.[14]

(41) a. Er   zit/*zitten twee liter wijn in de kaasfondue.
        there sit.SG/PL   two liter wine in the cheese.fondue
        'There are two liters of wine in the cheese fondue.'
   b. Er   *?zit/zitten twee glazen wijn in de kaasfondue.
        there   sit.SG/PL two glasses wine in the cheese.fondue
        'There are two glasses of wine in the cheese fondue.'
   (Doetjes 1997, 189–190)

Despite this singular marking with cardinals higher than 'one', both types of measure NPs can be argued to be plural. To see this, consider (42).

(42) een dikke/*dik vijf pond
     a   fat.AGR/fat five pound.N
     'a good five pounds'

The presence of the (singular) indefinite article is not a reliable diagnostic of singularity in Dutch, since in a number of constructions it also appears with plural NPs (see Bennis, Corver, and Den Dikken 1998), but the agreement marker on the adjective is. Since in Dutch attributive agreement surfaces if the NP is definite, plural, or common (Odijk 1992; Kester 1996), example (42), which is indefinite and has a neuter lexical noun, is most likely plural, despite the fact that the measure noun is singular.

The same conclusion can be drawn from definite cardinal-containing measure phrases. As the definite determiner and both demonstratives also have only two forms, the singular neuter versus all others, the form of the demonstrative in (43) is expected if the cardinal-containing measure NP is plural but requires an explanation otherwise.

(43) deze   / *dit   vijf pond   sterling/brood/bonen
     this.PL/C.SG   THIS.N.SG five pound.N.SG sterling/bread.N.SG/bean.PL
     'these five pounds sterling/five pounds of bread/five pounds of beans'

The explanation in question could potentially take the form of postulating in measure NPs a covert functional head that would be specified for the common gender. This seems unsatisfactory: setting aside the issue of the identity and semantics of this functional head, the question remains how to account for both the lack of plural marking on the nouns in Klooster's list in (35a) and its correlation with number marking on the predicate in standard Dutch.

To summarize, plural marking on the lexical NP and on the predicate in Dutch are both conditioned by individuation, which can be informally described as the degree to which different atoms in the denotation of a noun can be distinguished from each other. Assuming that measure-denoting nouns are generally treated as not individuated, making plural agreement conditional on the presence of the individuation feature correctly predicts that they are singular in cardinal-containing NPs and trigger singular agreement on the verb. Similar effects for individuation-conditioned NP-internal number marking obtain in Danish (Hankamer and Mikkelsen 2008) and in German (Grestenberger 2015).

If cardinals combined with semantically plural lexical NPs, neither the singular form of the lexical NP nor the correlation with singular agreement on the predicate would be expected.

4.3.3   An "Optional" Plural: The Case of Western Armenian

As discussed in section 4.2, the proposal by Bale, Gagnon, and Khanjian (2011) analyzes the singular marking in cardinal-containing NPs in languages such as Turkish and Hungarian as general number, rather than singular. On this proposal, in languages where cardinals combine with morphologically unmarked and therefore apparently singular NPs (44), cardinals are semantic modifiers and the apparent singular marking on the lexical NP actually indicates number neutrality. This subsective denotation is exemplified in (45) by a lexical entry for the cardinal *two* (adapted from Bale, Gagnon, and Khanjian 2011; $\Pi(Y)(x)$ means, as in chapter 2, that there exists a partition Y of x).

(44)   a.   három gyerek-(*ek)   Hungarian
          three   child-PL
          'three children'
          (Farkas and De Swart 2010)
       b.   iki çocuk-(*lar)   Turkish
          two boy-PL
          'two boys'
          (Bale, Gagnon, and Khanjian 2011)

(45) $\lambda P . \lambda x . x \in P \land \exists Y [\Pi(Y)(x) \land |Y| = 2 \land$ General number
          $\forall z [z \in Y \rightarrow \text{atom}(z)]],$
       where atom(x) is true if x is an atom.

In languages where cardinals combine with morphologically plural NPs, cardinals are treated as simple intersective modifiers, exemplified in (46) by another lexical entry for the cardinal *two* (adapted from Bale, Gagnon, and Khanjian 2011).

(46) $\lambda P_{PL} . \lambda x . x \in P_{PL} \land \exists Y [\Pi(Y)(x) \land |Y| = 2 \land$ Plural
          $\forall z [z \in Y \rightarrow z \in \text{min}(P_{PL})]],$
       where min(P) returns a set of the smallest possible Ps
       that do not share any parts with other members of P.

As a result, Bale, Gagnon, and Khanjian are able to propose that plural marking in cardinal-containing NPs is a reliable indicator of the semantic plurality of the lexical NP while its absence should not be analyzed as singularity (contra Ionin and Matushansky 2006). Cardinals can then be treated as systematically restrictive (rather than nonintersective, as in our analysis), and in addition, surface number marking is transparently linked to local semantic number.

Bale, Gagnon, and Khanjian then apply this analysis to number marking in Western Armenian, where cardinals can combine with both singular and plural lexical NPs, hypothesizing that both the denotations in (45) and (46) are available in that language. However, this analysis ignores the fact that Western Armenian plural marking is linked to specificity, as we now show.

A crucial assumption for the analysis defended by Bale, Gagnon, and Khanjian is number neutrality. In Western Armenian, a morphologically unmarked NP can serve as a predicate for either a singular or plural subject, while a plural NP can serve as a predicate only for plural subjects (47a–b). In argument positions a morphologically unmarked NP does not necessarily denote a single entity (47c). Therefore it is reasonable to conclude that morphologically unmarked (singular) NP predicates in Western Armenian have both singular and plural entities in their denotation, which makes them compatible with the subsective lexical entry for the cardinal *two* in (45).

(47) a. John-ə   dəgha-(*ner) e.   Western Armenian
        John.DEF boy-PL   is
        'John is a boy.'
        (Bale, Gagnon, and Khanjian 2011)
     b. John-ə   yev Brad-ə   dəgha-(ner) en.
        John.DEF and Brad.DEF boy-PL   are
        'John and Brad are boys.'
        (Bale, Gagnon, and Khanjian 2011)
     c. Maro-n   piʁ   desav.
        Maro.DEF elephant see.AOR.3SG
        'Maro saw an elephant/elephants.'
        (Sigler 1996, 38)

Bale, Gagnon, and Khanjian (2011) regard plural marking in Western Armenian cardinal-containing NPs as in (44b) as completely optional. However, as discussed by Donabédian (1993) and Sigler (1996), there are interpretational differences between singular and plural NPs in those contexts where both are possible; there are, furthermore, environments where plural marking is obligatory. These facts, presented below, are not easily accounted for on Bale, Gagnon, and Khanjian's analysis.[15]

Sigler (1996) argues that the presence of plural marking in Western Armenian NPs, whether they contain a cardinal on not, is connected to specificity:
for example, while a nonspecific, unmodified NP is infelicitous with plural
marking (48a), modification makes plural marking more felicitous, though
still optional (48b). According to Sigler, in (48a), the speaker is interested in
whether the hearer saw any elephants at the zoo, rather than in particular
elephants; modification, as in (48b), is one way of placing emphasis on individual elephants. When the NP is definite, carrying a definite suffix, plural
marking is obligatory for the expression of plurality (48c); plural marking is
likewise obligatory in the presence of an adjective like 'specific' (48d).[16]

(48) a. Gentanapanagan bardezin   meč pirʁ-(#er)   desak?
        zoological   garden.GEN.DEF in elephant-PL see.AOR.2PL
        'Did you see elephants at the zoo?'
     b. Gentanapanagan bardezin   meč həntgasdanen
        zoological   garden.GEN.DEF in Indian.CX.place.ABL.DEF
        pervad   pirʁ-(er)   desak?
        bring.PASS.PTCP elephant-PL see.AOR.2PL
        'At the zoo did you see elephants that were brought from India?'
     c. Gentanapanagan bardezin   meč pirʁ-*(er)-ə   desak?
        zoological   garden.GEN.DEF in elephant-PL-DEF see.AOR.2PL
        'Did you see the elephants at the zoo?'
     d. Masnavor/voroš   dup-*(er) barbecin.
        particular/specific box-PL empty.AOR.3PL
        'They emptied particular, specific boxes.'
        (Sigler 1996, 147–152)

The behavior of cardinal-containing NPs is quite similar. With unmodified
indefinites, plural marking is infelicitous (49a); in the presence of modification, which may bring out the specific reading, plural marking becomes
optional (49b); and in the presence of a definite suffix, plural marking is
obligatory (49c).

(49) a. Mer   dunə   kišerə   utə hyur-(#er) ge-c-av/an.
        1PL.GEN house.DEF night.DEF eight guest-PL   stay-AOR-3SG/3PL
        'Eight guests stayed overnight at our house.'
     b. Mer   dunə   kišerə   utə təram č-une-c-oʁ
        1PL.GEN house.DEF night.DEF eight money NEG-have-AOR-SR
        hyur-(er) ge-c-av/an.
        guest-PL stay-AOR-3SG/3PL
        'Eight guests who had no money stayed overnight at our house.'

c. Mer   utə hyur-*(er)-ə kišerə   mer   kovə   ge-c-an.
1PL.GEN eight guest-PL-DEF night.DEF 1PL.GEN side.DEF stay-AOR-3PL
'Our eight guests stayed overnight.'
(Sigler 1996, 147–152)

In order to account for the relationship between plural marking and specificity in Western Armenian, Bale, Gagnon, and Khanjian (2011) would need to
assume that the lexical entry in (46) obligatorily carries the feature [+specific],
whereas the lexical entry in (45) either bears the feature [−specific] or is
unspecified for specificity. However, this is nonexplanatory, and amounts to
little more than a restatement of the facts: there is no independent motivation
for linking the lexical entries and [±specific] features in this way and not the
other way around.

Number agreement on the predicate behaves the same, strongly suggesting
that it should be treated along the same lines. As shown in (49a–b), both singular and plural agreement on the verb are possible when the subject is an
indefinite cardinal-containing NP, whereas plural agreement on the verb is
required in the case of a definite (49c). It is not the case, however, that number
marking inside the subject and on the predicate are fixed independently, since
the latter crucially depends upon the former, as well as upon the transitivity
of the verb.

More specifically, Sigler (1996) makes a number of generalizations, illustrated in (50)–(51) (from Sigler 1996, 167–169, with glosses adjusted). In the
case of passive and unaccusative verbs ('fail' in Western Armenian is unaccusative), the verb must be singular if the lexical NP is singular (50a) and the
verb must be plural if the lexical NP is plural (50b); if the subject is definite,
number marking on the lexical NP and on the verb (50c) must match. In the
case of transitive verbs, plural marking is obligatory with semantically plural
cardinal-containing subjects, as in (51); according to Sigler (1996), some
speakers allow a singular lexical NP despite the plural marking on the verb,
whereas others prefer to use the plural marker on the NP (this is indicated by
square brackets around the plural marker in (51)).

(50) a. Kəsan usanoʁ kənuten-ə-mə   caxoʁe-c-av.
        twenty student exam-ABL-INDEF fail-AOR-3SG
        'There failed an exam twenty students.'
     b. Kəsan usanoʁ-ner kənuten-ə-mə   caxoʁe-c-an/*av.
        twenty student-PL exam-ABL-INDEF fail-AOR-3PL/3SG
        'Twenty students failed an exam.'

c. Kəsan usanoʁ-*(ner)-ə kənuten-ə-mə   caxoʁe-c-an.
   twenty student-PL-DEF exam-ABL-INDEF fail.AOR.3PL
   'The twenty students failed an exam.'
   (Sigler 1996, 167–168)

(51) Hink gin-[er]   surǰ   gəhəme-*(n)   gor.
     five   woman-PL coffee IMPF.drink-3PL PROG
     'Five women are drinking coffee.'
     (Sigler 1996, 169)

To summarize, the following holds in Western Armenian for plural marking on a lexical NP that appears with a cardinal above 'one' or that denotes a plurality. Plural marking is obligatory in the presence of a definiteness or specificity marker; optional in the absence of such markers, unless there is some emphasis on the individual or a possibility of a specific interpretation; and infelicitous for a bare, nonspecific indefinite interpretation. As for verbal agreement, an intransitive verb is singular if the lexical NP is singular and plural if the lexical NP is plural; transitive verbs are preferentially obligatorily plural, regardless of the NP form. In other words, number marking with cardinal-containing NPs, just like number marking on intransitive predicates combining with them, is conditioned by specificity. It can furthermore be demonstrated that importantly, the use of the singular for number neutrality in the absence of a cardinal does not have the same distribution as the singular in cardinal-containing NPs: as Sigler (1996, 156–158) demonstrates, in the subject position the use of the singular in bare NPs is restricted to semantically singular entities. Thus, Western Armenian can be taken as a further argument in favor of our view, stated in chapter 2, that a singular lexical NP appearing with a cardinal higher than 'one' is semantically singular.

4.3.4 Heterogeneous Number Marking with Cardinals

We next consider languages where the number on the lexical NP depends on the cardinal it appears with. We argue that these languages provide further evidence for our view that number marking on the lexical NP is agreement.

In Standard Arabic the lower simplex cardinals ('one' to 'ten') combine with a plural lexical NP (in the genitive case), as shown in (52), whereas the higher simplex cardinals combine with a singular lexical NP (in the accusative case for 'twenty', 'thirty', etc. and in the genitive case for the higher cardinals 'hundred', 'thousand', etc.), as shown in (53).

(52) a. ʔarbaʕ-at-u rijaal-in   Standard Arabic
        four-F-NOM man.PL-GEN
        'four men'

b. ?arbaʕ-u banaat-in
   four-NOM girl.PL-GEN
   'four girls'
   (Zabbal 2005)

(53) a. xamsuuna rajul-a
     fifty   man-ACC
     'fifty men'

   b. mi?at-u   rajul-in
      hundred-NOM man-GEN
      'a hundred men'
      (Zabbal 2005)

A similar pattern obtains in Scottish Gaelic (Greene 1992; see also Acquaviva 2006). While the cardinals 'one' and 'two' as well as the higher simplex cardinals ('twenty', 'hundred', etc.) combine with a singular lexical NP (54a), other lower simplex cardinals ('three' through 'ten') show a more complex pattern. Generally, they combine with a plural lexical NP (54b–c). However, when they merge with the higher cardinals *fichead* 'twenty' (55b), *ceud* 'hundred', and *mile* 'thousand', as well as the nouns *dusan* 'dozen', *duine* 'person', *latha* 'day', and *bliadhna* 'year' (55a), these remain singular.[17] (Compare (55b) to (55c), where 'twenty' is clearly singular.)[18]

(54) a. an dà leabhar/*leabhraichean   Scottish Gaelic
     the two book/book.PL
     'the two books'

   b. trì leabhraichean
      three book.PL
      'three books'

   c. ceithir leabhraichean mòra deug
      four   book.PL   big teen
      'fourteen big books'
      (Adger 2010, 341–342)

(55) a. sia bliadhna
     six year
     'six years'
     (Roibeard Ó Maolalaigh, pers. comm.)

   b. sia fichead
      six twenty
      'one hundred and twenty'
      (Greene 1992, 533)

    c. fichead cù
       twenty dog
       'twenty dogs'
       (Greene 1992, 532)

In Soninke (Mande, West Africa), as described by Creissels and Urmančieva (2017), a similar pattern obtains: while the cardinal *báané* 'one' behaves like an adjective and forms a compound with the noun, the lower cardinals 'two' through 'ten' follow the noun, which is then plural, and the higher cardinals precede the noun, which is singular and shows the syntax and phonology of a possessee.

In principle, it is possible to hypothesize a different semantics for those cardinals that combine with a singular lexical NP in Soninke, Standard Arabic, and Scottish Gaelic and those that combine with a plural one. Evidence against this as a general solution comes from the fact, noted above, that in Scottish Gaelic some lexical NPs occur in the singular even with those cardinals ('three' through 'ten') that normally occur with a plural NP. One could argue that those few nouns have a special functional status somehow precluding their being marked plural in cardinal-containing NPs. However, the nature of this status remains to be specified: treating them as classifiers would be unmotivated since, for one thing, 'person', 'day', and 'year' normally do not appear with a substance NP.

Instead, we suggest that all cardinals in Soninke, Standard Arabic, and Scottish Gaelic combine with semantically singular NPs and that the plural marking on the lexical NP that we see with some cardinals, as in (52) and (54b–c), is due to agreement. The exceptional behavior of the small class of Scottish Gaelic NPs that are always singular with cardinals (55a–b) is then attributed to agreement failure: number agreement is conditioned by some feature, which these nouns lack. As discussed in section 4.3.2, if plural agreement is conditional on the presence of the individuation feature, its absence on measure nouns is correctly predicted to cause agreement failure.

An example of the very same effect with discourse-independent features comes from Estonian Swedish (Rendahl 2001, 156; Koptjevskaja-Tamm and Wälchli 2001, 701), where only feminine nouns appear in the plural form in cardinal-containing NPs, while masculine and neuter nouns remain singular, as shown in (56).

(56) a. tri   mann   Estonian Swedish
       three man.SG
       'three men'

b. fem bärkiar
   five birch.PL
   'five birches'
(Koptjevskaja-Tamm and Wälchli 2001, 701)

The fact that plural marking in cardinal-containing NPs can be conditioned by the formal gender feature strongly argues against the hypothesis that cardinals combine with semantic plurals, as well as against the possibility that singular or plural marking on the lexical NP is determined by the semantics of the cardinal itself. Language-internal conditions on number marking with cardinals thus provide the strongest evidence for the hypothesis that this number marking is a result of agreement.

4.3.5 Plural Marking on the Lexical NP: Summary

In this section, we have examined three languages where number marking on the noun in cardinal-containing NPs is conditioned by the same factor as unquestionable cases of plural agreement. In Miya, both agreement and concord are conditioned by animacy (animate nouns can and must trigger agreement in number) and only animate nouns must agree in number in cardinal-containing NPs. In Dutch, a large class of measure nouns fails to be marked plural in cardinal-containing NPs and to trigger plural agreement on the verb, while still requiring plural agreement on determiners and modifying adjectives; that the relevant factor is individuation is shown by the fact that plural marking on measure nouns becomes obligatory if they denote individuated measures. In Western Armenian, both plural marking on the NP and plural marking on intransitive verbs are conditioned by specificity. Finally we saw further evidence, from heterogeneous number marking with cardinals cross-linguistically, that speaks in favor of treating plural marking on the lexical NP as agreement.

The hypothesis that number marking in cardinal-containing NPs is due to agreement with the entire cardinal-containing NP, rather than to the semantic plurality or singularity of the lexical NP that the cardinal combines with, can explain these correlations. The alternative hypothesis that the singular lexical-NP sister of a cardinal corresponds to general number is not supported by the facts: in Miya, number neutrality (general number) is possible but is not related to animacy, while Dutch has no general number at all.

Having demonstrated that plural marking on the lexical NP that occurs with cardinals is an instance of agreement rather than of true semantic plurality, we will next show that very similar agreement processes are at work with group nouns. Since a formal account of conditional agreement is not among the main aims of this book, we only point out the similarities here, without analyzing them; these observations will hopefully provide a springboard for future work in this area.

4.4 Semantic Number Agreement: Similarity between Group Nouns and Cardinals

Semantic number agreement is a well-known phenomenon in another area, namely, plural agreement with morphologically singular group nouns such as *team* and *government* (Munn 1999; Den Dikken 2001; and Sauerland and Elbourne 2002, among others). As we show below, the same subject NP may trigger singular or plural agreement as a function of its individuation, animacy, and so on, that is, the same factors that trigger plural marking with cardinal-containing NPs. We expect therefore that an examination of semantic agreement with group nouns will shed some light upon the behavior of cardinal-containing NPs.

4.4.1 Semantic Agreement on the Predicate with Group Nouns and Cardinals
In many dialects of English, including British English, group nouns may take either singular or plural agreement, as shown in (57) (from Sauerland and Elbourne 2002).

(57) a. The Government is ruining this country.
　　 b. The Government are ruining this country.

Intuitively, the plural option is possible because of the lexical semantics of group nouns, which imply a plural entity. Importantly, Munn (1999) and Sauerland and Elbourne (2002) claim that plural agreement with group nouns is impossible in *there* constructions, as shown by (58), from Sauerland and Elbourne 2002.

(58) a. A committee was/were holding a meeting in here.
　　 b. There was/*were a committee holding a meeting in here.

While Munn (1999) and Sauerland and Elbourne (2002) explain the above contrast by means of formal syntactic proposals, we argue that it is driven by individuation (i.e., the degree to which different atoms in the denotation of a noun can be distinguished from each other). Support for this view comes from the behavior of group nouns with complements. While the availability of plural agreement for group nouns with no complement appears to be specific to British English, the presence of an *of* PP makes plural agreement possible in American English as well. The examples in (59) show that group nouns with *of* PP complements—that is, pseudopartitives—may trigger both types of agreement (cf. Akmajian and Lehrer 1976).

(59) a. A large herd of zebras *was/were* discovered last year.
      b. A group of tourists *was/were* visiting the museum.

The occurrence of plural marking on the predicate in (59) could, at first glance, be attributed to agreement attraction by the intervening plural NP (*zebras, tourists*) (Bock and Miller 1991 and Eberhard, Cutting, and Bock 2005, among others). However, two facts argue against this. First, agreement-attraction errors are relatively infrequent: Eberhard, Cutting, and Bock (2005, table 3, row 5) report that across studies, the rate of production of plural marking on the verb with a singular head noun and a plural local noun (e.g., *the key to the cabinets*) varies from 1 to 30 percent, with an average of 13 percent. In contrast, a quick search of the Corpus of Contemporary American English (COCA; Davies 2008–) for *a group of [n] was* and *a group of [n] were* (where *[n]* stands for a wildcard noun) yielded fifty-one hits each: that is, we find a 50 percent rate of plural agreement when the NP is headed by a collective noun, well above what would be expected if this were due to agreement attraction.

Second, plural agreement inside a *there* construction, too, is much facilitated when an *of* complement is present, even though agreement attraction should not be relevant here (since the plural local noun comes after the verb). A web search has turned up a multitude of examples of the form *there were a team/ herd/band*, nearly all of them followed by an *of* PP. Some examples are given in (60).

(60) a. Sure enough, there were a herd of elk running down into the valley
          to our left.
          (http://wildlife.state.co.us/hunt/Youth/YouthTellOfTheHunt.asp)
      b. There were a team of doctors coming behind here to give the care
          and treatment.
          (http://www.leaderu.com/marshill/mhr03/peter1.html)
      c. Also there were a group of Australian visitors trying to sell their
          wares.
          (http://funtime.comics.org.nz/ftcpresents/iconz1995.html)

As always with corpus examples, the question arises of whether they represent a genuinely available grammatical option or a performance error. In this case, a COCA search suggests that plural agreement in examples such as (60) is indeed fully grammatical: the sequence *there were a group of* yielded twenty-four hits, compared to sixty-seven hits for *there was a group of,* indicating that plural agreement is not particularly rare (26 percent), though the rate is lower than what we find when the NP is in subject position (see above).[19]

On the other hand, in the absence of an *of* PP, plural agreement seems to be unavailable: a sentence-initial *The group were* yielded zero hits, as opposed to 131 hits for *The group was*. These numbers suggest that American English allows plural agreement for collective nouns only when the noun is followed by an *of* PP, and that while plural agreement is dispreferred with *there* insertion, it is still available.

We propose informally that the presence of an *of* PP, which specifies the members of the group denoted by the collective noun, makes the individuated reading of the NP more likely, and hence facilitates plural agreement. We furthermore speculate that the *there* construction (where the associate is typically a nonspecific indefinite) is less readily compatible with the individuated reading than the subject position.

Coming back to cardinals, we see both similarities and differences with group nouns. On the one hand, crosslinguistically, both group NPs and cardinal-containing NPs can trigger both singular and plural agreement on the predicate, with plural agreement potentially conditioned by individuation (see the Dutch facts in section 4.3.2). However, these two NP types do not necessarily behave the same within a language. In English, cardinal-containing NPs normally require plural agreement on the predicate, as seen in (61a); singular agreement on the predicate is possible when the lexical NP is a measure NP, as in (61b). When the measure NP in (61b) is used with a singular verb, we obtain the normal, nonindividuated reading (the entire twenty pounds is for sale for five dollars); when it is used with a plural verb, we obtain the somewhat pragmatically odd individuated reading (i.e., each pound of apples is for sale for five dollars; this reading is facilitated if each pound is packaged individually).

(61) a. Ten students are/*is studying for the exam.
     b. Twenty pounds of apples is/?are for sale for five dollars.

To sum up, both group and cardinal-containing NPs in English exhibit sensitivity to individuation, yet their behavior is not entirely parallel (while singular agreement is preferred for group nouns, especially in American English, plural agreement is required for most cardinal-containing NPs). We cannot explain these differences without a formal analysis of semantic agreement, which lies beyond the scope of this work.

4.4.2 Semantic Agreement on the Determiner with Group Nouns and Cardinals

An additional factor that needs to be taken into account is determiner agreement. While *team, committee, band,* and the like can trigger plural verbal agreement in British English, they cannot take plural determiners (see Corbett 1979; Barlow 1992; Munn 1999; Sauerland and Elbourne 2002), as shown in (62).[20] Cardinal-containing NPs show the opposite pattern: plural determiners are required in most cases (63a), while both singular and plural determiners are possible with measure nouns (63b).

(62) a. *These band are going to be playing.
    b. This band are going to be playing.
    c. This band is going to be playing.
    (Munn 1999)

(63) a. These/*this five students failed the exam.
    b. That/those five gallons of milk came in handy.
    (Gawron 2002)

Thus, we observe another difference between group nouns and cardinal-containing NPs, but we note again that the availability of singular versus plural agreement with cardinal-containing NPs is conditioned by individuation: singular agreement is possible only with nonindividuated measure nouns (intuitively, *that five gallons of milk* denotes a single container of milk, while *those five gallons of milk* is compatible with multiple containers).

To summarize, plural agreement with group nouns requires a mechanism for semantic agreement (Corbett 1979, 1983, 2006; Pollard and Sag 1994; Wechsler and Zlatić 2003, 2012; Wechsler 2011; etc.), which may also be extended to agreement with and within cardinal-containing NPs. Such a unification of cardinals and group nouns with respect to semantic agreement is desirable because higher cardinals crosslinguistically are often historically derived from group and measure nouns and synchronically there often exist some null-derived cardinal-to-measure nouns such as *thousand* (cf. *a thousand books* vs. *a thousand of books*). In the absence of a formal theory of semantic versus syntactic agreement, we cannot propose an analysis of the differences between cardinals and group nouns.[21]

4.5 Summary

In this chapter, we have discussed three distinct but interrelated phenomena: plural marking (or lack thereof) on the lexical NP that occurs with a cardinal;

plural versus singular marking on the predicate, when a cardinal-containing NP is in subject position; and plural versus singular marking on determiners and adjectives that modify a cardinal-containing NP. While number marking on determiners/adjectives and predicates is commonly analyzed as agreement, we have argued that plural marking on the lexical NP is also an instance of agreement. Evidence for this proposal comes from the fact that, crosslinguistically, plural marking on the lexical NP (as well as plural marking on predicates, determiners, and adjectives) is conditioned by such factors as individuation, specificity, and animacy. We have also pointed to similarities in the behavior of cardinal-containing NPs and pseudopartitives headed by group nouns with regard to semantic agreement. However, a full theory of semantic agreement lies beyond the scope of this chapter.

5   Additive Complex Cardinals

In chapters 2 and 3, we discussed the semantics and syntax of complex cardinals involving multiplication, such as *two hundred*. In this chapter, we take up complex cardinals involving addition, such as *thirty-six, two hundred and five*, and *three thousand seven hundred and fifty-six*. In the majority of familiar languages the construction of complex cardinals involving addition (henceforth, *additive complex cardinals*) is achieved either by simple juxtaposition of the required number of multiplicands (*two thousand twenty-five*) or by coordination (*one hundred and one*). In Ionin and Matushansky 2006 we argue that the former case should be treated as an instance of asyndetic coordination. A broader crosslinguistic examination, however, yields yet another way of constructing additive complex cardinals, namely using an adposition; adpositions are also used in the construction of subtractive complex cardinals.

The goal of this chapter is to provide a syntactic and semantic analysis of additive complex cardinals that is compatible with the semantics and syntax that we proposed in chapters 2 and 3. Our basic proposal is that a cardinal-containing NP such as *twenty-two books* has the underlying structure of *twenty books and two books*, and that the semantics of this coordinated NP is derived by an additive rather than an intersective interpretation of *and*.

We start, in section 5.1, by providing an overview of different strategies found crosslinguistically for building additive complex cardinals. In section 5.2, we analyze the most common type of additive complex cardinals, that involving coordination, as a case of NP deletion of the lexical NP in one of the conjuncts. We show that there is direct evidence in favor of the NP-deletion analysis from languages such as Russian and Welsh, while other languages, such as Polish, Romanian, Inari Sami, and Greek, require the additional stipulation that the lower cardinal inside an additive cardinal becomes more nominal as a result of agreement. This proposal is laid out in section 5.3. Section 5.4 considers alternative analyses of the syntax of additive complex cardinals and shows that they have no advantages over our proposal. Finally, section 5.5

considers the role of extralinguistic strategies in the formation of complex cardinals involving coordination.

In section 5.6, we shift our focus to complex cardinals built by means of locative and comitative adpositions, further supporting our proposal that complex cardinals are built by independently available syntactic means. Section 5.7 provides an interim summary of the syntax of complex cardinals before we pass on to their semantics in section 5.8.

We show that the standard intersective interpretation of *and* does not derive the correct interpretation of additive complex cardinals. The required additive interpretation of *and* can be either postulated, as in Heycock and Zamparelli 2000, 2005, or derived from the intersective meaning, as in Champollion 2016. Finally, we show that neither the intersective nor the additive interpretation of *and* can account for the full range of NP-internal conjunctions, both with and without cardinals. We propose that another, disjunctive interpretation of *and* is required to account for those cases, and we sketch a way of deriving this interpretation from the basic intersective one.

5.1 Different Strategies for Encoding Addition

Crosslinguistically, three different strategies can be used for additive complex cardinals. As discussed above, one is coordination, illustrated in (1). Asyndetic coordination (e.g., *twenty-two* in English) is included in this category as well. A second option is the use of a spatial adposition, as in Biblical Welsh, illustrated in (2) (see also chapter 7 for more on Biblical Welsh). And the third option is the use of a comitative adposition, as in (3).

(1) *Conjunction*
    a. three hundred and five   English
    b. twee-en-twintig   Dutch
       two-and-twenty
       'twenty-two'
    c. laba iyo   toban   Somali
       two CONJ ten
       'twelve'
       (Saeed 1999)

(2) *Locative preposition:* ar *'on'*
    a. un   ar ddeg   Biblical Welsh
       one on ten
       'eleven'

b. pedwar ar bymtheg
   four   on fifteen
   'nineteen'
c. naw ar hugain
   nine on twenty
   'twenty-nine'
   (Hurford 1975, 163–164)

(3) *Comitative preposition:* ǝd *'with'*
   a. mæráw médd-æn ǝd   sæmmos   Tamashek (Tuareg of Mali)
      ten.M   man-M.PL with five
      'fifteen men'
   b. mærɑw-æt ḍeḍ-en   ǝd   sæmmós-æt
      ten-F   woman-F.PL with five-F.PL
      'fifteen women'
      (Heath 2005, 251–252)

The latter two strategies permit not only addition but also subtraction, as illustrated in (4) and (5).

(4) *Locative preposition:* de *'from'*
   un-de-viginti   Latin
   one-from-twenty
   'nineteen'

(5) *Caritative preposition:* bi *'without'*
   a. deš bi-jekh   Welsh Romani
      ten without-one
      'nine'
   b. bi-trin-engiro   trianda   Russian Romani
      without-three-GEN thirty
      'twenty-seven'
      (Elšik and Matras 2006, 167)

In this chapter, we will mostly focus on coordination, but other additive strategies will also be discussed.

5.1.1 Multiplication Is Not Addition

It is easy to see that the cascading structure that we proposed for complex cardinals involving multiplication (see (3) in chapter 3) cannot be extended to those involving addition. From the morphological point of view, complex cardinals that involve addition often contain an overt conjunction (1) or a preposition ((2) and (3)). On the semantic side, the interpretation of a sequence of two cardinals when multiplication versus addition is involved is naturally not the same. Since our structure for complex cardinals involving multiplication does not contain any functional heads, it is simply impossible to get a different interpretation for additive complex cardinals using the same structure.

On the syntactic side, there is no case assignment or number agreement within additive complex cardinals, unlike those involving multiplication. This is illustrated in (6) for Russian.

(6) a. pjat'   sot   Russian: multiplicative
       five.NOM/ACC hundred.PL.GEN
       'five hundred'
    b. tysjača   sto   Russian: additive
       thousand.NOM hundred.NOM/ACC
       'one thousand one hundred'

Further evidence for two distinct structures for additive and multiplicative complex cardinals comes from the Russian-specific phenomenon of approximative inversion (Mel'čuk 1985; Fowler 1987; Franks 1994, 1995; Billings 1995; Isakadze 1998; Yadroff and Billings 1998; Pereltsvaig 2006a,b; Zaroukian 2012; Matushansky 2015; Rothstein and Khrizman 2015), illustrated in (7). Approximative inversion reverses the normal linear order between a cardinal and a noun, with the semantic effect of imprecision.

(7) a. pjat' časov
       five hour.PL.GEN
       'five hours'
    b. časov   pjat'
       hour.PL.GEN five
       'about five hours'

As discussed by Matushansky (2015), in an NP containing a complex cardinal that involves multiplication, approximative inversion actually fronts the first multiplicand, as shown in (8b) (the inverted version of (8a)), rather than the lexical noun, as in (8c). However, approximative inversion cannot similarly target an addend (the part being added) in a complex cardinal involving addition, as shown in (9b).

(8) a. sorok tysjač   mašin
        forty thousand.PL.GEN car.PL.GEN
        'forty thousand cars'
    b. tysjač   sorok mašin
        thousand.PL.GEN forty car.PL.GEN
        'some forty thousand cars'

    c. *mašin   sorok tysjač
       cars.GEN forty thousand.PL.GEN

(9) a. million   tysjača   mašin
       million.NOM thousand.NOM car.PL.GEN
       'one million one thousand cars'

    b. *tysjača   million   mašin
       thousand.NOM million.NOM car.PL.GEN

It is not the case that the word for 'thousand' simply cannot be fronted (see (8b)), and it is also not the case that the simple presence of addition blocks fronting. Taking the complex cardinal that involves addition in (10a), we see that fronting of the lexical NP is possible (10b), but fronting of 'thousand' is ungrammatical (10c). In this example, 'thousand' is a multiplicand of only one addend and fronting it would violate the Coordinate-Structure Constraint (Ross 1967). In contrast, fronting 'thousand' is perfectly fine as long as it is a shared multiplicand of both addends, as in (10d).

(10) a. million   sto   tysjač   raz
       million.NOM hundred.NOM thousand.PL.GEN time.PL.GEN
       'one million one hundred thousand times'

    b. raz_i ètak million   sto   tysjač t_i
       time.PL.GEN so   million.NOM hundred.NOM thousand
       'some one million one hundred thousand times'

    c. *$tysjač_i$ ètak million   sto t_i raz
       thousand.PL.GEN so   million.NOM hundred.NOM time.PL.GEN

    d. $tysjač_i$ ètak sto   pjat'desjat t_i soldat   i
       thousand.PL.GEN so   hundred.NOM fifty.NOM   soldier.PL.GEN and
       oficerov
       officer.PL.GEN
       'some one hundred and fifty thousand soldiers and officers.'

These data provide strong evidence in favor of analyzing additive complex cardinals as coordination: the impossibility of extraction out of only one addend is explained as a Coordinate-Structure Constraint violation.

The remaining issue is the contrast between (8c), where the lexical NP cannot be fronted by approximative inversion, and (10b), where it can. We hypothesize that the Across-the-Board fronting of the noun becomes possible in (10b) precisely because the higher multiplicands cannot be so fronted. In (8c) *tysjač* is an intervener and so *mašin* cannot be targeted. In (10b), on the other hand, the lexical noun *raz* is the first nominal that can be fronted and so it is.

We now lay out the details of the coordination treatment of complex cardinals like *one hundred and two* and *twenty-two* and discuss the problems that arise from this treatment.

5.1.2 Asyndetic Coordination

It is quite common crosslinguistically for complex cardinals involving addition to contain no overt conjunction, as illustrated in (11). We consider this to be an instance of asyndetic coordination, which is defined as coordination in the absence of an overt conjunction (we are aware of no instances of asyndetic disjunction). The phenomenon is in fact attested in different parts of language and in different languages. For instance, adjectival predicates are often conjoined with no overt *and* even in English, as illustrated by (12a), which is truth-conditionally equivalent to (12b).

(11) a. thirty-five   English
       b. dvadcat' dva   Russian
          twenty   two
          'twenty-two'

(12) a. a tall dark stranger
       b. a tall and dark stranger

In addition, coordination without an overt conjunction is attested crosslinguistically, as noted by Payne (1985; cited in Winter 1995), Stassen (2000), and Drellishak (2004):

(13) Ñe   niyo'jɵ   nipita   ni'ɵ.   Andoke
      be.PAST her.brother her.aunt her.sister   (Macro-Carib, Witotoan)
      'It was her aunt, her brother, and her sister.'
      (Stassen 2000, 5; glosses adjusted for consistency)

Gawron (2002) notes that the conjunction *and* is optional in the domain of measurements, as illustrated by the truth-conditionally equivalent (14a) and (14b). Examples (15a–b) illustrate the effect for monetary units, where an overt conjunction can even be ungrammatical; on the other hand, in (15c), the overt conjunction is required.

(14) a. six feet six inches of finest silk
       b. six feet and six inches of finest silk

(15) a. two dollars (and) seventy-five cents
       b. two dollars (*and) seventy-five
       c. five gigabytes ??/*(and) twenty megabytes of data

The behavior of conjunction in measure expressions seen above illustrates that, depending on the construction, the presence of an overt conjunction can be obligatory, optional, or disallowed. The same holds for additive complex cardinals: in some languages (e.g., Russian), numerical expressions never contain an overt conjunction, while in others (e.g., Arabic), an overt conjunction is obligatory for addition (Zabbal 2005; see also chapter 7).

5.1.3   Ordering Constraints on Asyndetic Coordination

In chapter 3, we discussed the fact that multiplicative and additive complex cardinals typically differ with regard to word order. While multiplicative complex cardinals typically exhibit the order *low-high*, the opposite is usually the case for additive complex cardinals. Hurford (2003) shows that in the case of addition, in the absence of an overt conjunction, the high cardinal nearly always precedes the low cardinal: thus, the number 22 is pronounced *twenty-two*, not **two-twenty*, and the number 102 is pronounced *a hundred two*, not **two hundred*. On the other hand, examples like *two and twenty*, where the overt conjunction unambiguously signals addition, are attested crosslinguistically, for example, in German, as examples like (16) show.

(16) zwei-*und*-zwanzig   German
       two-and-twenty
       'twenty-two'

Some languages in which the cardinals 'eleven' through 'nineteen' exhibit the order low-high (despite the absence of an overt conjunction) are English, Hebrew (see chapter 7), and Scottish Gaelic (*ceithir deug*, literally 'four ten', i.e., 'fourteen'; see Hurford 2003). In some of these languages (e.g., English and possibly Hebrew), these cardinals form one word, and the order low-high is word-internal (e.g., *four-teen*). The fact that teens are an exception to the usual ordering for additive complex cardinals may be due to morphology, if *-teen* is a suffix. (In other words, we can assume that teens are not derived via conjunction; see chapter 7 for a specific proposal with regard to the '-teen' forms in Hebrew). While this solution does not extend to other Germanic languages, such as German and Dutch, where the low-high order in addition is mandatory below 'hundred', overt conjunction unambiguously signals addition in those cases, as illustrated in (16).

We do not believe that the above ordering constraints need be or can be linguistically encoded. We draw a parallel with proper names: whereas in the majority of languages the order is obligatorily *first name, last name*, the reverse holds for Japanese and for Sinhala. Likewise, in the use of titles such as *Mr.* and *Mrs.*, a married couple may be referred to as *Mr. and Mrs. Smith*, but not as *Mrs. and Mr. Smith*. Another example is the use of dates: in English, we say *on Sunday, August first*, not *on August first, Sunday*. For all these cases, the compulsory ordering does not seem to have any deep linguistic significance.

We next move on to implementing addition via coordination in the syntax.

5.2   Addition as NP-Level Coordination

Given the syntax and semantics that we have proposed, in a complex NP such as *three hundred (and) forty students*, each coordinated cardinal must contain an instance of the lexical NP *students*: *three hundred students and forty students*.[1] In Ionin and Matushansky 2006, we discuss two ways in which *three hundred students and forty students* can be converted into *three hundred (and) forty students*: (i) right-node raising of the lexical NP and (ii) PF deletion of the lexical NP in the first conjunct.[2] Our proposal here is different: we argue that only NP deletion (ii) is necessary for deriving additive complex cardinals. Postponing our argument against right-node raising until section 5.4, we first present evidence in favor of the presence of multiple NPs in complex cardinals involving addition; then we show how additive complex cardinals in various languages can be derived via NP deletion.

5.2.1   Evidence for Multiple Copies

Our proposal that coordinated cardinals are derived from coordinated NPs finds support in the Bantu language Luvale (Zweig 2006, citing Horton 1949), in Biblical Hebrew, in Biblical Welsh (Hurford 1975; see chapter 7 for extensive discussion), and in archaic Russian. As observed in Ionin and Matushansky 2006, in these languages the lexical NP may appear in all conjuncts of a coordinated cardinal, as shown in (17)–(20) (see also Corbett 1978).

(17) *mikoko* makumi atanu na-*mikoko* vatanu   Luvale
     sheep   ten   five   and-sheep five
     'fifty-five sheep'
     (Horton 1949)

(18) a. təšaʕ mēʔôt   *šānā* u-šlōšîm *šānā*   Biblical Hebrew
        nine hundred.PL year and-thirty year
        'nine hundred and thirty years'
        (Genesis 5:5)
     b. šalôš *šānîm* wə- ʔarbaʕ mēʔôt   *šānā*
        three year.PL and four   hundred.PL year
        'four hundred and three years'
        (Genesis 11:15)

    c. tēšaʕ *šānîm* u- mātayim   *šānā*
       nine year.PL and hundred.DU year
       'two hundred and nine years'
       (Genesis 11:19)

(19) a. saith *mlynedd* ac wyth gan   *mlynedd*   Biblical Welsh
       seven year.PL and eight hundred year.PL
       'eight hundred and seven years'
    b. naw can   *mlynedd* a   deng *mlynedd* a
       nine hundred year.PL and ten   year.PL and
       de-ugain   *mlynedd*
       two-twenty year.PL
       'nine hundred and fifty years'
    (Hurford 1975, 175, 176)

(20) tridcat' *let*   i   tri   *goda*   Archaic Russian
    thirty   year.PL.GEN and three year.PAUC
    'thirty-three years'

    Next we will show that the process responsible for the disappearance of all lexical NPs but one in (most cases of) additive complex cardinals is NP deletion.

5.2.2 Direct Evidence for NP Deletion

The tree in (21) shows the derivation of an additive complex cardinal via PF deletion of the lexical NP in one of the conjuncts.

(21)

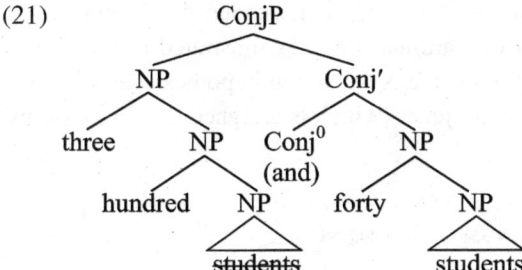

    While in Ionin and Matushansky 2006 we argue that both NP deletion and right-node raising are crosslinguistically attested, here we will opt for a more economical proposal allowing only NP deletion (see section 5.4.1 for arguments against right-node raising).

    Support for the NP-deletion view comes from situations where coordinated cardinals do not assign the same case yet case conflict does not seem to arise. In Russian, for instance, the case that surfaces on the lexical NP is that assigned in the right conjunct. This is illustrated in (22): the case on the noun in 'twenty-two books' (22c) depends on 'two' rather than on 'twenty' (see chapter 6 for more extensive discussion of Russian).

(22) a. dva   šagá   Russian
       TWO.NOM step.PAUC
       'two steps'
   b. dvadcat'   šagóv
       twenty.NOM step.PL.GEN
       'twenty steps'
   c. dvadcat'   dva   šagá
       twenty.NOM TWO.NOM step.PAUC
       'twenty-two steps'

The facts in (22) are compatible with the NP-deletion view. On this analysis, (22c) is derived from (23), via asyndetic coordination plus NP deletion of the first lexical NP. The case on the second NP is then naturally that assigned by its sister, the second cardinal.

(23) [dvadcat' ~~šagov~~]   i   [dva šagá]
     twenty   step.PL.GEN and two step.PAUC

Even stronger support for the NP-deletion view comes from the behavior of complex cardinals containing the adjectival *odin* 'one'. As discussed in chapter 3, *odin* is clearly adjectival, agreeing in number and gender with the lexical NP. It is the only Russian cardinal that does not assign case. With all complex cardinals ending in 'one' in Russian ('twenty-one', 'hundred and one', etc.), the lexical NP receives the same case, gender, and number as it would with the simplex adjectival cardinal 'one', as illustrated in (24a). The easiest way to account for this is via the NP-deletion hypothesis, as in (24b), where the lexical NP in the first conjunct (which is assigned genitive case by 'twenty') is deleted.

(24) a. dvadcat'   odna   kniga
       twenty.NOM one.F.SG.NOM book.F.SG.NOM
       'twenty-one books'
   b. [dvadcat'   ~~knig~~]   i   [odna   kniga]
       twenty.NOM book.PL.GEN and one.F.SG.NOM book.F.SG.NOM

Polish paucal cardinals provide further evidence for the same conclusion: the additive complex cardinal in (25c) behaves like the paucal cardinal in (25b) rather than the higher cardinal in (25a), both with respect to the case of the lexical NP (which is nominative rather than genitive plural) and with respect to verbal agreement (which must be plural rather than default).[3]

(25) a. Tu było   dwadzieścia stołów.   Polish
       here was.N.SG twenty.NV   tables.GEN
       'Twenty tables were here.'
       (Feldstein 2001, 66)

   b. Dwa   stoły   rozbiły   się.
       two.NV.NOM table.M.PL.NOM break.PAST.NV.PL REFL
       'Two tables broke.'
       (Klockmann and Šarić 2015)

   c. Dwadzieścia trzy   koty   bawiły   się.
       twenty.NV   three.NV.NOM cat.PL.NOM play.NV.PL REFL
       'Twenty-three cats were playing.'
       (Swan 2002, 199)

Yet another potential piece of evidence in favor of the NP-deletion view comes from languages like Biblical Welsh and Scottish Gaelic, where the lexical NP does not have to appear after the last conjunct of an additive cardinal. The examples in (26) show that in Scottish Gaelic the lexical NP can occur inside the complex cardinal. Exactly the same phenomenon is found in Biblical Welsh (Hurford 1975, 175–177), as shown in (27), where several but not all copies of the lexical NP are preserved (for more discussion of Modern and Biblical Welsh, see chapter 7). It is also found in archaic English: compare the archaic (28a) with the more contemporary (28b).[4]

(26) a. tri *fear* dheug 's da fhichead   Scottish Gaelic
       three man ten   and two twenty
       'fifty-three men'

   b. da fhichead *fear* 's a   tri-deug
       two twenty   man and PRT three-ten
       'fifty-three men'
       (Hurford 2003; attributed to Paterson 1968)

(27) a. ddwy *flynedd* a   thri ugain a   chan   Biblical Welsh
       two   years   and three twenty and hundred
       mlynedd
       years
       'one hundred sixty-two years'
       (Genesis 5:18)

   b. can   *mlynedd* a   phymtheng *mlynedd* a
       hundred year.ADN and fifteen   year.ADN and
       thri   ugain
       three twenty
       'one hundred seventy-five years'
       (Genesis 25:7)

(28) a. threescore *years* and ten (Psalm 90)
     b. fourscore and seven *years* ago (the Gettysburg Address)

Right-node raising (across-the-board right extraposition of the lexical NP) cannot account for the distribution of the lexical NPs inside the complex cardinals in (26) and (27), but these cases are easy to derive via NP deletion, as shown in (29) for the Scottish Gaelic examples (for ease of exposition, we have added the unpronounced conjunction *'s* 'and' between each pair of adjacent conjuncts).

(29) a. [[tri fear] ~~'s~~ [dheug ~~fear~~] 's [da [fhichead ~~fear~~]]]
     b. [[da [fhichead fear]] 's [a tri ~~fear~~] ~~'s~~ [deug ~~fear~~]]

The right-node-raising hypothesis seems incompatible with the facts above. Right-node raising is normally impossible when the common NP is assigned different cases in the two conjuncts (Borsley 1983; Dyła 1984; Franks 1993; etc.), as occurs in (22)–(25), and the preservation of the leftmost rather than the rightmost lexical NP is also unexpected. However, the NP-deletion approach would seem to be incompatible with the more frequent pattern of case and number marking in additive complex cardinals, where the only surfacing lexical NP is systematically plural and assigned the case determined by the higher cardinals. We turn to this next.

5.2.3 Unexpected Case or Number on the Lexical NP

If additive complex cardinals are derived by NP deletion, we expect that it is always the cardinal adjacent to the overt lexical NP that determines its case and number. However, this is not always the case. In Germanic languages (30) and in Modern Greek (31), for example, the lexical NP in additive complex cardinals ending in 'one' is plural rather than singular.

(30) a. one hundred and one books/*book   English
     b. honderd(en)één   kinderen   Dutch
        hundred.and.one child.PL
        'one hundred and one children'

(31) a. ikosi   enas   anðres/*anðras   Modern Greek
        twenty one.M.SG man.PL/SG
        'twenty-one men'
     b. ikosi   mia   ɣinekes/*ɣineka
        twenty one.F.SG woman.PL/SG
        'twenty-one women'
     (Savu 2012, 18)

To derive the patterns in (30) and (31), Ionin and Matushansky (2006) hypothesize that they are constructed via right-node raising on the assumption that number agreement takes place after right-node raising. However, this view cannot account for the more complex patterns in Inari Sami and Finnish.[5]

In Inari Sami, the lower cardinals 'two' through 'six' assign accusative case to the lexical NP, as in (32a). All higher cardinals, including coordinated cardinals that contain a lower cardinal, assign partitive case to the lexical NP. The lexical NP is assigned partitive case both when the lower cardinal comes before the higher cardinal, as in (32b), and when it comes after the higher cardinal, as in (32c). The same pattern is observed in additive complex cardinals containing 'one' (33), which does not assign any case when simplex (34).

(32) a. kyehti/kulmâ/nelji/vittâ/kuttâ päärni   Inari Sami
         two/three/four/five/six   child.ACC
         'two/three/four/five/six children'
      b. nelji-nubáloh pärnid
         four-teen   child.PART
         'fourteen children'
      c. kyeht-lov-nelji pärnid
         two-ten-four   child.PART
         'twenty-four children'
      (Nelson and Toivonen 2000)

(33) Mun uáinám kyeht-lov-ohtâ poccud.
      I   see   two-ten-one   reindeer.PART
      'I see twenty-one reindeer.'
      (Nelson and Toivonen 2000)

(34) a. Mun uáinám ovdâ   kieðâ.
         I.NOM see.1SG one.ACC hand.ACC.SG
         'I see one hand.'
      b. Ohtâ   kietâ   lii   tobbeen.
         one.NOM hand.NOM.SG is.3SG there
         'One hand is there.'
      (Nelson and Toivonen 2000)

The surface strings in (32b–c) and in (33) cannot be derived either by right-node raising or by NP deletion: neither accounts for the partitive case on the lexical NP. The same issue arises in Finnish, where as a simplex cardinal 'one' does not assign case (35a), yet when it appears as part of a complex cardinal it combines with the lexical NP in the partitive (35b–c).

(35) a. yksi kirja   Finnish
         one book.NOM
         'one book'

b. yksi-toista kirjaa
   one-teen   book.PART
   'eleven books'
c. kaksi-kymmentä-yksi kirjaa
   two-ten.PART-one   book.PART
   'twenty-one books'

It is altogether implausible to suggest that case assignment can follow right-node raising because, as mentioned above, right-node raising is sensitive to the cases assigned inside the conjuncts. This means that neither a right-node-raising account nor the NP-deletion one can explain the above facts without an additional stipulation: namely, that the lowest cardinal does not behave in addition (within an additive complex cardinal) as it does in isolation. That is, 'one' inside 'twenty-one' has different morphosyntactic properties than 'one' occurring on its own: inside 'twenty-one', it assigns the same case as higher cardinals (in Inari Sami and Finnish) and triggers plurality (in Modern Greek and Germanic).

If for some reason, the additive 'one' in (33) and (35b–c) and the additive lower cardinals in (32b–c) assign partitive, then the NP-deletion account makes the correct prediction, as sketched in (36). Likewise, if in Modern Greek and in Germanic the additive cardinal 'one' triggers plural rather than singular agreement inside a complex cardinal, the plural lexical NP in (30) and (31) would follow.

(36)   ConjP   NP deletion

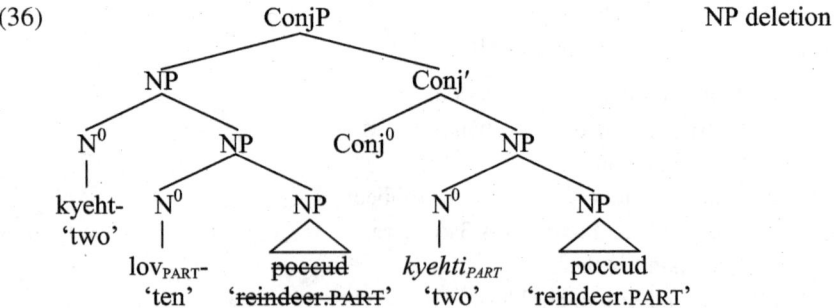

These facts could also potentially be accounted for under the right-node-raising account, with the same stipulation that we need for the NP-deletion account. However, given that we already have independent evidence for NP deletion in additive complex cardinals (see section 5.2.2), the NP-deletion account is to be preferred over the right-node-raising account.

But the solution seems to be altogether ad hoc: why should the morphosyntactic properties of lower cardinals change when they form part of an additive complex cardinal? In section 5.3 we will provide independent evidence that lower cardinals can indeed change their morphosyntax. We will demonstrate that the direction of this change is towards higher nominality and speculate why such more-nominal lower cardinals do not appear in isolation. We will then, in section 5.4, argue against both the right-node-raising approach and the most likely alternative hypothesis, namely that the cascading structure with subsequent NP coordination we have been assuming is not suitable for additive complex cardinals, and that the complex cardinal should be treated as a constituent to the exclusion of the lexical NP.

5.3 Syntactic Changes Triggered by Coordination

Setting aside number marking, Finnish and Inari Sami additive complex cardinals show that inside a complex cardinal, lower cardinals become more nominal: instead of behaving like adjectives that do not affect their sister, as they do in isolation, they assign to it the case that the higher cardinals do. While multiple explanations can be envisaged for these facts, we will argue that they arise as a result of the changed morphosyntactic specification of the lower cardinal. In what follows we will examine a number of languages where it is the lower cardinal itself that undergoes some subtle phonological or morphological change when appearing inside a complex cardinal. We will show that, while in each of these languages an alternative explanation can be suggested, only the hypothesis that the lower cardinal becomes more nominal inside an additive complex cardinal explains the repeated pattern across multiple languages.

5.3.1 Romanian Additive Complex Cardinals
As is easy to see, lower simplex cardinals in Romanian have some adjectival properties. Thus the lowest cardinals, 'one' and 'two', agree with the lexical NP in gender (37). 'One' through 'nineteen' combine with the lexical NP directly (38a), whereas 'twenty' through 'hundred' require the preposition *de* between the cardinal and the lexical NP (38b).[6]

(37) a. doi   băieți   Romanian
       two.M boy.PL
       'two boys'
     b. două fete
       two.F girl.PL
       'two girls'
       (Savu 2012, 19)

(38) a. şapte case
        seven house.PL
        'seven houses'
     b. şaizeci   de minute
        six-ten.PL of minute.PL
        'sixty minutes'

Turning to additive complex cardinals, we see that the preposition *de* is always required, even if the cardinal closest to the lexical NP is one of those that do not occur with *de* on their own, such as 'two' in (39) (cf. (37); the same facts hold for 'three' through 'nine'). The presence of an overt genitive preposition suggests a more nominal character, although, importantly, the lower cardinal here does not become fully nominal, as shown by the fact that it still agrees with the lexical NP in gender. The same is true for additive cardinals that end in 'one', which combine with plural rather than singular lexical NPs (40).

(39) a. două-zeci   şi   doi   *de* băieţi
        two.F-ten.PL and two.M of boy.PL
        'twenty-two boys'
     b. două-zeci   şi   două   *de* fete
        two.F-ten.PL and two.F of girl.PL
        'twenty-two girls'
     (Savu 2012, 19)

(40) a. două-zeci   şi   unu   *de* băieţi/*băiat
        two.F-ten.PL and one.M of boy.PL/SG
        'twenty-one boys'
     b. două-zeci   şi   una   *de* fete/*fată
        two.F-ten.PL and one.F of girl.PL/SG
        'twenty-one girls'
     (Savu 2012, 19)

The Romanian 'one' provides us with some morphological evidence that points to the lower cardinal itself as the locus of change, rather than the additive complex cardinal as a whole or the cardinal-containing NP as a whole. The form that 'one' takes inside an additive cardinal (40) differs from its form when functioning as a simplex cardinal (41). Evidence that the reduced forms *un* and *o*, which also function as indefinite articles in Romanian, truly are cardinals in (41) comes from (42). The fact that 'one' is in focus in (42) contrasting with other possible numbers shows that it cannot be the indefinite article (for semantic reasons) or a form that is reduced only for phonological reasons.

(41) a. un/*unu băiat
     one.M   boy
     'one boy'
  b. o/*una fată
     one.F   girl
     'one girl'
    (Savu 2012)

(42) numai o/*una carte
    only   one.F book
    'only one book'

The fact that the cardinal 'one' takes a different phonological shape when appearing inside an additive complex cardinal can be taken as indication that it is not entirely the same lexical item as the simplex cardinal 'one'. However, the form appearing in additive complex cardinals must be used in contexts of NP ellipsis: while the short form *o* is required in (43a), only the long form is possible in (43b), which is an answer to the question "Did you buy some books?"

(43) a. Am cumparat o/*una carte
    I   bought   one.F book
    'I bought one/a book.'
  b. Am cumparat una/*o.
    I   bought   one.F
    'I bought one.'

We propose that the change in the phonological shape of the cardinal 'one' in Romanian is indicative of a change in its featural specification. To be more specific: We assume—with Munn (1987, 1993), Woolford (1987), and Zoerner (1995), among others (see Progovac 1998 for an overview)—an asymmetric structure for coordination, as in (21) above, with the conjunction treated as the head. It can therefore agree with both its complement and its specifier (the two conjuncts) and therefore the nominal features of one conjunct can be transmitted to the other conjunct. More adjectival cardinals such as 'one' through 'nine' in Romanian thus become more nominal inside additive complex cardinals.

Other explanations can be envisaged. For Romanian, one possibility is that the lexical NP undergoes right-node raising (and the longer form is used before the gap or the trace left by movement); however, this explanation cannot be extended to Polish, which we will discuss now.

5.3.2 Polish Additive Complex Cardinals

Postponing until chapter 6 a full discussion of the extremely complex Polish cardinal system, we observe here that the lower cardinals 'one' through 'four' agree in gender and case with their lexical NP. The examples in (44)–(45) illustrate this.

(44) a. Jeden   chłopiec przyszedł.   Polish
         one.M.NOM boy.NOM came.M.SG
         'One boy came.'
     b. Jedna   dziewczyna przyszła.
         one.F.NOM girl.NOM   came.F.SG
         'One girl came.'
     c. Jedno   dziecko   przyszło.
         one.N.NOM child.NOM came.N.SG
         'One child came.'

(45) a. Dwie   dziewczyny pływały.
         two.F.NOM girl.PL.NOM swam.NV.PL
         'Two girls swam.'
     b. Dwa   krzesła/stoły   rozbiły   się.
         two.NV.NOM chair.N.PL.NOM/table.M.PL.NOM break.PAST.NV.PL REFL
         'Two chairs/tables broke.'
     (Klockmann and Šarić 2015)

When appearing inside an additive complex cardinal, however, the cardinal 'one' loses its ability to agree.[7] The lexical NP that follows it becomes plural and appears in the genitive case, as in (46). The paucal cardinals do not change their behavior, as (47) shows.

(46) a. Czterdzieści *jeden* dziewczyn uczestniczyło.
         forty.NV   one   girl.PL.GEN participate.PAST.N.SG
         'Forty-one girls took part.'
     b. Czterdziestu *jeden* mężczyzn   czekało.
         forty.V   one   man.PL.GEN wait.PAST.N.SG
         'Forty-one men were waiting.'
     (Swan 2002, 199)

(47) Dwadzieścia trzy   koty   bawiły   się.
      twenty.NV   three.NV.NOM cat.PL.NOM play.NV.PL REFL
      'Twenty-three cats were playing.'
      (Swan 2002, 199)

The right-node-raising analysis cannot be adopted for Polish, as it would produce incorrect results for (47). Here the lexical NP continues to agree in case and number with the paucal cardinal 'three' (i.e., with the rightmost cardinal rather than with the entire complex cardinal or with any other simplex cardinal: 'twenty' as a simplex cardinal requires the lexical NP to be genitive plural). Furthermore, the genitive case assigned to the lexical NP by additive complex cardinals, as in (46), patterns with the genitive case assigned by other cardinals (and differs from the genitive case otherwise assigned inside a noun phrase) in that in oblique-case contexts like those in (48) it turns into the oblique case assigned to the entire numeral NP.

(48) a. autobus z   dwudziestu jeden *pasażerami*
         bus   with twenty.INS one   passengers.INS
         'a bus with twenty-one passengers'
         (Swan 2002, 199)

   b. z   tysiącem   siedmdiuset czterdziestu ośmiu   *osobami*
      with thousand.INS seventy.INS forty.INS   eight.INS persons.INS
      'with one thousand seven hundred and forty-eight persons'
      (Bielec 1999, 246)

We explain the above facts by hypothesizing that inside additive complex cardinals, lower cardinals become more nominal, inheriting the nominal features of the higher cardinal, as we suggested for Romanian. The outcome for 'one' and the paucal cardinals in Polish is not identical because of their different underlying featural specifications. As discussed in chapter 6, while singulars in Polish are specified for the feature [αanimate], plurals are only distinguished by the feature [αhuman]. It seems reasonable to assume that the paucal cardinals, being plural, are specified for the feature [αhuman], and that 'one', being singular, is not. The indeclinability of 'one' inside an additive cardinal can therefore be connected to vocabulary-insertion failure in the absence of the feature [αhuman] on a nominal element.

Alternatively, the Polish facts might suggest that the lower cardinal in an additive complex cardinal involves the arithmetical use (see section 2.4), which would explain why in Polish a 'one' in that context does not agree anymore and becomes indeclinable. This hypothesis, however, is not borne out. On the one hand, the paucal cardinals in Polish, just like the cardinal 'one' in Romanian, continue to agree in gender, which does not happen in the arithmetical use. On the other hand, even though the cardinal 'one' changes its behavior in an additive complex cardinal to allow genitive plural, the paucal cardinals continue to determine the case marking on the lexical NP and verbal agreement (47); thus cardinals do not become syntactically uniform inside additive complex cardinals, which in their arithmetical use they do.

5.3.3 Hebrew Additive Complex Cardinals

While the full range of the syntactic and morphological peculiarities of Hebrew cardinals will be examined in chapter 7, we will briefly discuss here the behavior of the adjectival lower cardinals 'one' and 'two' in Hebrew to obtain some further indication of the nature of the change that lower cardinals undergo inside additive complex cardinals. The Hebrew 'one' is postnominal when simplex (49a) and prenominal when part of an additive complex cardinal (49b). The cardinal 'two' has a bound form and unbound form; it takes the bound form only when simplex (50a), not inside an additive complex cardinal (50b). The number of the lexical NP in complex cardinals is systematically plural.

(49) a. delet   exad   Hebrew
        door.F one.F
        'one door'
     b. esrim   ve-exad   dlatot
        twenty and-one.F door.F.PL
        'twenty-one doors'

(50) a. štey/*štayim   dlatot
        two.F_{BOUND}/two.F door.F.PL
        'two doors'
     b. esrim   ve-štayim/*štey   dlatot
        twenty and-two.F/two.F_{BOUND} door.F.PL
        'twenty-two doors'

Once again we note the same general tendency: the lower adjectival cardinals become more nominal, without becoming fully nominal. More specifically, the lower adjectival cardinals retain their ability to agree with the lexical NP, while losing certain peculiarities of behavior. With regard to the bound–unbound alternation in (50), adjectives and nouns do not differ in Hebrew in their ability to take the bound form, but the lack of such a form is characteristic of measure nouns, suggesting that it is this subclass of nouns that additive cardinals approximate.

5.3.4 German: Multiple Options

German presents a particularly interesting case, in that with any complex cardinal higher than 'one hundred' that ends in 'one', the lexical NP can be either singular, as in (51a), or plural, as in (51b–c). If the lexical NP is singular, the cardinal *ein* 'one' is inflected for case and gender (51a). If it is plural, the cardinal *ein* either takes the uninflected form (51b) or appears as *eins* (51c), the form that it normally takes under NP ellipsis. The latter option is one that is becoming more common in German. (The examples in (51) are from Fabricius-Hansen et al. 2009; the translations were provided, along with comments, by Klaus von Heusinger, pers. comm.)

(51) a. tausend-und-eine   Nacht   German
        thousand-and-one.F night
        'a thousand and one nights'
     b. tausend-und-ein   Ideen
        thousand-and-one idea.PL
        'a thousand and one ideas'
     c. tausend-und-eins   duftende Kräuterschätze
        thousand-and-one.INFL fragrant herb.PL
        'a thousand and one fragrant herbs'

These data show that German allows all three options discussed so far. Example (51a) is a case of simple NP deletion with the overt NP agreeing with 'one', as in Russian (section 5.2.2). Example (51b) behaves like Polish (section 5.3.2), with the default, uninflected form of 'one'. And in (51c), we see a special form of 'one' appearing inside a complex cardinal, similar to what happens in Romanian (section 5.3.1).

5.3.5   Interim Summary

To sum up, we have seen evidence from four different languages that lower cardinals change their properties when occurring inside an additive complex cardinal. In Romanian, lower cardinals inside an additive complex cardinal require the preposition *de* 'of'; in Polish, they assign the same case as higher, more nominal, cardinals; and in Hebrew, they occur prenominally and disallow the bound form. These apparently disparate phenomena all have two things in common: the lower cardinal inside an additive complex cardinal takes on the properties of higher cardinals, and these properties characterize the cardinal as less adjectival and more nominal. The analysis according to which the more nominal form of the lower cardinal results from agreement with a higher cardinal also explains why these more nominal forms do not occur as simplex cardinals (e.g., why the Hebrew 'one' is not prenominal when it is simplex).

In other languages the same pattern is observed. In Inari Sami and Finnish, lower cardinals inside additive complex cardinals become more nominal and as a result assign the same case as higher cardinals, and likewise in English, the cardinal 'one' inside additive complex cardinals triggers plural marking as a result of having become more nominal, even though the connection between plural marking and nominality is not direct. The agreement of the lower cardinal with the higher, combined with the NP-deletion mechanism, allows us to derive additive complex cardinals crosslinguistically.

5.4 Alternative Takes on the Formation of Additive Complex Cardinals

In the previous sections, we have proposed that the syntax of additive complex cardinals can be accounted for by the assumption that the lower cardinal used inside an additive complex cardinal may have a different featural makeup from its equivalent used in isolation. We have shown that crosslinguistically, two patterns emerge with regard to the syntax of additive complex cardinals: the lexical NP can behave in the way determined by the cardinal it is adjacent to (Russian) or as if it were combining with a higher cardinal (Hebrew, Inari Sami, Finnish, English, etc.). In addition, some languages (Polish) fall in between, with behavior varying as a function of the cardinal. We have argued that agreement between the lower and higher cardinal, followed by NP deletion of the lexical NP in one of the conjuncts, can account for all of these patterns.

We now consider whether the patterns that we observe can be accounted for in a different way: either (a) by using right-node raising to explain why only one instance of the lexical NP is present in the surface representation or (b) by using a totally different syntactic structure, namely adjunction of the complex cardinal to the NP. We will attempt to show now that neither of these alternatives provides a better solution.

5.4.1 Right-Node Raising

The tree in (52) shows how to derive additive complex cardinals via right-node raising (for analyses of right-node raising as rightward movement, see Ross 1967; Postal 1974; Abbott 1976; Grosu 1976; and Sabbagh 2003, among others; for alternatives see Wilder 1997, 1999; Hartmann 2000; and Bachrach and Katzir 2009, among others).

(52)

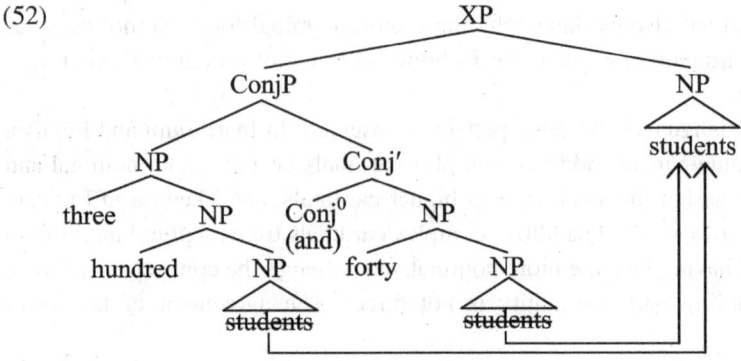

As discussed in section 5.2.3, right-node raising should fail if the lexical NPs in the conjuncts are not the same. Once we assume, however, that the properties of the lower cardinals can change to match the properties of higher cardinals, the right-node-raising analysis becomes applicable. Our reasons against adopting the right-node-raising approach are as follows. First, additive complex cardinals can also be constructed with an adposition (section 5.6), in which case, the right-node-raising approach is inapplicable. Second, right-node raising cannot derive additive complex cardinals in languages such as Biblical Welsh and Scottish Gaelic, where the lexical NP can occur in either conjunct (see section 5.2.2). Given that we already need NP deletion to account for such languages, positing right-node raising for other languages is superfluous: it is conceptually simpler to have only one strategy for deriving additive complex cardinals, and we have shown that NP deletion can account for additive complex cardinals in all the languages under consideration.

5.4.2 NP Adjunction

The fact that additive complex cardinals in languages such as Polish, Romanian, Finnish, and Inari Sami are incompatible with the analysis involving coordinated numeral NPs without modifications to account for the unexpected case and number marking found in these languages may suggest that the entire analysis is on the wrong track. While there is, as we have shown in previous chapters, abundant evidence that simplex cardinals function as heads and that multiplication involves complementation, perhaps the facts discussed in section 5.2.3 suggest that the lexical NP should be treated as a sister to the entire complex cardinal, as in (53). To achieve compositionality it then becomes necessary to either type shift the lowest cardinal in each conjunct ($\langle\langle e,t\rangle,\langle e,t\rangle\rangle \rightarrow \langle e,t\rangle$), as in (53a), or to hypothesize a null nominal pro-form or a null classifier in these positions, as in (53b).

(53) a.   NP   Type shift

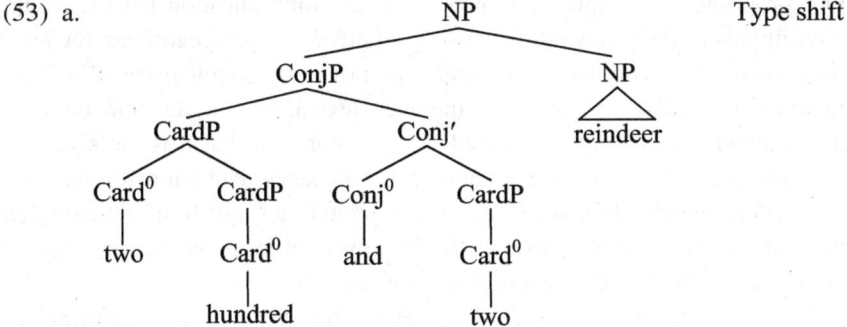

b.

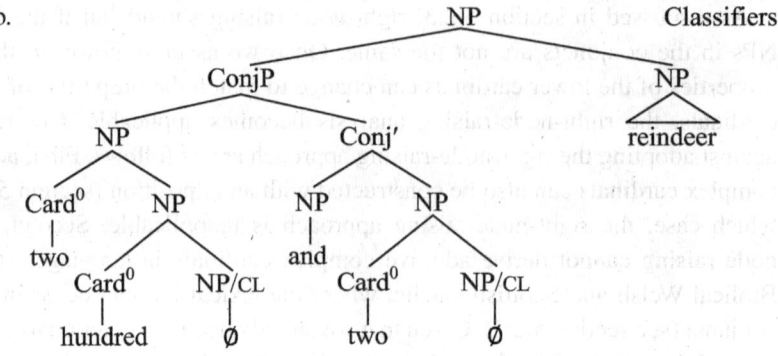

Both structures in (53) run into a problem with languages like Polish or Russian, where the case and number marking on the lexical NP depends on the choice of the lowest cardinal, as it is expected in both of these structures that the case and number on the NP sister of the complex cardinal will be independent of the internal composition of the complex cardinal. Not only does it have to be assumed that the featural specification of the rightmost cardinal is somehow transmitted to the complex cardinal as a whole, it must also be assumed, to account for the divergence between 'one' and other paucal cardinals in Polish ((46) versus (47)), that 'one' changes its properties when used as part of a complex cardinal while other paucal cardinals do not. While the hypothetical type shift in (53a) provides some motivation for why the properties of the cardinal should change, once it is assumed that the properties of the cardinal can change, there is no more need to appeal to a type shift as well, and the coordination-cum-deletion structure discussed above becomes preferable for reasons of theoretical simplicity.

5.4.3 Interim Summary

Our proposal for complex cardinals involving multiplication led us, as did crosslinguistic data, towards analyzing additive complex cardinals (at least those involving coordination) as conjunction of cardinal-containing NPs. This hypothesis predicts, however, that the overt lexical NP should show the case and number morphology associated with the cardinal that it is the sister of. This prediction is correct for a number of languages, including Russian, yet wrong for a much larger set of languages, including English, where case and number marking on the lexical NP becomes uniform across all additive complex cardinals, regardless of their internal composition.

To account for the differences between the behavior of a lower cardinal in isolation and inside an additive complex cardinal, we proposed that inside an additive complex cardinal it becomes "more nominal," without yet specifying what this corresponds to on either the semantic side or the morphosyntactic one. We further observed that any approach attempting to derive additive complex cardinals from simplex cardinals faces the challenge of explaining the specific changes in behavior of certain lower cardinals when occurring inside additive complex cardinals in various languages. We then showed that once the mechanism driving this change is assumed to exist, alternatives to an approach involving NP deletion do not provide any advantage. Meanwhile, NP deletion itself has to be hypothesized in any case to explain languages such as Russian, where the entire additive complex cardinal behaves like the last simplex cardinal in it does when in isolation.

5.5   The Role of Convention in Coordination

The above account of the syntax of additive cardinals overgenerates: it predicts both nonstandard coordination (e.g., *twenty-fifteen* to mean 'twenty plus fifteen', i.e., 'thirty-five') and free appearance of both asyndetic and overt coordination (e.g., *thirty and nine, a million two*, etc.). In this section, we argue that such overgeneration is prevented by extralinguistic constraints, just as we argued for extralinguistic constraints on ordering inside additive complex cardinals (section 5.1.3).

5.5.1   Nonstandard Coordination

We propose that cardinal coordination is not constrained semantically but rather that extralinguistic, arithmetical, constraints are involved: for instance, languages that use a base-ten system typically disallow the coordination of two simplex cardinals whose value is at least ten but below one hundred (hence the impossibility of *thirty-fifteen, twenty-ten*, etc.), though there are exceptions (e.g., the French *soixante-dix*, lit. 'sixty-ten', for 'seventy').

As already noted in section 5.1.3 the presence of an overt conjunction alleviates some of the constraints on complex cardinals. As shown in (54), the presence of *and* makes unconventional coordination at least marginally possible.

(54) a. *twenty-seventeen books   Asyndetic coordination
       b. #twenty and seventeen books   Overt conjunction

As indicated by the grammaticality judgments, in the presence of an overt conjunction, convention can sometimes be overridden. A convention-overriding conjunction occurs in (55), a line from a children's poem by the Russian poet Teffi.

(55) tridcat' tri   *i*   dva kota   i   četyre koški   Russian
     thirty   three and two cat.M.PL and four   cat.F.PL
     'Thirty-three and two tomcats and four tabbycats.'

The possibility of (55) indicates that the constraints on coordination can
be overridden when there is some reason to separate two pluralities, and
this separation requires an overt coordination. As discussed in Ionin and
Matushansky 2006, measure phrases exhibit a parallel behavior. Normally, the
larger unit (*feet*) has to appear before the smaller unit (*inches*), as shown in
(56a); with asyndetic coordination, the reverse order is ungrammatical (56b).
However, when an overt conjunction is used, convention can be overridden:
both orders are possible, as indicated by (56c–d). For (56d) to be acceptable, it is helpful to clearly separate the five inches of silk from the five feet
of silk in space and/or time. This is parallel to what happens with cardinals
in (55).

(56) a.   I bought five feet five inches of blue silk.
     b. ??/*I bought five inches five feet of blue silk.
     c.   I bought five feet and five inches of blue silk.
     d.   I bought five inches and (then) five feet of blue silk.

It is possible that an overt conjunction changes the prosodic properties
of the cardinal-containing NP and makes it compatible with the new
interpretation.

To sum up, we suggest that extralinguistic, arithmetic constraints determine
which cardinals are coordinated, and in what order. We draw a parallel between
these arithmetical constraints and the constraints on coordination of titles. As
already noted (section 5.1.3), in English, a married couple may be referred to
as *Mr. and Mrs. Smith* but not as *Mrs. and Mr. Smith*. In addition to this ordering constraint, there is a constraint on which titles can go in which position:
Dr. and Mrs. Smith is completely acceptable, but in many social circles, *Mr.
and Dr. Smith* is not an acceptable way of referring to a woman with a doctor-
ate and her husband. This is a purely social constraint, not a linguistic one (as
evidenced by the fact that *Mr. and Dr. Smith* is becoming acceptable as
women's social roles change). We believe that the constraint against *twenty-
seventeen* is similarly traceable to extralinguistic (in this case, arithmetic)
conventions.

5.5.2 Crosslinguistic and Intralinguistic Variation in the Realization of Conjunction

As discussed in Ionin and Matushansky 2006, cardinals differ as to whether
they use asyndetic or overt coordination, both within the same language and

among different languages. For instance, in English, *fifty-three* disallows an overt *and* while *one thousand and one* requires it (at least for some speakers), and in *three hundred (and) fifty* it is optional. Similar differences exist in the coordination of other phenomena, such as times. For example, overt conjunction is optional for the coordination of minutes and seconds (57a) but obligatory, or at least strongly preferred, for the coordination of months and days (57b). Thus, coordinated cardinals in which an overt conjunction is obligatory versus optional have a parallel in other types of coordinated structures.

(57) a. She swam across the pool in three minutes (and) fifteen seconds.
  b. Their vacation lasted exactly two months ??/*(and) twelve days.

The constraints on coordination of cardinals have to be explained by any theory of complex cardinals. For example, if complex cardinals are treated as morphological compounds, arithmetic constraints on compound formation have to be assumed. This is no more explanatory than assuming that such constraints are in the syntax, as we do.

Case marking inside an additive complex cardinal can be taken as evidence against our NP-conjunction approach. As illustrated in (58)—from Miljan and Cann 2013, with glosses adjusted (see also Õispuu 1999, 127)—in Estonian only the last cardinal in an additive complex cardinal is case-marked by the case assigned from the outside; all prior conjuncts are marked genitive.

(58) Kõigil   *saja*   *kolmekümne* viiel   Estonian
    all.SG.ADE hundred.SG.GEN thirty.SG.GEN five.SG.ADE
    õpilasel   oli   pilet   olemas.
    student.SG.ADE be.PAST.3SG ticket.SG.NOM be.INF.INE
    'All one hundred and thirty-five students had a ticket.'

In section 6.7 we demonstrate on the basis of Russian that the situation is actually more complex: it can be that some but not all nonrightmost cardinals inside a complex one fail to be marked with the externally assigned case, and this shows that case-marking failure cannot be taken as evidence that the complex cardinal forms a constituent to the exclusion of the lexical NP.

Another phenomenon that can be attributed to a conventionalized arithmetic constraint is the *und* ('and') reduction found with complex cardinals in German. As shown by Wiese (2003), *und* may be phonologically reduced to *'n* when occurring inside a complex cardinal (*drei-'n'-achtzig*, lit. 'three-'n-eighty'), but (unlike the English *and*) it may not be phonologically reduced outside of the domain of cardinals (e.g., with proper names: **Kai 'n Achim* to mean 'Kai and Achim'). We believe that this again is due to convention, rather than to a special status of cardinals.

5.6 Adpositional Strategies inside Complex Cardinals

As discussed in section 5.1, in addition to coordination (both overt and asyndetic), languages use a number of adpositional strategies for forming nonmultiplicative complex cardinals. These include adpositional addition ((2) and (3)), and adpositional subtraction, ((4) and (5)). We address these two strategies in more detail below. Other rare strategies, including overcounting and division, are discussed by Comrie (2005a), see also note 8.

5.6.1 Adpositional Strategies in Encoding Addition

Languages use two adpositional strategies for encoding addition: the locative strategy (59a) and the comitative strategy (59b).[8]

(59) a. *Locative preposition:* ar *'on'*
　　　[ddeng mlynedd [_{PP}ar hugain]] a　　chant　　　　　Biblical Welsh
　　　ten　year.PL　　on twenty　and hundred
　　　'a hundred and thirty years'
　　　(Genesis 5:2; Hurford 1975, 175)

　　b. *Comitative preposition:* ə̀d *'with'*
　　　mæráw médd-æn　ə̀d　　sæmmos　　　　Tamashek (Tuareg of Mali)
　　　ten.M　man-M.PL with five
　　　'fifteen men'
　　　(Heath 2005, 251–252)

From the syntactic standpoint the two strategies in (59) do not differ: both involve what looks most like adjunction of a PP to an NP. We will retain the label PP for both prepositional phrases, as in (59), and postpositional phrases, as in (60), from Mano (a Mande language of Liberia and Guinea).

(60) a. ɓɔ́ɔ́ ŋwū　　pèèlɛ̄ là vṳ yììsɛ̄ wɛ́lɛ́ yààkā　　　　　　Mano
　　　bag hundred two　on ten four unit three
　　　'two hundred forty-three bags'
　　b. wáá　　pèèlɛ̄ (là) ŋwū　　yààkā là vṳ sɔ́ɔ́lī wɛ́lɛ́ sáládō
　　　thousand two　on hundred three　on ten five　unit six
　　　'two thousand three hundred fifty-six'
　　　(Khachaturyan 2015, 58)

However, because of the pragmatics of the adpositions involved, we expect an asymmetry between spatial and comitative adpositions in the ordering of higher versus lower cardinals. With locative adpositions, the ground (i.e., the complement of the adposition) is normally bigger than the figure: in the absence of special context, (61a) is a more pragmatically felicitous statement than (61b). Conversely, with comitative adpositions, it is the figure that is normally bigger than the ground, as shown in (61c–d).

(61) a.   the bottle on the desk
      b.   #the desk under the bottle
      c.   the cup with the handle
      d.   #the handle with the cup

The ordering of higher versus lower cardinals reflects this asymmetry. In (59a), with a locative preposition, the lower cardinal is the figure and the higher cardinal is the ground. In (59b), with a comitative preposition, the opposite is the case. In (60), with a locative postposition, even though the order is the opposite of that in (59a), the structure is the same: it is the higher cardinal that functions as the complement of the adposition. In other words, adpositions used inside complex cardinals appear to retain their original pragmatic properties. One difference is that in the case of cardinals, the order is grammaticalized rather than only preferred. The same holds for coordination, as discussed in section 5.1.3.

Given the above, we propose that adpositional complex cardinals have the same structure as other PPs in the same language, as schematized in (62).

(62)

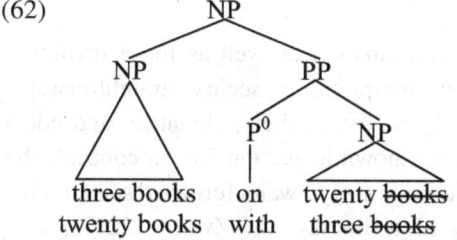

The tree in (62) has the lexical NP appearing overtly in the figure rather than the ground. However, crosslinguistically, an overt lexical NP can show up in either figure or ground. In Biblical Welsh (from Hurford 1975), the lexical NP can appear overtly in the figure, as in (63a), or in both figure and ground, as in (63b) (we have been unable to find examples where it appears in the ground only). It can also be omitted in both figure and ground, provided that it is present elsewhere in the complex cardinal, as in (63c): as this complex cardinal is built using additive coordination as well as adpositional coordination, the lexical NP is provided in the conjunct that does not involve adposition. (For more discussion of complex cardinals in Biblical Welsh, please see chapter 7.)

(63) a. un   mlynedd ar [ddeg ar hugain]
         one year   on ten   on twenty
         'thirty-one years'
         (2 Kings 22:1, 2 Chronicles 34:1; Hurford 1975, 198)
     b. [bymtheng mlynedd a   phedwar ugain] ac wyth gan   mlynedd
         fifteen   year   and four   twenty on eight hundred year
         'eight hundred and ninety-five years'
         (Genesis 5:16; Hurford 1975, 182)
     c. can   mlwydd a   [deg ar hugain]
         hundred year   and ten on twenty
         'one hundred and thirty years'
         (2 Chronicles 24:15; Hurford 1975, 183)

No theory that complex cardinals are built extralinguistically or in the lexicon can successfully explain why syntactic mechanisms are used in their construction or why the pragmatic properties of adpositions are retained. The use of coordination (especially asyndetic coordination) in complex cardinals could be viewed as a quirk of lexical composition. However, the availability of other strategies, all independently attested in the syntax of the languages where they are used, can only be accounted for by a syntactic approach to complex cardinals.

For complex cardinals involving coordination as well as those involving comitative adpositions, the additive interpretation seems straightforward: comitative prepositions are additive by virtue of their semantics, and coordination too can be additive, as will be shown in section 5.8. In contrast, the compositional semantics for the locative strategy awaits further research. One promising direction is provided by Matushansky and Zwarts (2017), who explore locative prepositions in the domain of measurement (see section 2.3).

5.6.2 Adpositional Strategies in Encoding Subtraction
One of the minor strategies noted by Hurford (1975) for constructing complex cardinals is subtraction, which can be encoded in two ways, by a caritative preposition or a directional one. The first option is illustrated in (64), from Biblical Welsh, involving the caritative preposition *onid* 'if not, unless, except' (Pughe 1832, 389). The second option is exemplified by (65), from Latin, with the ablative preposition *de* 'from', and by (66), from Ainu, with the allative particle *e* 'to' (Batchelor 1905).

(64) onid   un mlwydd   cant   Welsh
      unless one year.of.age hundred
      'ninety-nine years'
      (Hurford 1975, 198)

(65) Un-de-viginti   pueri duo-de-quadraginta manus   habent. Latin
     one-from-twenty boy.PL two-from-forty   hand.PL.ACC have.3PL
     'Nineteen boys have thirty-eight hands.'

(66) wan e ine hot ne   Ainu
     ten to four twenty
     'seventy'
     (Batchelor 1905, 97)

Assuming our proposal, both the minuend (the part being diminished) and the subtrahend (the part being subtracted) must contain a copy of the lexical NP, as schematized in (67) and in (68a). However, caritative and ablative subtractions do not have the same internal structure. Caritative subtraction clearly has the same structure as addition: the subtrahend functions as the complement of the adposition, and the minuend is the NP modified by the resulting PP, as schematized in (67) for the Welsh example in (64). Ablative subtraction seems different. If the structure in (67) were used for (68a), it would give rise to an incorrect semantics: *de* is an ablative preposition requiring a complement that can denote a location, which 'twenty boys' cannot.[9] In fact, ablative subtraction strongly resembles degree modification of locative PPs, schematized in (68b). If, however, the subtrahend is a degree modifier of the minuend, then the minuend and indeed the entire cardinal-containing NP in (68a) is a PP.

(67) [NP [PP onid [NP un mlwydd]] [NP cant ~~mlwydd~~]]

(68) a. un ~~puer~~ de viginti pueri
     b. [PP [NP two feet] [from [NP the house]]]

The complexity of adpositional addition and subtraction requires developing a theory of how locative adpositions function in nonspatial domains. Once again, we refer the reader to section 2.3 and the work of Matushansky and Zwarts (2017), who propose an approach to the compositional semantics of prepositional measures such as *under twenty feet*. Their approach seems a potential starting point for such a theory of adpositional addition and subtraction, given the similarity between cardinals and measures explored elsewhere in this work. Indeed, adpositional addition and subtraction are attested with measure NPs, in particular in the domain of time measurement. Examples (69a–b) illustrate locative subtraction and addition, respectively, while (70) is a case of caritative subtraction.

(69) a. tien (minuten) voor   zes   Dutch
       ten minute.PL in front of six
       'ten minutes to six'

b. tien (minuten)   over half zes
   ten   minute.PL over half six
   'twenty minutes to six'

(70) bez   pjati   (minut)   desjat' časov   Russian
     without five.GEN minute.PL.GEN ten   hour.PL.GEN
     'five minutes to ten'

We leave a fully worked-out implementation of the semantics of adpositional addition and subtraction for future research.

5.7 Summary: The Syntax of Nonmultiplicative Complex Cardinals

In this chapter, we have shown that additive complex cardinals can be constructed by means of conjunction (asyndetic or overt) as well as locative and comitative adpositions. These facts strengthen our proposal that complex cardinals are built by syntactic means used elsewhere in the language. We have provided an analysis of the syntax of additive complex cardinals that makes use of NP deletion as a means of removing all instances but one of the lexical NP. We have also shown that lower cardinals undergo morphosyntactic changes inside additive complex cardinals, and we have attributed this to agreement. Finally, we have discussed the role of extralinguistic constraints that determine which cardinals can combine inside an additive complex cardinal, and in which order. We have drawn parallels with other domains to show that such extralinguistic constraints are not specific to cardinals.

Now that we have provided an analysis of the syntax of additive cardinals, we move on to a consideration of their semantics. In the next section, we provide a semantic analysis of complex cardinals involving coordination, leaving the semantics of adpositional complex cardinals for future research.

5.8 The Semantics of Coordinated Cardinals

On our analysis, an NP like *twenty-two books* involves the coordination of *twenty books* and *two books*, where both are predicates over semantically plural individuals. It is easy to show that the standard Boolean semantics of *and* in (71) (Partee and Rooth 1983), type-lifted to apply to predicates, does not yield the expected result for this NP, as shown in (72).

(71) $[\![\text{and}]\!] = \lambda f \in D_{\langle e,t \rangle} . \lambda g \in D_{\langle e,t \rangle} . \lambda x \in D_e . f(x) \wedge g(x)$

(72) $[\![\text{and}]\!] ([\![\text{two books}]\!])([\![\text{twenty books}]\!]) \approx \lambda x . [\![\text{twenty books}]\!] (x) \wedge [\![\text{two books}]\!] (x)$

As discussed in Ionin and Matushansky 2006, the reading in (72) is not available for cardinal-containing NPs involving addition: this is expected for pragmatic reasons, since nothing can be simultaneously twenty books and two books.[10] Instead, we need an analysis that can successfully capture the actual meaning of *twenty-two books*, which is that of a set of plural individuals each of which consists of a sum of twenty books and two books.

As discussed above, we hypothesize that addition is achieved via asyndetic coordination; however, the standard Boolean meaning of *and* does not yield the right truth conditions. In the remainder of this chapter, we will discuss the collective interpretation of *and* (Link 1983, 1984; Krifka 1990a; Lasersohn 1995, chap. 14; and Le Bruyn and De Swart 2014, among others) and how to derive it from the standard intersective (Boolean) interpretation. We will first present the solution proposed by Heycock and Zamparelli (2000, 2005) for plural predicates inside DPs. In (73), two different meanings are available, as shown in (73a–b), but only (73a) can be achieved by the standard Boolean coordination.

(73) His friends and colleagues came to the party.
    a. A set of people each of whom is his friend *and* his colleague came to the party.
    b. A set of people each of whom is his friend *or* his colleague came to the party.

We will then show that when applied to additive complex cardinals, Heycock and Zamparelli's analysis runs into problems, one of which is its incompatibility with adpositional cardinals (section 5.6). We will present Champollion's alternative to Heycock and Zamparelli's approach, which has the advantage of deriving the additive interpretation of *and* from its standard Boolean meaning by manipulating the interpretation of the conjuncts. We will show that Champollion's analysis can be extended to adpositional cardinals. Finally, we will present evidence that *and* can be interpreted disjunctively with cardinals just as elsewhere, and that neither of the two analyses mentioned above can account for this interpretation. We will sketch a proposal for how to analyze such disjunctive conjunction.

5.8.1 Set Product

The fact that the intersective meaning of *and* in (71) cannot account for all types of conjunction has already been discussed in the literature with respect to conjoined entities, as in (74), and conjoined predicates, as in (75).

(74) Mary and John slept.

(75) a. The books are old and new.

    b. These men and women met in the park.

As shown by Krifka (1990a), Lasersohn (1995), and Winter (1996, 1998, 2001b), the standard view of Boolean coordination leads to problems with such examples: under standard assumptions, the proper names in (74) are not predicates; (75a) does not mean that each book is old and new simultaneously; and in (75b), the intersection of the set of men and the set of women is empty. All of these cases require an additive interpretation of *and*.

With respect to additively interpreted conjoined predicates (75), one solution that has been proposed is to hypothesize an additional meaning for *and*, that of set product (Heycock and Zamparelli 2000, 2005). We will first describe this proposal, then show how we adopted it in Ionin and Matushansky 2006 for cardinal-containing NPs involving addition.

Heycock and Zamparelli (2000, 2003, 2005) discuss the fact that conjoined NPs can have either intersective or additive readings (which they term the *joint* and *split* readings, respectively).

(76) a. That liar and cheat is not to be trusted.   Intersective

    b. My doctor and dentist are both sailors.   Additive

    c. The men and women sat down to lunch.   Additive

    d. The soldier and sailor came home from the voyage.   Ambiguous

In (76a), the conjoined NP has an intersective reading: the same individual is both a liar and a cheat. In contrast, (76b–c) are examples of additive readings: in (76b), the properties of 'doctor' and 'dentist' are borne by two different individuals, and similarly in (76c), some individuals are men and some are women. While in (76a–b) agreement marking on the verb disambiguates in favor of either the intersective or the additive reading, in (76c) the intersective reading is ruled out by pragmatic considerations (since no individual can be both a man and a woman at once). Sentences like (76d) are ambiguous between the two readings: this sentence can mean either that a person who is both a soldier and a sailor came home (77a) or that two people, one a soldier and the other sailor, came home (77b).

(77) a. $\lambda x . \text{soldier}(x) \wedge \text{sailor}(x)$ Intersective

    b. $\lambda X . X = y \oplus z \wedge \text{soldier}(y) \wedge \text{sailor}(z)$ Additive

Heycock and Zamparelli (2000, 2003, 2005) derive both the intersective and additive readings of *and* by assuming that *and* returns a set product, as defined in (78). Heycock and Zamparelli (2000, 2003, 2005) define set product in terms of sets; in (79), we redefine it in terms of plural individuals, given that our analysis is principally couched in these terms.

(78) *Set product (SP)*

$\mathrm{SP}(A_1,...,A_n) =_{\mathrm{def}} \{X : X = a_1 \cup ... \cup a_n, a_1 \in A_1, ... , a_n \in A_n\}$

(79) *Set product (SP)*

$\mathrm{SP}(A_1,...,A_n) =_{\mathrm{def}} \{X : X = a_1 \oplus ... \oplus a_n, a_1 \in A_1, ... , a_n \in A_n\}$

To illustrate how set product works, consider the three sample scenarios in (80) (based on Heycock and Zamparelli 2000, 2003, 2005) for the singular-NP coordination *soldier and sailor*.

(80) a. $[\![\text{soldier}]\!] = \{a,b,c\}$, $[\![\text{sailor}]\!] = \{m,n,o\}$

$[\![\text{soldier and sailor}]\!] = \mathrm{SP}([\![\text{soldier}]\!], [\![\text{sailor}]\!])$

$= \{a \oplus m, a \oplus n, a \oplus o, b \oplus m, b \oplus n, b \oplus o, c \oplus m, c \oplus n, c \oplus o\}$

  b. $[\![\text{soldier}]\!] = \{a,b,c\}$, $[\![\text{sailor}]\!] = \{b,c,d\}$

$[\![\text{soldier and sailor}]\!] = \mathrm{SP}([\![\text{soldier}]\!], [\![\text{sailor}]\!])$

$= \{b,c,a \oplus b,a \oplus c,a \oplus d,b \oplus c,b \oplus d,c \oplus d\}$

  c. $[\![\text{soldier}]\!] = \{a,b,c\}$, $[\![\text{sailor}]\!] = \{a,b,c\}$

$[\![\text{ soldier and sailor}]\!] = \mathrm{SP}([\![\text{soldier}]\!], [\![\text{sailor}]\!])$

$= \{a,b,c,a \oplus b,a \oplus c,b \oplus c\}$

The scenario in (80a) represents the default case, where the set of soldiers and the set of sailors are completely disjoint; the set-product operation therefore results in *soldier and sailor* denoting a set of plural individuals (pairs) where one member is a soldier and the other a sailor. Thus, *the soldier and sailor* will necessarily denote a pair of individuals. There are, however, some special cases, where the two sets are not disjoint in the model. For instance, in (80b), the two sets denoted by *soldier* and *sailor* are not disjoint, representing a case of partial overlap. Therefore, *soldier and sailor* now denotes a set consisting both of single individuals (each of whom is a soldier and a sailor at once) and of plural individuals (pairs where one member is a soldier and the other a sailor). As a result, *the soldier and sailor* could, in this scenario, denote either a single individual or a pair of individuals. Finally, (80c) represents a case of complete overlap between the two sets (the intersective reading).

As shown above, Heycock and Zamparelli derive intersective readings (where *a soldier and sailor* denotes one individual, like in (80c)) and additive readings (where it denotes a plural individual, like in (80a)) via the same semantic operation (set product). However, there is reason to believe that these are in fact two different readings, with the intersective reading derived by the standard Boolean *and*. Evidence comes from intensional contexts such as (81), which can be used to express a hiring request for a single position; hiring two people, one a coordinator and the other an administrator, would not satisfy this hiring request. This indicates that the intersective and additive readings are truth-conditionally distinct. We therefore have to assume that the intersective meaning of *and* is available.

(81) We need to hire a coordinator and administrator for this language program.

For Heycock and Zamparelli, coordination of plural NPs (e.g., *soldiers and sailors*) is assumed to involve first set-product formation followed by application of the pluralizing operator (closure under sum), as illustrated in (82). Note that plurals of set products are simple sets of plural individuals. In other words, Heycock and Zamparelli (2000, 2003, 2005) assume, like us, that the plural marking on the NP does not correspond to the star operator but rather reflects agreement with a higher star operator.[11]

(82) a. $[\![\text{soldier}]\!] = \{\{a\},\{b\}\}$, $[\![\text{sailor}]\!] = \{\{m\},\{n\}\}$
  b. $[\![\text{soldier and sailor}]\!] = \text{SP}([\![\text{soldier}]\!], [\![\text{sailor}]\!])$
  $= \{a \oplus m, a \oplus n, b \oplus m, b \oplus n\}$
  c. $*[\![\text{soldier and sailor}]\!] = *\{a \oplus m, a \oplus n, b \oplus m, b \oplus n\}$
  $= \{a \oplus m, a \oplus n, b \oplus m, b \oplus n, a \oplus m \oplus n, a \oplus b \oplus m,$
  $a \oplus b \oplus n, b \oplus m \oplus n, a \oplus b \oplus m \oplus n\}$

Assuming the intersective analysis of cardinals, Heycock and Zamparelli (2000, 2003, 2005) derive the meaning of, for instance, *three soldiers and sailors* by picking out all sets with the cardinality 3 from the extension of *soldiers and sailors* given in (82c). As we will show in section 5.8.4, with our semantics for cardinals, set product does not derive the right interpretation of *three soldiers and sailors*. For now, however, we turn to additive complex cardinals like *thirty-three sailors*, for which the set-product analysis does work.

5.8.2 The Application of Set Product to Additive Complex Cardinals

In Ionin and Matushansky 2006 we use the set-product operation to derive the meaning of additive complex cardinals, such as *twenty-two books*. On our assumptions, *twenty-two books* is underlyingly *twenty books and two books*. Both conjuncts (*X* and *Y*) are sets of plural individuals (consisting *X* of twenty books and *Y* of two books), and nothing prevents the set-product operation from applying. *Twenty books and two books* will then denote a set of plural individuals each combining a twenty-book plural individual with a two-book plural individual and thus consisting of twenty-two books in total. In the same manner, the set-product operation can successfully derive the meaning of any two conjoined cardinal-containing NPs, such as *five soldiers and seven sailors*.

However, as discussed above (see (80b–c)), set product allows for the possibility of overlap: the same individual student could in principle be part of both the twenty-student plural individual and the two-student plural individual whose set product is denoted by *twenty-two students*, with the result that twenty- and twenty-one-student plural individuals would be included in its meaning. Yet *twenty-two students* does not have this meaning. So why is overlap impossible for cardinal-containing predicates, while it is allowed elsewhere?

By our hypothesis, (83a) is derived from (83b) by NP deletion. To the extent that (83b) is felicitous, overlap is equally impossible in it. Furthermore, the same lack of overlap occurs even when the lexical NPs in the two conjuncts are different: only a mathematician set on devising a puzzle would treat (83c) as being about twenty people, even though teaching assistants are nearly always graduate students. Finally, the situation is not specific to NPs containing cardinals: we see the same lack of overlap in (83d).

(83) a. Twenty-two students came to the party.
     b. Twenty students and two students came to the party.
     c. Twenty graduate students and five teaching assistants came to the party.
     d. All graduate students and several teaching assistants came to the party.

The lack of overlap is also seen when measurements or money are considered: (84a) cannot be about a mere six feet of silk, with the six inches included in the six feet, and (84b) cannot be about five dollars, with the seventy-five cents included in the five dollars.

(84) a. I bought six feet (and) six inches of finest silk.
     b. This cost five dollars (and) seventy-five cents.

We propose that the lack of overlap is pragmatic: the whole purpose of measuring and counting is to achieve the maximal precision given the context and the speaker's knowledge. Treating overlap as a possible option is expressly contrary to this purpose. Some deviations are possible, as in riddles (85a), but only when expressly signaled, as by overt conjunction. While it is difficult to construct corresponding examples with cardinal-containing NPs, it is not impossible, as shown by (85b). Overlap is only possible with an overt conjunction (85b) and is impossible with asyndetic coordination (85c). This parallels the effect that we observe with *respectively* (see note 1).

(85) a. The guests were two syntacticians and two phonologists. That is, Noam Chomsky, Morris Halle, and Paul Kiparsky.

b. Looking at my lists of required course books, I see that I need one
hundred, twenty, and five books. But actually, there is overlap
among my lists, so the total number of books I need to buy is only
one hundred and two.

c. Looking at my lists of required course books, I see that I need one
hundred twenty-five books. #But actually, there is overlap among
my lists, so the total number of books I need to buy is only one
hundred and two.

In Ionin and Matushansky 2006, we proposed that the effect of an overt conjunction is pragmatic in nature. The conventional way of talking about 125
books is to use asyndetic coordination; to use the unconventional *one hundred,
twenty, and five books* instead, one needs to have a good reason, such as sepa-
rating the books in space and/or time (e.g., into separate lists, as in (85b)),
which in turn creates the possibility of overlap. Another pragmatic reason to
use an overt conjunction has to do with using different guises for the same
individual, as we discussed in Ionin and Matushansky 2006.

Overlap with additive readings should be distinguished from intersective
readings, as discussed above (see (72)). The reason that (86a) lacks the intersective reading is that no plural individual can simultaneously be twenty
professors and five professors. The same holds for the conjunction of two
cardinal-containing NPs, for the same reason: (86b) lacks the reading on which
the same plural individual is both six professors and two deans at once.

(86) a. We need twenty-five professors for a quorum.
  b. We need six professors and two deans for a quorum.

While the intersective reading is simply logically impossible in (86), it
is both logical and possible in the examples in (87) (due to an anonymous
reviewer), where the conjoined NPs are in predicate position, and in (88),
where the intersective reading in argument position is more difficult to obtain
but still possible.[12]

(87) a. Barack Obama and Joe Biden are two powerful world leaders and
    two caring family men.
  b. Isaac Newton and Robert Hooke were two incredibly successful
    scientists but also two bitter rivals.

(88) a. ?These two powerful world leaders and two caring family men
    disagree on many issues.
  b. ?These two incredibly successful scientists and two bitter rivals
    fought continuously.

Since additive complex cardinals never involve the same cardinality in the two conjuncts (i.e., there are no complex cardinals of the form *two-two* or *thirty-thirty*), intersective readings of additive complex cardinals do not exist.

5.8.3 Deriving Set Product

A major issue for accounts of the nonintersective, additive meaning of *and* is how to relate it to the intersective reading. Some accounts (e.g., Link 1983, 1984; Hoeksema 1988; Le Bruyn and De Swart 2014) assume lexical ambiguity between the two interpretations; however, this fails to capture the crosslinguistic fact that the same lexical item has both readings. Heycock and Zamparelli (2000, 2005) attempt to solve this problem by positing that set product is the only interpretation that *and* has (see (80)). As pointed out above, this approach fails to capture the fact that intersective and additive readings are truth-conditionally distinct (see (81)).

Champollion (2016) provides a solution to this issue by deriving the additive reading from the intersective reading. The solution that he proposes relies on shifting the types of the two conjuncts (cf. Winter 1998, 2001a), so as to allow the intersective *and* to apply, resulting in an additive interpretation. More precisely, he proposes the operation of Existential Raising, which turns both conjoined predicates into generalized quantifiers:

(89) $[\![\text{ER}]\!] = \lambda P_{\langle \tau, t \rangle} . \lambda P_{\langle \tau, t \rangle} . \exists x_\tau [x \in (P \cap Q)]$

In *man and woman*, *and* has the standard intersective semantics, while both conjuncts are assumed to contain a covert instance of the ER operator in (89). When applied to *man*, ER returns the set of all sets containing a man and potentially something else. When this set is intersected (via *and*) with the set denoted by ER(woman), which denotes the set of all sets containing a woman and potentially something else, the result is a set of sets each of which contains a man, a woman, and potentially something else. This derivation is given in (90). As the final step, Champollion postulates a covert minimizing operator MIN, which selects from this set of sets only the minimal sets, that is, those sets which contain a man and a woman and nothing else, as in (91).

(90) $[\![\text{ER(man) and ER(woman)}]\!]$
    a. $= [\![\text{and}]\!] (\text{ER}([\![\text{man}]\!]))(\text{ER}([\![\text{woman}]\!]))$
    b. $= [\lambda P . \exists x [\text{man}(x) \wedge P(x)]] \cap [\lambda P . \exists y [\text{woman}(y) \wedge P(y)]]$
    c. $= [\lambda P . \exists x \exists y [\text{man}(x) \wedge \text{woman}(y) \wedge P(x) \wedge P(y)]]$

(91) $[\![\text{MIN}]\!] = \lambda Q_{\langle \tau t, t \rangle} . \lambda P_{\langle \tau, t \rangle} . P \in Q \wedge \forall P' [P' \subset P \to \neg (P' \in Q)]$

Crucially, Champollion, following Winter (2001a), does not distinguish between plural individuals and sets. In other words, the set containing Ann and Beth is for him the same as the plural individual Ann $\oplus$ Beth. If we accept this premise, we can derive the additive interpretation of predicate conjunction, that is, set product.

The advantage of Champollion's proposal over that of Heycock and Zamparelli (2000, 2005) is not restricted to its parsimony. Crucially, the first step in this proposal (application of the ER operator) makes the complement of *and* a generalized quantifier. This is a possible semantic type for the complement of an adposition. Given that additive complex cardinals can be constructed not only via coordination but also by means of adpositions (see section 5.6), Champollion's proposal potentially allows us, as long as we assume an appropriately modified meaning for the relevant locative adpositions, to generalize this type of derivation to adpositional additive cardinals. This is something that cannot be achieved in Heycock and Zamparelli's approach unless the denotation of a set-product operator (79) is given to locative adpositions.

5.8.4   The Disjunctive Meaning of *And*

The additive interpretation of conjunction discussed above (whether implemented via set product or according to Champollion's approach) cannot account for all types of conjunction. While, as we showed in section 5.8.2, set product successfully derives the interpretation of *twenty-two books*, it cannot derive the additive interpretation of *and* when a cardinal combines with a conjoined NP, as in *four soldiers and sailors*. Assuming the set-product interpretation of *and*, *soldier and sailor* denotes a set of pairs, each consisting of a soldier and a sailor. This means that, given our semantics for cardinals, *four soldiers and sailors* would correspond to a set of plural individuals each of which consists of four nonintersecting soldier-sailor pairs. This would amount to eight individuals in total, four soldiers and four sailors (models where the intersection of the set of soldiers and the set of sailors is not empty are even less compatible with our analysis). For our semantics to work, it is necessary instead for *soldier and sailor* to denote a disjunction, *soldier or sailor*, so that *four soldiers and sailors* denotes a set of plural individuals each of which consists of four elements, where each element is a soldier or a sailor.

Importantly, this disjunctive interpretation is also required for conjunction in some downward-entailing environments, which do not involve cardinals. The disjunctive reading of *and* is available to singular NPs with *every*, for example, as in (92).

(92) a. Every boy and girl received a prize.
      b. Every boy and girl who enters the room receives a prize.

The sentence in (92a) can be paraphrased as *Every individual who is either a boy or a girl received a prize*—precisely the reading derived by disjunctive *and.* Set product would instead derive the additive reading, on which every boy-girl pair received a prize; this reading is also available to (92a), but requires some indication that boys and girls are paired off. Even more clearly, (92b) means that every individual boy or girl who enters the room receives a prize, and not that every boy-girl pair receives a prize. We observe the same effect in some other downward-entailing contexts, as in (93), where the discussion concerns individuals each of whom is a cat or a dog.

(93) a. John doubts that cats and dogs are intelligent.
    b. Few cats and dogs are as intelligent as humans.

To sum up, the need for a disjunctive analysis of *and* is not a result of our semantic analysis of cardinals; it would also resolve conjunction in cases like (92) and (93). We will now sketch an analysis that can derive this disjunctive interpretation of *and* from its default Boolean interpretation, by assuming a number of independently motivated type shifts. As a first step, we will adopt the proposal of Winter (2001a) deriving the additive *and* for entities from the Boolean *and*. We will then propose that the entities composed by this *and* are kinds constructed on the basis of the predicate denotation of the two conjuncts. Finally, we will discuss the realizations of the plural kind individual that is formed this way and argue that depending on what we take to be instantiations of plural kinds, we can obtain the additive and disjunctive meanings of *and.*

With regard to the additive interpretation of *and*, Winter (1996) provides a straightforward way of reducing the additive *and* in conjunctions like *Mary and John* to the Boolean *and*: as proposed by Partee and Rooth (1983), an NP denoting an entity (type e) can be shifted to the generalized-quantifier type $\langle\langle e,t\rangle,t\rangle$ via the operation of "Montague Raising," as in (94a). If both type e conjuncts are lifted via "Montague Raising" into the generalized-quantifier type, the Boolean *and* can combine with them, with the derivation in (94b) (based on Winter 1996, 353).

(94) a. $[\![\text{Mary}]\!] = m \rightarrow \lambda P \,.\, P(m)$
    b. $[\![\text{Mary and John}]\!] = [\![\text{and}]\!](\lambda P \,.\, P(j))(\lambda R \,.\, R(m))$
       $= \lambda Q \lambda P \,.\, P(j)(Q) \wedge \lambda R \,.\, R(m)(Q)$

The constituent *Mary and John* denotes then the set of all sets that contain both Mary and John. Applying the MIN operator in (91) to this set yields the minimal set that contains only Mary and John. Maintaining the assumption that plural individuals are sets gives us the plurality of John and Mary, that is, their sum.

The same reasoning can be used for the interpretation of examples like (95), where a cardinal combines with a conjoined NP. Per our hypothesis that plural marking on an NP sister of a cardinal is due to agreement (see chapter 4 for details) and does not entail the presence of the pluralizing operator on that NP, the underlying semantic representation of (95) involves the coordination of two singulars, as shown in (96).

(95) The seven men and women are smiling for the camera.

(96)

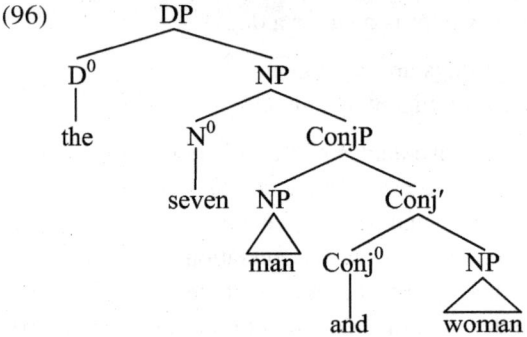

To derive (95), we start with the coordination of two singular kinds: the kind man and the kind woman, derived via the *down* operator (Chierchia 1984, 1998), which shifts properties to kinds.[13] As is easy to see, if a singular noun can be treated as a kind and therefore as having the type of an entity (Chierchia 1984, 1998), its denotation can then be shifted to the generalized-quantifier type. The same reasoning applies as before, yielding an additive interpretation: a "plural kind" generalized quantifier, as in (97a). We should now map this plural kind back into a predicate denotation, that is, the set of instantiations of this kind. Following Chierchia (1984) and Partee (1986), we use the operation *pred* for this purpose, that is, to pass from a conjunction of kinds to the set of its instantiations. How is a plural kind instantiated? Two possibilities seem the most likely: (i) a set of plural individuals consisting of one member of one kind and one member of the other kind (97b), that is, Heycock and Zamparelli's set product,[14] and (ii) a set of individuals each of which is a member of at least one of the two kinds (97c), that is, set union. This sets the stage for the possibility that under certain conditions, *and* can be interpreted disjunctively.

(97) a. ⟦man and woman⟧ = $man_{kind} \oplus woman_{kind}$
    b. $\lambda x . x = y \oplus z \wedge man(y) \wedge woman(z)$ Set product
    c. $\lambda x . man(x) \vee woman(x)$ Set union

We propose that both set-product (additive) and set-union (disjunctive) interpretations of *and* are in fact available. The reason is that it is possible to extract two kinds of atoms from a conjunction of kinds: absolute atoms (the set of all men and all women), which gives us the set-union interpretation (97c), and derived atoms (plural individuals), each consisting of two absolute atoms from the two kinds, which gives us the set-product reading (97b).[15]

Assuming that the set-union interpretation is available for the conjunction of kinds, we now illustrate how it works with conjunction under a cardinal, as in *seven men and women*. We start with the set-union definition in (97c), on which *man and woman* denotes a set of individuals each of whom is a member of the kind man or a member of the kind woman—that is, each of whom is a man or a woman.

The NP *man and woman* can now combine with the cardinal *seven*, resulting in a set of plural individuals, each of which consists of seven nonintersecting parts, where each part is either a man or a woman, as shown in (98a). Combining the definite determiner, *the seven men and women* has the semantics in (98b), paraphrased informally in (98c).

(98) a. $\lambda x \in D_e \,.\, \exists S \in D_{\langle e,t \rangle} \, [\Pi(S)(x) \wedge |S| = 7 \wedge$
$$\forall s \in S \, [[[man]](s) \vee [[woman]](s)]]$$

b. $\iota x \,.\, \exists S \in D_{\langle e,t \rangle} \, [\Pi(S)(x) \wedge |S| = 7 \wedge$
$$\forall s \in S \, [[[man]](s) \vee [[woman]](s)]]$$

c. the unique plural individual that is divisible into seven nonintersecting parts each of whom is a man or a woman

It initially seems that the truth conditions that we have achieved are not sufficiently precise: it has now become possible for *seven men and women* to denote a set of seven individuals all of whom are men or all of whom are women. Contrary to the initial impression, however, this is actually a welcome result. In a situation where it is known that we are talking about seven women only, it is indeed pragmatically odd to talk about seven men and women. However, under other circumstances, *seven men and women* may indeed refer to seven women only. Suppose that a presenter on a weekly TV program randomly selects seven men and women to interview each time. It is possible to comment on this with (99). Even if on some occasions only men or only women are interviewed, (99) can still be used, showing that our truth conditions are indeed correct.

(99) And each time she finds something interesting to tell about each of these seven men and women!

Other examples of this type are found in (100). The sentence in (100a) conveys that seeing just men, or just women, or both men and women, working at desks would all lead to the conclusion that the location is an office. Similarly, in (100b), the earning of twenty cents follows if ten individuals, regardless of their gender, complete the necessary steps.

(100) a. If you see men and women working at desks with computers, you
       can assume the location is an office.
       (http://vickyzhu.blogspot.com/2010/06/how-to-analyze-pictures
       -in-doing-toeic.html)
    b. Imagine if ten men and women register under you. If they review
       all 4 ads, it's $0.20 (twenty cents) for you.
       (clickbanktrafficwarrior.net/neobux-guide-and-strategy/)

The reading on which *men and women* or *ten men and women* is allowed
to consist exclusively of men or exclusively of women cannot be derived via
Heycock and Zamparelli's set product or Champollion's *and*: a set product of
men and women necessarily consists of sets each of which contains at least
one man and at least one woman. Thus, our application of disjunctive *and* to
plural-NP conjunctions has the welcome advantage of deriving a reading not
derived by the set-product operation. The same mechanism should be used to
derive the disjunctive readings of the conjoined NPs in (92) and (93).

5.8.5   Summary: Different Readings of *And*

In this section, we have shown that NP coordination comes in three flavors:
the (Boolean) intersective reading; the additive reading (derived via either set
product or Champollion's *and*); and the novel disjunctive reading. We have
suggested that the latter two readings of *and* are both derived from the intersective reading as a result of multiple type-shifting operations on the coordinated NPs.

We note that the disjunctive reading of *and* is not in complementary distribution with the additive reading. Recall that (92a), *Every boy and girl received
a prize*, is ambiguous between the disjunctive reading (every boy or girl
receives a prize) and the additive one (every boy-girl pair receives a prize). In
the same manner, a conjoined bare plural such as *men and women* normally
has an additive interpretation, denoting a set of plural individuals each consisting of both men and women, but can also have a disjunctive interpretation, as
in (100a). This is further illustrated in (101), where the two readings are truth-
conditionally distinct. While (101a) has the additive reading (the speaker will
only eat his or her hat if both cats and dogs are found to be sufficiently intelligent), (101b) allows for the disjunctive reading (seeing goats only would be
sufficient for making the relevant conclusion).

(101) a. If cats and dogs are as intelligent as humans, I will eat my hat.
     b. If you see goats and sheep through your car window, you must be
        in the country.

We do not know whether downward-entailing environments like in (101) are the only environments where the disjunctive reading of conjoined bare plurals arises or whether this reading is simply undetectable in other environments. In contrast, with conjunction under a cardinal, such as *five men and women*, the disjunctive reading is the only one available, as shown above. Conversely, for additive complex cardinals (*twenty-two students*) and for conjunction of cardinal-containing NPs more generally (*five men and seven women*), the additive reading is the only one available: *five men and seven women* can never mean, disjunctively, a plural individual consisting of either five men or seven women. We have no explanation for why the different meanings of *and* arise in different syntactic and semantic configurations, but there is compelling evidence for the existence of both.

5.9   Conclusion

In this chapter, we have provided both syntactic and semantic analyses of additive complex cardinals, such as *twenty-two books*. Based on our analysis of simplex cardinals, we proposed that *twenty-two books* is underlyingly *twenty books and two books*, and therefore that it receives the same syntactic and semantic analysis as any coordination of two cardinal-containing NPs, for example, *five books and two magazines*. Supporting evidence for this analysis comes from the fact that in some languages the lexical NP can be overt in both conjuncts.

On the semantic side, we have proposed that coordination of two cardinal-containing NPs (*five men and two women* as well as *twenty-two books*) involves the additive meaning of *and*, derived by the set-product operation of Heycock and Zamparelli (2000, 2005), or the semantics for *and* provided by Champollion (2016). On the other hand, NP coordination under a cardinal (*five men and women*) cannot be derived via set product, and we have proposed that it is obtained via the disjunctive interpretation of *and*. We have shown that both non-Boolean types of *and*, additive and disjunctive, are available to other types of coordinated structures besides those involving cardinals, and we have attempted to derive both types from the standard Boolean meaning.

6 Cardinals in Slavic Languages

As discussed in chapters 3 and 5, cardinals in Russian and other Slavic languages provide compelling evidence that complex cardinals involving multiplication and addition are formed in the syntax rather than the morphology. In particular, we have shown that Russian cardinals 'five' and up exhibit many properties of nouns, including the ability to assign case, and that case assignment within complex cardinals involving multiplication has the same characteristics as case assignment between a cardinal and a lexical NP. In this chapter, we take a more in-depth look at the properties of cardinals (as well as fractions) in Slavic languages. The main theoretical contributions of this chapter are as follows. We provide evidence that simplex cardinals do not form a unified morphosyntactic or lexical category, but in fact form a cline, continuum, or "squish" (Ross 1972). We furthermore show that the gradation from more adjectival to more nominal cardinals is determined by their degree of φ-completeness detectable from their ability to trigger number and gender agreement and to show heterogeneous patterns of case marking in oblique cases. Finally, we discuss the properties of the count (adnumerative) form across Slavic languages.

The first part of this chapter focuses on the question of the lexical category of cardinals in Russian. In section 6.1, we argue that most cardinals in Russian fall on a continuum between nouns and adjectives, rather than behaving just like nouns or just like adjectives. In section 6.2, we show that while cardinals 'five' and higher behave more like nouns than adjectives (just as we argued in chapter 3), most of these cardinals are defective nouns when it comes to properties such as case assignment in oblique cases and inherent gender. In section 6.3, we consider vague cardinals (*many*, *several*) as well as the cardinals 'two' through 'four'; we show that while such cardinals are best analyzed as adjectival, they also do not behave completely like regular adjectives. Section 6.3 also takes up the question of the paucal case assigned by lower cardinals and argues for the view that lexical NPs marked with paucal case are plural rather than singular, despite the surface similarity of the paucal form to the genitive-singular form. In section 6.4, we extend our analysis of cardinals to lexical fractions, showing that, just like simplex cardinals, fractions in Russian fall on a continuum from fully nominal to defective nouns. In section 6.5, we take a brief detour into predicate agreement with cardinal-containing NPs, highlighting yet another similarity between cardinals and measure nouns. Finally, section 6.6 provides further evidence that cardinals are distinct from regular nouns by showing that in both Russian and Bulgarian, the lexical NP sister of (some) cardinals occurs in a special count form that is not attested elsewhere.

While the bulk of this chapter is concerned with the behavior of simplex cardinals in Slavic, section 6.7 takes up the question of Russian complex cardinals. We show that their declensional patterns support our proposal (chapters 2 and 3) that a complex cardinal does not form a unit to the exclusion of the lexical NP.

Most of the discussion in this chapter focuses on Russian, with data from Bulgarian, Ukrainian, and Belarusian included where relevant. In section 6.8, however, we switch the focus to cardinals in Polish, which exhibit a number of interesting characteristics not shared by cardinals in Russian. Our explanation of the behavior of Polish cardinals rests on the view that they, just like Russian cardinals, are ϕ-defective nouns.

6.1   The Cardinal "Squish" in Russian

Hurford (1975, 1987) demonstrates that crosslinguistically lower cardinals are usually similar to adjectives while the highest cardinals are nouns. In accordance with this generalization, the Russian 'one' is clearly an adjective and the highest cardinals ('million', 'billion', etc.) are clearly nouns. The cardinals between 'one' and 'million', though, are on a continuum from adjectives to nouns, exhibiting some but not all of the properties of regular adjectives and nouns. Russian thus provides compelling evidence that (most) cardinals are neither fully nominal nor fully adjectival, and that the binary divide between nouns and adjectives should be replaced by a more fine-grained analysis in terms of feature specification.

In order to establish the lexical category of a given Russian cardinal, we first need some diagnostics for nominal versus adjectival status. We consider four diagnostics of nouns: nominal-declension pattern, genitive-case assignment, gender, and animacy. We consider two diagnostics of adjectives: adjectival-declension pattern and agreement in number and gender with the modified noun. For each diagnostic, we will compare the behavior of cardinals to that of "regular" (noncardinal) nouns (section 6.2) and adjectives (section 6.3); we will also consider how lexical fractions behave on these diagnostics (section 6.4).

Table 6.1 summarizes the behavior of cardinals, vague cardinals such as *many/few*, and fractions with respect to the properties examined in the next three subsections. We will show that the different factors listed in table 6.1 are responsible for the internal variation in the morphology and syntax of the lexical-semantic class of Russian simplex cardinals.

6.2 Nominal Cardinals in Russian

We first consider the behavior of Russian higher cardinals, 'five' and above, and argue that they are nominal. We show that on some diagnostics (namely, declension patterns, case assignment in direct cases, and animacy), most higher cardinals behave exactly like other nouns. On other diagnostics (namely, case marking under oblique case assignment and gender agreement vs. plural agreement), the behavior of most higher cardinals diverges from that of regular nouns, but it also does not characterize higher cardinals as adjectival.

6.2.1 Nominal-Declension Patterns

Table 6.2 is a partial illustration of the Russian declensional pattern in the singular; it includes only those declension-gender combinations that—as we will show—are exhibited by cardinals. As shown in table 6.2, all three declension-gender combinations are exhibited by both animate and inanimate nouns, but only in the case of second-declension nouns does animacy affect the declensional pattern in the singular: inanimate second-declension nouns show a nominative-accusative syncretism, whereas animate second-declension nouns show a genitive-accusative syncretism.

Table 6.3 provides evidence that, morphologically, most simplex cardinals in Russian behave like singular nouns. The words for 'five' through 'nineteen', as well as 'twenty', 'thirty', 'fifty', 'sixty', 'seventy', and 'eighty', all exhibit the third-declension pattern, as illustrated in the table with *šest'* 'six'. (For discussion of cardinals below 'five', see section 6.3.) The word for 'thousand' exhibits the first-declension pattern (but with two forms in the instrumental—more on this below), while the words for 'million', 'billion', and so on exhibit the pattern of inanimate second-declension nouns.

Four cardinals, *pol-* 'half', *sorok* 'forty', *devjanosto* 'ninety', and *sto* 'hundred', have only two forms in the singular: the direct one (the masculine surface -∅- for *pol-* and *sorok*, the neuter underlying [o] for the latter two)

Table 6.1
The Russian cardinal "squish"

	Nouns	'million'	'one-fourth'	'thousand'	'five' to 'hundred'	'many'/ 'few'	'one-half' (*pol*)	'three', 'four'	'two'	'one'
Animacy distinction	+									
Inherent gender	+	+	+	+						
Case assignment	+	+	+	+	+	+	+	+	+	
Homogeneous pattern				±	+	+	+	+	+	
Paucal form								+	+	
Number and animacy agreement								+	+	+
Gender agreement in the plural									+	
φ-agreement in the singular										+

Table 6.2
Russian declensional patterns for singular nouns (partial pattern)

	First declension (F)		Second declension (M)		Third declension (F)	
	+animate	−animate	+animate	−animate	+animate	−animate
	'girl'	'rain cloud'	'boy'	'table'	'mouse'	'wool'
NOM	devočk-a	tuč-a	mal'čik-∅	stol-∅	myš-∅	šerst'-∅
ACC	devočk-u	tuč-u	mal'čik-a	stol-∅	myš-∅	šerst'-∅
GEN	devočk-i	tuč-i	mal'čik-a	stol-a	myš-i	šerst'-i
DAT	devočk-e	tuč-e	mal'čik-u	stol-u	myš-i	šerst'-i
LOC	devočk-e	tuč-e	mal'čik-e	stol-e	myš-i	šerst'-i
INS	devočk-oj	tuč-ej	mal'čik-om	stol-om	myš-ju	šerst'-ju

Table 6.3
Russian declensional patterns for nominal cardinals

	First declension (F)	Second declension (M)	Third declension (F)
	'thousand'	'million'	'six'
NOM	tysjač-a	million-∅	šest'-∅
ACC	tysjač-u	million-∅	šest'-∅
GEN	tysjač-i	million-a	šest'-i
DAT	tysjač-e	million-u	šest'-i
LOC	tysjač-e	million-e	šest'-i
INS	tysjač-ej/tysjač-ju	million-om	šest'-ju

and the oblique one (*polu-*, *soroka*, *devjanosta*, *sta*), permitting their analysis as defective second-declension nouns.

Although most cardinals have no plural form, the higher multiplicands *sto* 'hundred', *tysjača* 'thousand', *million* 'million', and so on can appear as complements to other cardinals and in consequence may be marked plural.[1] The plural declensional pattern is illustrated in table 6.4.

As shown in table 6.4, the cardinals *tysjača* 'thousand' and *million* 'million' have a complete declensional paradigm in the plural.[2]

Thus, 'thousand' and 'million' behave just like regular nouns in terms of their declensional pattern, in both the singular and the plural. The fact that *sto* 'hundred' too has a nominal, albeit somewhat deficient, plural paradigm when appearing in complex cardinals (Vinogradov 1952, vol. 1, 376–377; Garde 1998, 245–246; see also note 1) provides evidence that it too is nominal.

Table 6.4
Russian declensional patterns for nouns and nominal cardinals, plural

	'thousand'	'hundred'	'rain cloud'	'million'	'dream'
NOM	tysjač-i	–	tuč-i	million-y	sn-y
ACC	tysjač-i	–	tuč-i	million-y	sn-y
GEN	tysjač-∅	sot-∅	tuč-∅	million-ov	sn-ov
DAT	tysjač-am	st-am	tuč-am	million-am	sn-am
LOC	tysjač-ax	st-ax	tuč-ax	million-ax	sn-ax
INS	tysjač-ami	st-ami	tuč-ami	million-ami	sn-ami

6.2.2 Case Assignment in Direct Cases

Many nouns in Russian appear with a genitive-marked sister NP, as shown in (1). Simplex cardinals above 'four', when appearing in a direct case (nominative or accusative), also take a genitive-marked sister NP (2a), which is obligatorily plural, exactly as with group nouns (2b) and nominal quantifiers (2c). Thus, cardinals in Russian behave like nouns and unlike, say, demonstratives, which must bear the same case as the lexical NP (2d).

(1) a. syn Leny   Russian
       son Lena.GEN
       'a/the son of Lena'
    b. miska kaši
       bowl cereal.GEN
       'a/the bowl of cereal'

(2) a. šest'   / tysjača   / million   knig   /
       six.NOM/ACC thousand.NOM million.NOM/ACC book.PL.GEN
       *knigi
       book.PL.NOM/ACC
       'six/a thousand/a million books'
    b. para   knig   / *knigi
       pair.NOM book.PL.GEN   book.PL.NOM
       'a couple of books'
    c. bol'šinstvo   knig   / *knigi
       majority.NOM/ACC book.PL.GEN   book.PL.NOM/ACC
       'the majority of books/most books'
    d. èti   knigi   / *knig
       these.NOM/ACC book.PL.NOM/ACC   book.PL.GEN
       'these books'

The hypothesis that simplex cardinals 'five' and higher are nominal is supported by the fact that in direct cases they can be the only element inside a cardinal-containing NP that is marked with the case assigned to that NP, as shown in (3a–b). This again parallels the behavior of regular nouns, as shown in (3c–d).

(3) a. My posetili tusjaču   muzeev.
       we visited thousand.ACC museum.PL.GEN
       'We have visited a thousand museums.'

   b. Rabota v sto   raz   važnee   ljubvi.
      work   in hundred.ACC/NOM times.GEN more.important love.GEN
      'Work is a hundred times more important than love.'

   c. My posetili kvartiru   pisatelja.
      we visited apartment.ACC writer.GEN
      'We visited the apartment of the/a writer.'

   d. Položi   saxar   v čašku   čaja.
      put.IMPER sugar.ACC in cup.ACC tea.GEN
      'Put sugar in a cup of tea.'

Although both the syntax and the morphology of case assignment support the hypothesis that higher cardinals are nominal, it can also be demonstrated that as nouns, they are defective. We turn to this next.

6.2.3 Case Assignment in Oblique Cases: Homogeneous versus Heterogeneous Patterns

When a cardinal-containing NP is in a direct case (nominative or accusative), the sister of the cardinal, if it is higher than 'four', obligatorily bears genitive case, as shown in the previous section. When a cardinal-containing NP is assigned an oblique case, however, this case is assigned to the lexical NP as well as to the cardinal, as shown in (4a) and (5a). This pattern, the *homogeneous* pattern of oblique-case marking (Babby 1987), is exhibited by the cardinals 'five' through 'hundred' (as well as 'two' through 'four', which assign paucal case, as discussed in section 6.3.2). In contrast, with regular nouns, the *heterogeneous* pattern is found: only the head noun is assigned the oblique case, while the second NP retains genitive-case marking, as shown in (4b–c) and (5b–c).

(4) a. s   pjat'ju knigami/*knig
       with five.INS book.PL.INS/PL.GEN
       'with five books'

   b. s   bol'šinstvom *knigami/knig
      with majority.INS   book.PL.INS/PL.GEN
      'with the majority of books/with most books'

c. s   korobkoj *knigami/knig
   with box.INS   book.PL.INS/PL.GEN
   'with a box of books'

(5) a. k pjati   knigam/*knig
       to five.DAT book.PL.DAT/PL.GEN
       'to five books'
    b. k bol'šinstvu   *knigam/knig
       to majority.DAT   book.PL.DAT/PL.GEN
       'to the majority of books/to most books'
    c. k korobke *knigam/knig
       to box.DAT   book.PL.DAT/PL.GEN
       'to a box of books'

Multiplicands higher than 'thousand' ('million', 'billion', etc.) exhibit the
heterogeneous pattern, taking a genitive-marked sister NP even in an oblique
case, as in (6b); compare the lower cardinal in (6a). In this respect they behave
just like group nouns and nominal quantifiers (6c–d).

(6) a. Ona zavedovala   pjat'ju   otdelami/*otdelov
       she manage.PAST million.INS department.PL.INS/GEN
       'She managed five departments.'
    b. Ona zavedovala   millionom *otdelami/otdelov
       she manage.PAST million.INS department.PL.INS/GEN
       'She managed a million departments.'
    c. Ona zavedovala   desjatkom   *otdelami/otdelov
       she manage.PAST group.of.ten.INS department.PL.INS/GEN
       'She managed some ten departments.'
    d. Ona zavedovala   bol'šinstvom *otdelami/otdelov
       she manage.PAST majority.INS   department.PL.INS/GEN
       'She managed the majority of departments/most departments.'

Tysjača 'thousand' appears to alternate between the two patterns, with the
heterogeneous pattern (7a) and the homogeneous pattern (7b) both attested
(Mel'čuk 1985, 289–294; Timberlake 2004), although the latter appears
dispreferred.

(7) a. čelovek s   tysjačju   lic
       man.NOM with thousand.INS face.PL.GEN
       'a man with a thousand faces'
    b. geroj   s   tysjačju   licami
       hero.NOM with thousand.INS face.PL.INS
       'a hero with a thousand faces'
    (Timberlake 2004, 191)

Babby (1987) and following him Franks (1994, 1995) and Rappaport (2002) propose that cardinals assign *structural* genitive case while other nouns assign *inherent* genitive case. Inherent case is stronger than structural case, so when an inherent oblique case is assigned (by a verb or a preposition), it overrides a structural case, though not another inherent case.

In analyzing the variation between homogeneous and heterogeneous case marking, we would like to follow the same intuition (namely that cases can be overridden by other cases), but only in spirit, because the structural–inherent divide does not, so to speak, cut the cake properly. The relevant evidence comes from accusative-case assignment. Accusative is assigned not only to the direct object of a transitive verb (i.e., in Spec, vP) but also to complements of many prepositions, such as *pro* 'about' and *za* 'for; behind'. This fact obliges us to hypothesize that both structural and inherent versions of this case are available, on the accepted view that prepositions assign inherent case. However, as is easy to see, for example in (8), the accusative case assigned by a preposition does not override the genitive (or the paucal) assigned by the cardinal.

(8) a. pro   pjat'   knig
       about five.NOM/ACC book.PL.GEN
       'about/on five books'
   b. za četyre   dnja
       for four.NOM/ACC day.PAUC/SG.GEN
       'before/in four days'

For this reason we depart from the structural–inherent-case distinction as the factor determining the morphological case that the lexical NP surfaces in. We hypothesize instead, with Matushansky (2008b, 2010, 2012) and Arkadiev (2014), that the surface case represents a realization of all case features assigned to an NP. Oblique cases being more richly specified than direct cases, they must precede direct cases in vocabulary-insertion rules and therefore override them.

Unlike the genitive case assigned by (most) cardinals, the genitive case assigned by nouns (and the highest cardinals; see above) cannot be overridden by the case assigned from outside. To account for this fact, we assume that the surface genitive morphology can correspond to more than one underlying feature bundle and that the feature bundle assigned by nouns contains features not present in the feature bundle assigned by cardinals. Given that cardinals, unlike nouns and adjectives, are featurally defective, it is altogether reasonable to assume that they assign a smaller set of case features, one that does not intervene in the realization of the inherent case assigned from outside.

6.2.4 Animacy

The grammatical feature of animacy (also taken to be a subgender as opposed to an independent grammatical feature: see Corbett 1991, 165–168) manifests itself in the realization of underlying accusative case in certain declensions. While for most animate nouns accusative surfaces as genitive (9b), for most inanimate nouns, including group nouns, accusative is syncretic with nominative (9a).[3]

(9) a. My narisovali nebo   / sneg   / lan'   /
       we drew   sky.ACC/NOM snow.ACC/NOM doe.ACC/NOM
       lesa.
       forests.ACC/NOM
       'We drew the sky/snow/a doe/forests.'
    b. My narisovali studenta   / studentov.
       we drew   student.M.ACC/GEN student.PL.ACC/GEN
       'We drew a student/the student/(the) students.'

Unlike nouns, adjectives have no inherent φ-features; Russian adjectives (including demonstratives) exhibit gender, number, and animacy agreement with the nouns they modify. Although feminine adjectives have a dedicated accusative-case form, accusative case on masculine, neuter, and plural adjectives is syncretic with either nominative (as in (10a)) or genitive (as in (10b)) according to the animacy of the nouns they modify.

(10) a. My narisovali ètot   belyj   sneg.
        we drew   this.ACC/NOM white.ACC/NOM snow.ACC/NOM
        'We drew this white snow.'
     b. My narisovali ètogo   vysokogo   studenta.
        we drew   this.ACC/GEN tall.ACC/GEN student.ACC/GEN
        'We drew this tall student.'

It is easy to see that cardinals higher than 'four' behave like nouns rather than like adjectives in this regard: with the exception of the first-declension cardinal *tysjača* 'thousand', which has a dedicated accusative-case form, cardinals higher than 'four' show the accusative-nominative syncretism (11a–c).

(11) a. My narisovali šest'   / *šesti   zvězd/devušek.
        we drew   six.ACC/NOM   six.GEN star.PL.GEN/girl.PL.GEN
        'We drew six stars/girls.'
     b. My priglasili sorok   / *soroka   studentov.
        we invited   forty.ACC/NOM forty.GEN student.PL.GEN
        'We invited forty students.'

c. My priglasili million   / *milliona   studentov.
   we invited   million.ACC/NOM   million.GEN student.PL.GEN
   'We invited a million students.'
d. My priglasili otrjad   / *otrjada   soldat.
   we invited   detachment.ACC/NOM   detachment.GEN soldier.PL.GEN
   'We invited a detachment of soldiers.'

In other words, cardinals behave just like group nouns (11d), which decline like inanimate nouns irrespective of their complement, and unlike adjectives, which agree in animacy with nouns they modify, as in (10).

6.2.5 Gender versus Number Agreement

We now consider gender as a diagnostic of nominal status. Regular singular nouns in Russian, including group nouns, obligatorily trigger gender agreement on the determiner (12a–b).[4] In contrast, cardinals higher than 'four', despite declining morphologically like singular nouns, do not exhibit gender agreement with the determiner, triggering plural agreement instead (12c).

(12) a. ètot/*èta/*èti   otrjad   soldat
        this.M.SG/F.SG/PL detachment.NOM soldier.PL.GEN
        'this detachment of soldiers'
     b. èta/*ètot/*èti   čast'   soldat
        this.F.SG/M.SG/PL military unit.NOM soldier.PL.GEN
        'this detachment of soldiers'
     c. èti/*ètot/*èta   sorok/desjat' studentov
        this.PL/M.SG/F.SG forty/ten.NOM student.PL.GEN
        'these forty/ten students'

With regard to agreement on the verb, cardinals higher than 'four' trigger either plural agreement (13a) or default agreement, which in Russian corresponds to the neuter singular (13b), but not gender agreement, neither feminine nor masculine (13c) (see section 6.5 for more discussion). In this, cardinals again differ from regular nouns—including group nouns—which obligatorily trigger gender agreement on the verb (14).

(13) a. K nam prišli   pjat' krasivyx   devušek.
        to us   arrived.PL five beautiful.PL.GEN girl.PL.GEN
        'Five beautiful girls arrived at our place.'
     b. K nam prišlo   pjat' krasivyx   devušek.
        to us   arrived.N.SG five beautiful.PL.GEN girl.PL.GEN
        'Five beautiful girls arrived at our place.'
     c. *K nam prišla/prišël   pjat' krasivyx   devušek.
        to us   arrived.F.SG/M.SG five beautiful.PL.GEN girl.PL.GEN

(14) a. *Na stanciju pribyli   otrjad   soldat.
        in station arrived.PL   detachment.NOM soldier.PL.GEN
     b. *Na stanciju pribylo   otrjad   soldat.
        in station arrived.N.SG detachment.NOM soldier.PL.GEN
     c. Na stanciju pribyl   otrjad   soldat.
        in station arrived.M.SG detachment.NOM soldier.PL.GEN
        'A detachment of soldiers arrived at the station.'

The highest cardinals ('thousand', 'million', 'billion', etc.) show mixed behavior with respect to gender agreement: the cardinal-containing NP can trigger plural agreement (15a) or gender agreement (15b–c), both on the determiner and on the verb.[5] The default agreement on the verb (neuter singular) is always possible as well (16). Nouns that denote groups of a precise quantity, such as *djužina* 'dozen', *desjatok* 'group of ten', and *sotnja* 'group of one hundred', also allow plural as well as gender agreement on the determiner, although gender agreement is preferred (17).

(15) a. Otkuda   voz'mutsja   èti   million/tysjača
        wherefrom take.3PL.REFL this.PL million.NOM/thousand.NOM
        čelovek?
        person.PL.GEN
        'Where would one get these one million/thousand people from?'
     b. Neponjatno, čem   pitalsja   ètot   million
        unclear   what.INS fed.M.SG.REFL this.M.SG million.NOM
        čelovek.
        person.PL.GEN
        'It is unclear what this million people fed on.'
     c. Neponjatno, čem   pitalas'   èta   tysjača
        unclear   what.INS fed.F.SG.REFL this.F.SG thousand.NOM
        čelovek.
        person.PL.GEN
        'It is unclear what this thousand people fed on.'

(16) a. Prišlo   tysjača   štrafov.
        arrived.N.SG thousand.NOM fine.PL.GEN
        'A thousand fines arrived.'
     b. Ušlo   million   rublej.
        left.N.SG thousand.NOM ruble.PL.GEN
        'A million rubles was gone.'

(17) a. èta/??èti   djuzhina   studentov
        this.F.SG/PL dozen.NOM student.PL.GEN
        'this dozen students'

b. ètot/??èti desjatok studentov
   this.M.SG/PL ten.NOM student.PL.GEN
   'this group of ten students'
c. èta/?èti sotnja studentov
   this.F.SG/PL hundred.NOM student.PL.GEN
   'this group of one hundred students'

The fact that cardinals from 'five' to 'hundred' do not trigger gender agreement on the determiner might be taken to mean that these cardinals are adjectival modifiers of the lexical NP, and that the plural determiner is agreeing with the plural lexical noun. Arguments against this view are the nominal declensional pattern of these cardinals (see table 6.3) and the fact that they do not exhibit animacy, gender, or number agreement with the lexical noun, as true adjectives do. Rather, we believe that these cardinals are best analyzed as defective nouns—specifically, nouns that are defective with respect to the gender feature.

6.2.6 Cardinal or Lexical Noun?

As discussed above, the cardinals 'five' through 'hundred' resemble regular nouns in some but not all respects. Like regular nouns, they assign surface genitive case to their sister NP, and they do not agree with the lexical NP in gender or number, as adjectives do; they also exhibit nominal declensional patterns. Yet unlike regular nouns, these cardinals do not have inherent gender (as evidenced by their inability to trigger gender agreement) and exhibit the homogeneous pattern of case assignment in oblique cases. The homogeneous pattern indicates that the case-feature bundle assigned from outside the NP comprises the case-feature bundle assigned by the cardinal. We conclude that these cardinals are ϕ-defective nouns (cf. Klockmann 2012, 2013).

The cardinals 'thousand' and 'million' in Russian have more in common with regular nouns: namely, an ability to trigger gender agreement on the determiner and the heterogeneous case-assignment pattern in oblique cases (although in the case of 'thousand', the homogeneous pattern is also possible). At the same time, two properties set 'thousand' and 'million' apart from (most) lexical nouns. First, as shown above, they can trigger plural agreement on the determiner (18a) (see also (15a)); however, so do some group nouns, as shown in (17) above (and note that group nouns have no problems triggering plural agreement in other languages, e.g., *committee* in British English—see chapter 4). Second, 'thousand' and 'million' (like other cardinals) can combine with a (plural) premodifier in the genitive case, like *čelyx* 'a whole.GEN', *dobryx* 'a good.GEN', or *kakix-to* 'some.GEN' (Worth 1959; Mel'čuk 1985, 322), as shown in (18b).[6] This ability is not available to group nouns (18c).

(18) a. I   protekli   očerednye million   let.
    and passed.PL next.PL   million.NOM years.GEN
    'And then another million years passed.'
  b. celyx   million/tysjača/sorok   čelovek
    whole.PL.GEN million/thousand/forty.NOM person.PL.GEN
    'a whole million/thousand/forty people'
  c. celaja/*celye/*celyx   djužina/kuča
    whole.F.NOM/PL.NOM/PL.GEN dozen/bunch.NOM
    čelovek/ljudej
    person.PL.GEN/people.GEN
    'a whole dozen/bunch of people'

We propose that Russian words for 'thousand' and 'million' are in fact
ambiguous between a true noun reading and a cardinal reading. On the noun
reading, 'thousand' and 'million' behave just like 'dozen' and other measure
nouns: they trigger gender and number agreement on determiners and adjectives, and they obligatorily exhibit the heterogeneous pattern of case assignment in oblique cases, just like other nouns. On the cardinal reading, they
can trigger either gender agreement or plural agreement on determiners and
adjectives, and they vary in the pattern of case assignment. In the case of
'million', the noun and cardinal are completely homophonous and cannot
be teased apart. But for 'thousand', it is possible to tease them apart. As a
noun (19a), *tysjača* declines like a regular first-declension feminine noun
(see table 6.3); as a cardinal (19b), it exhibits a different declensional pattern.
The declensional pattern correlates with the pattern of case assignment: the
nominal 'thousand' obligatorily shows the heterogeneous pattern, but the cardinal 'thousand' allows both heterogeneous and homogeneous patterns. The
nominal 'thousand' obligatorily triggers gender and number agreement on
the determiner (20a), whereas the cardinal 'thousand' allows both singular
and plural agreement (20b–c)—but only with plural agreement (20b) is the
homogeneous pattern of case assignment possible.

(19) a. Ja ušël   s   raboty s   tysjačej   bumažek/*bumažkami,
    I leave.PAST from work   with thousand.INS paper.DIM.GEN/INS
    'I left work with a thousand pieces of paper.'
  b. Ja ušël   s   raboty s   tysjačju   bumažek/?bumažkami.
    I leave.PAST from work   with thousand.INS paper.DIM.GEN/INS
    'I left work with a thousand pieces of paper.'

(20) a. s   ètoj/*ètimi   tysjačej   studentov/*studentami
    with this.F.SG.INS/PL.INS thousand.INS student.PL.GEN/PL.INS
    'with this thousand students'

b. s   ètimi   tysjačju   studentov/studentami
with this.PL.INS thousand.INS student.PL.GEN/PL.INS

c. s   ètoj   tysjačju   studentov/*studentami
with this.F.SG.INS thousand.INS student.PL.GEN/PL.INS

Example (20c) suggests a correlation between the featural composition of the cardinal and its case-assigning properties. The declensional pattern is still that of a cardinal, yet agreement on the determiner shows the presence of the gender feature, which makes this instance of 'thousand' opaque to the case assigned from outside, unlike the 'thousand' in (20b).

6.2.7 Summary: The Diagnostics of Cardinals versus Nouns

In this section, we have shown that the Russian cardinals 'five' and up are on a continuum with respect to which properties of nouns they possess, from the least nominal ('five' through 'hundred') to the most nominal ('million'). The lower cardinals in this group ('five' through 'hundred') exhibit the homogeneous case pattern in oblique cases, which means that the case-feature bundle that they assign is properly contained in the one assigned from outside. In contrast, 'thousand' and 'million' exhibit the heterogeneous case pattern, which means that they behave like regular nouns, with a bigger case-feature bundle to assign. We have furthermore seen that the case pattern correlates to some extent with the ability to trigger gender/number agreement, lending support to the intuition that cardinals can be more or less ϕ-impoverished compared to regular nouns. We turn next to those cardinals that are clearly adjectival, namely the lower cardinals.

6.3 Adjectival Cardinals and Paucal Case in Russian

In the previous section, we focused on the cardinals 'five' and up in Russian and showed that they exhibit many (though not all) of the properties of Russian nouns. We now turn our attention to adjectival cardinals. As discussed in the previous section, the two diagnostics that we consider for the adjectival status of cardinals are the adjectival declension pattern and agreement in number and gender with the modified noun.

In agreement with Hurford's (1975, 1987) crosslinguistic generalization that lower cardinals are more adjectival, the cardinal *odin* 'one' in Russian is fully adjectival: it agrees with its sister in number, gender, and case, and it does not assign case. The examples in (21) illustrate this.

(21) a. odin   student   / odna   studentka
one.M.NOM student.M.NOM one.F.NOM student.F.NOM
'one male student/one female student'

b. Ja videla   odnogo   studenta   / odnu
   I see.PAST one.M.ACC/GEN student.M.ACC/GEN one.F.ACC
   studentku.
   student.F.ACC
   'I saw one male student/one female student.'
c. ob   odnix   nožnicax
   about one.PL.LOC scissors.PL.LOC
   'about one pair of scissors'

The declensional pattern of *odin* 'one' is that of an irregular adjective: it appears with the nominal endings in the direct cases and in addition, like many other functional adjectives, it undergoes a readjustment rule in the plural (Halle and Matushansky 2006).

In contrast, the cardinals 'two' through 'four', as well as vague cardinals, show a mix of nominal and adjectival properties, as detailed below.

6.3.1 Adjectival Vague Cardinals

In Russian, cardinals differ from lexical nouns and adjectives in their ability to assign a case that is only detectable in the direct cases. This property, however, is not restricted to words denoting precise quantities. The vague cardinals (words that name vague quantities) *neskol'ko* 'several', *skol'ko* 'how many', *stol'ko* 'so many', *mnogo* 'many', and *nemnogo* 'not many' exhibit exactly the same behavior, as (22) illustrates.[7]

(22) a. Ja znaju   mnogo   / neskol'ko   učënyx.
        I know.1SG many.ACC/NOM several.ACC/NOM scientist.PL.GEN
        'I know many/several scientists.'
     b. Ja znakoma so   mnogimi   / s   neskol'kimi   učënymi.
        I familiar.F with many.PL.INS with several.PL.INS scientist.PL.INS
        'I am familiar with many/several scientists.'

Though exhibiting the mixed declensional pattern of possessives, demonstratives, and similar functional adjectives, these vague cardinals combine with a genitive-plural NP, just like the more nominal cardinals 'five' to 'thousand' (compare (23c–d) to (23a–b) below). Importantly, just like with those cardinals the homogeneous pattern of case marking resurfaces in oblique cases ((22b); see Franks 1995).

The declensional patterns of vague cardinals place them somewhere between adjectives and nouns, or more precisely, between the higher cardinals 'five' to 'hundred' and the more adjectival paucal cardinals (see section 6.3.2). On the one hand, they don't show animacy agreement with the lexical NP. As shown in table 6.2, for animate nouns in the second declension, the accusative form is syncretic with the genitive, while for inanimate nouns, it is syncretic with the nominative. As shown in (23a–b), attributive adjectives obligatorily agree with the noun in animacy. The vague cardinals do not follow this pattern, as illustrated in (23c–d).

(23) a. Ja kupila   krasnye   stoly.
      I buy.PAST red.PL.ACC/PL.NOM table.PL.ACC/PL.NOM
      'I bought (some) red tables.'

    b. Ja kupila   seryx   slonov.
      I buy.PAST gray.PL.ACC/PL.GEN elephant.PL.ACC/PL.GEN
      'I bought (some) gray elephants.'

    c. Ja kupila   mnogo   stolov.
      I buy.PAST many.ACC/NOM table.PL.ACC/PL.GEN
      'I bought many tables.'

    d. Ja kupila   mnogo   slonov.
      I buy.PAST many.ACC/NOM elephant.PL.ACC/PL.GEN
      'I bought many elephants.'

On the other hand, as shown in table 6.5, in the oblique cases the vague cardinals show the declensional patterns of plural adjectives, such as possessive adjectives. As shown in (24), vague cardinals, just like attributive adjectives, agree with the plural noun that they modify in number. In this, they behave differently from the nominal cardinals discussed in the previous section.

(24) a. k   maminym/mnogim   sosedjam
      to   mother's.DAT/many.DAT neighbor.PL.DAT
      'to mother's neighbors/to many neighbors'

    b. o   maminyx/mnogix   podrugax
      about mother's.LOC/many.LOC friend.F.PL.LOC
      'about mother's friends/about many friends'

Table 6.5
Declensional pattern for vague cardinals in Russian

	'many'	Regular possessive adjective ('mother's')	Underlying forms for adjectives
NOM	mnog-o	mam-in-y	(-yj-)-ɨ
ACC			
GEN	mnog-i-x	mam-in-yx	(-yj-)-xɨ
DAT	mnog-i-m	mam-in-ym	(-yj-)-mɨ
LOC	mnog-i-x	mam-in-yx	(-yj-)-xɨ
INS	mnog-i-mi	mam-in-ymi	(-yj-)-mi

6.3.2 Adjectival Paucal Cardinals

We next consider the so-called paucal cardinals, 'two' through 'four', which, as discussed in chapter 3, assign paucal case to the lexical NP. Paucal cardinals differ both from vague cardinals and from the nominal cardinals 'five' and up with respect to declensional patterns as well as case-assignment properties. First, paucal cardinals exhibit the declensional pattern of a plural irregular adjective rather than that of a noun or a regular adjective, as shown in table 6.6 (which is the same as table 3.6 in chapter 3).[8] Second, while most of the nominal cardinals combine with the lexical NP in the genitive plural, the lower cardinals *dva/dve* 'two.M/F', *tri* 'three', and *četyre* 'four', as well as the quantifier *oba/obe* 'both.M/F', all combine with the form known as paucal, as shown again in (25a–b). The paucal form is visible in direct cases only: when the entire cardinal-containing NP is in an oblique case, the lexical NP also bears the oblique case (and furthermore, is plural), as shown in (25c–d)—in other words, like the cardinals 'five' to 'hundred', the lower cardinals show the homogeneous pattern of case assignment in oblique cases.

(25) a. četyre sestry
        four   sister.PAUC/SG.GEN
        'four sisters'

    b. četyre brata
        four   brother.PAUC/SG.GEN
        'four brothers'

    c. s   četyr'mja sëstrami   / *sestry
        with four.INS   sister.PL.INS   sister.PAUC/SG.GEN
        'with four sisters'

    d. s   četyr'mja brat'jami   / *brata
        with four.INS   brother.PL.INS   brother.PAUC/SG.GEN
        'with four brothers'

Table 6.6
Declensional pattern for paucal cardinals in Russian

	'two'	'three'	'four'
NOM	dv-á/dv-é	tr-í	četyr-e
GEN	dv-ú-x	tr-ё-x	četyr-ё-x
DAT	dv-ú-m	tr-ё-m	četyr-ё-m
LOC	dv-ú-x	tr-ё-x	četyr-ё-x
INS	dv-u-mjá	tr-e-mjá	četyr'-mjá

Note: Accusative case is not included in the table, because it is syncretic with the nominative when the lexical NP is inanimate and syncretic with the genitive when the lexical NP is animate.

The differences within this class of paucals are minimal and concern the nature of the vowel in the theme suffix (see Halle and Matushansky 2006). Paucal cardinals exhibit two properties of adjectives: (i) *dva/dve* 'two.M/F' and *oba/obe* 'both.M/F' show gender agreement with their sister (26a–b); and (ii) all the paucal cardinals exhibit animacy agreement with the lexical NP, as shown by the fact that in an accusative-case position they take a nominative-case ending with an inanimate NP (26c) and a genitive-case ending with an animate one (26d).

(26) a. dve/obe   sestry
       two/both.F.NOM sister.PAUC/SG.GEN
       'two/both sisters'
     b. dva/oba   brata
       two/both.M.NOM brother.PAUC/SG.GEN
       'two/both brothers'
     c. Ja kupila   tri   stola.
       I buy.F.PAST three.ACC/NOM table.PAUC/SG.GEN
       'I bought three tables.'
     d. Ja kupila   trëx   slonov.
       I buy.F.PAST three.ACC/GEN elephant.PL.ACC/PL.GEN
       'I bought three elephants.'

The agreement patterns that the paucal cardinals and 'both' exhibit strongly suggest (cf. Neidle 1988) that they are adjectives. However, a problem arises: adjectives are not known for assigning case to NPs they modify. The semantic type of cardinals ($\langle\langle e, t\rangle, \langle e, t\rangle\rangle$) is such that they can be taken to combine with their sisters in a head-complement relation (which is what we have been assuming for more nominal cardinals, which assign case to their sisters) or as adjuncts/specifiers (in which case it is unsurprising that they should agree like adjectives do). Neither configuration would seem compatible with the syntax of paucal cardinals: if they are merged as heads, their case-assigning properties can be explained, but not their adjectival behavior (the projection of an adjective should behave as an adjective rather than as an NP, so they cannot be assumed to be adjectival); if they are merged as adjuncts or specifiers, on the other hand, we do not expect them to assign case. We propose therefore that the adjectival nature of certain cardinals is reflected in their feature specification (i.e., they bear uninterpretable ϕ-features) and that their status in bare phrase structure as simultaneously heads and maximal projections makes it possible for them to exhibit mixed behavior. The configuration that they appear in, a head and its sister, makes case assignment as possible as agreement, as detailed in the next section.

6.3.3 Paucal Cardinals as Both Maximal Projections and Heads

As shown above (26a–b), the cardinal *dva/dve* 'two.M/F' and the quantifier *óba/óbe* 'both.M/F' show gender agreement with the lexical NP they combine with. The same holds for *poltorá/poltorý* 'one and a half.M/F'. We will refer to all three as dual Qs.

The gender agreement shown by the dual Qs is the only case in Russian where gender agreement takes place in plural NPs. However, on our hypothesis, the lexical NP that appears with the cardinal is semantically singular, despite plural morphology (a case of number agreement), so gender agreement with this NP is not so surprising, given that singular NPs in Russian trigger gender agreement. The question nonetheless arises of the mechanism of the two-way number-gender agreement seen here (the cardinal agreeing with the NP in gender while the NP agrees with the cardinal in number), and of how this agreement interacts with case assignment.

Assuming the same cascading structure for cardinals as always, the dual Q is merged as a sister to the lexical NP, as in (27) (the Card^0 notation is for the sake of convenience only). Whichever of the three dual Qs it is, it is syntactically a head and therefore, we hypothesize, can probe into its sister. If the dual Q bears an uninterpretable gender feature and an interpretable number feature, whereas the lexical NP that it combines with is specified, conversely, for an interpretable gender feature and an uninterpretable number feature, then valuation succeeds in both directions. The paucal case is assigned as part of the same process.

(27)   CardP   'two'

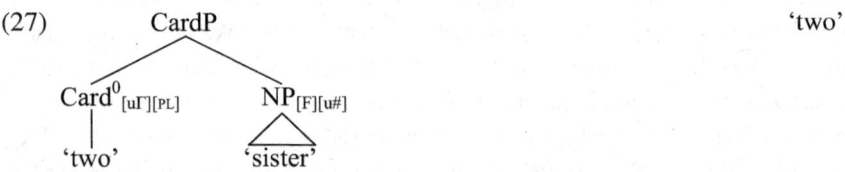

Card^0$_{[uΓ][PL]}$ NP$_{[F][u\#]}$

'two'   'sister'

The hypothesis that the dual Qs are specified for an interpretable number feature allows us to explain as well why the declensional pattern of 'two' and 'both' fits that of *plural* irregular adjectives:[9] assuming that adjectival inflection does not distinguish between interpretable and uninterpretable specifications of the number feature, plural morphology is obtained in the absence of agreement, as is presumably the case for pluralia tantum nouns.[10]

It is clear from the preceding discussion that the dual Qs cannot be treated as either adjectives or nouns. They are distinguished from adjectives by the presence of an interpretable number feature, whereas their gender feature, unlike that of nouns, is uninterpretable and unvalued. They therefore provide additional evidence for rejecting the binary distinction between nouns and

adjectives in favor of a more fine-grained analysis in terms of featural specification. On this view, cardinals illustrate the existence of intermediate sub-
types within each lexical category (similar to the existence of two types of
adjectives in Japanese: see Dixon 1977; Miyagawa 1987; Kubo 1992; Nishiyama 1999; Baker 2003; Backhouse 2004; etc.) or intermediate categories
(similar to the nominal adjectives in Kabyle (Djemai 2008) or Eastern Riffian
(Oomen 2012)).

6.3.4 Paucal-Case Marking

In the previous subsections, we discussed the so-called paucal cardinals, which
assign paucal case to their lexical NP sister. In this section, we will further
discuss the status of the paucal form. We will argue against the hypothesis that
the paucal form in Russian corresponds to the paucal number and demonstrate
that paucal cardinals differ from other cardinals in that they assign an impoverished version of structural genitive, the paucal case. In this subsection, we
provide evidence that paucal is distinct from the genitive singular; the next
two subsections take up the question of whether paucal should be analyzed as
number or case, arguing in favor of the latter option.

For most nouns, the paucal form is syncretic with the genitive singular, as
was shown in (25a–b). However, there are two classes of nouns for which the
paucal is distinct from the genitive singular. The first of these are feminine
surnames derived with the possessive suffixes -ov- and -in- (Franks 1995, 52),
as illustrated in (28).[11] A last name such as *Petrov* is derived from a first name
such as *Petr* 'Peter' through the addition of the possessive -ov- suffix; the
feminine version of this surname requires a feminine ending, -a in the nominative, as shown in (28a). When this feminine last name combines with a paucal
cardinal, as in (28b), it obligatorily surfaces in the paucal, not the genitive
singular (compare to (28c)).

(28) a. Marija   Petr-ov-a
        Maria.NOM Peter-POSS-F.SG.NOM
        'Maria Petrova'
     b. dve   Petrovy   /* Petrovoj
        two.F.NOM Petrov.F.PAUC/PL.NOM   Petrov.F.SG.GEN
        'two (female) Petrovs'
     c. brat   Petrovoj
        brother.M.NOM Petrov.F.SG.GEN
        'Petrova's brother'

The second class of nouns in which the paucal and genitive-singular forms
are distinct consists of five nouns in the second declension: *šag* 'step', *sled*

'footprint', *rjad* 'row', *čas* 'hour', and *šar* 'ball' (see Mel'čuk 1985). As shown in (29) and (30), the genitive-singular form and the paucal form have different stress patterns, and it is the paucal form that must follow the cardinal: see (29a–b) and (30a).[12] The genitive-singular form is used in other contexts, such as (29c–d) and (30b).

(29) a. četyre časá/*čása   Paucal form after cardinal
       four   hour.PAUC/SG.GEN
       'four hours'

    b. četyre šará/*šára
       four   balloon.PAUC/SG.GEN
       'four balloons'

    c. okolo čása/*časá   Genitive-singular form after preposition
       near hour.SG.GEN/PAUC
       'about an hour'

    d. cvet šára/*šará
       color balloon.SG.GEN/PAUC
       'the color of the balloon'

(30) a. četyre šagá
       four   step.PAUC
       'four steps'

    b. zvuk šága
       sound step.SG.GEN
       'the sound of a step'

In order to convey when we are dealing with the paucal form rather than the genitive singular, we will use the lexical NP *šag* 'step' in all relevant Russian examples from now on.

If the cardinal and the lexical noun are not adjacent, there is a strong preference for the paucal form to become identical to the genitive singular (Mel'čuk 1985, 432–434), as (31) shows.[13]

(31) a. dva   širokix   šága/*šagá
       two.M.NOM broad.PL.GEN step.SG.GEN/PAUC
       'two broad steps'

    b. dva   s   polovinoj šága/*šagá
       two.M.NOM with half.INS   step.SG.GEN/PAUC
       'two and a half steps'

Just like the simplex cardinals 'five' to 'hundred', the paucal cardinals, as shown above in (25c–d), trigger the homogeneous case-marking pattern in oblique cases. Additional examples are in (32).

(32) a. bez četyrëx šarov/*šará
without four.PL.GEN balloon.PL.GEN/PAUC
'without four balloons'

b. s dvumja rjadami/*rjadá
with two.INS row.PL.INS/PAUC
'with two rows'

The distinction between the genitive-singular form and the paucal form is more systematic in Ukrainian, where with the cardinals *dva/dvi/dvoe* 'two', *tri* 'three', and *čotyry* 'four', as well as the quantifier *obidva/obidvi* 'both', many masculine nouns appear in the so-called dual form, which has the nominative-plural ending but the stress pattern of the genitive singular (Mel'čuk 1985, 436; Korunec 2003, 203). This is shown in (33) and (34).

(33) a. tyn 'fence.SG.NOM', týnu 'fence.SG.GEN', tyný   Ukrainian
'fence.PL.NOM'

b. dva (čornyx) týny
two black.PL.GEN fence.DU
'two black fences'
(Mel'čuk 1985, 436)

(34) a. vórog 'enemy.SG.NOM', vóroga 'enemy.SG.GEN',
vorogí 'enemy.PL.NOM'

b. tri vórogi
three enemies.DU
'three enemies'
(Mel'čuk 1985, 436)

Likewise, in Belarusian the nominative-plural form and the paucal form of neuter nouns are distinguished by stress (Akiner 1983), as (35) and (36) show.

(35) a. sióły 'village.PL.NOM', siałá 'village.SG.GEN'   Belarusian

b. dva/try/čatyry siałý
two.N/three/four village.PAUC
'two/three/four villages'

(36) a. píśmy 'letter.PL.NOM', piśmá 'letter.SG.GEN'

b. dva/try/čatyry piśmý
two.N/three/four letter.PAUC
'two/three/four letters'

While some researchers have argued that the paucal form in Russian (and, by a natural extension, in Ukrainian and Belarusian) corresponds to a particular number (Yadroff 1999; Rakhlin 2003; Bailyn and Nevins 2008; etc.), others have claimed that it is a particular case (Mel'čuk 1985; Franks 1994, 1995; Rappaport 2002, 2003a,b).[14] It is the latter hypothesis that we will defend here. To do so we will first examine the hypothesis that the paucal form corresponds to a number specification. If we assume this hypothesis, the question then arises which case is assigned to the lexical NP. Two answers are equally possible: (i) the paucal cardinals behave like the cardinal 'one' and assign no case, and (ii) the paucal cardinals behave like the higher cardinals and assign genitive case. We will demonstrate that both possible answers fail to satisfactorily account for the syntax of paucal-cardinal-containing NPs, in particular for the case marking on attributive adjectives and the plural marking on the paucal cardinal itself. On the other hand, the hypothesis that the paucal form corresponds to a particular case marking can deal with these issues quite straightforwardly.

6.3.5 Paucal Number, No Case

As Corbett (2000, 23fn.) observes, since the paucal form in Russian (as well as in Ukrainian and Belarusian) cannot be used to denote a small number (two to four) in the absence of a cardinal, it cannot be part of the grammatical-number system. In other words, the paucal morphology on the noun does not seem to have any semantic import, unlike the real paucal number. This conclusion is also required by our assumption (chapter 2) that the lexical NP that cardinals combine with is necessarily semantically singular.

It is possible, however, that the paucal-number marking on the lexical NP is uninterpretable and arises as a result of agreement with the paucal cardinal. In fact, this is more or less the analysis that we advanced to account for the plural marking on the noun in a cardinal-containing NP in Russian, English, and many other languages (chapter 4).

One problem for the view that paucal marking reflects number agreement rather than case is the surface case marking on attributive adjectives in lexical NPs combining with paucal cardinals. As the examples in (37) illustrate, when the entire cardinal-containing NP appears in a direct case, adjectives inside the lexical NP (unlike those higher than the cardinal) must appear in the genitive plural, with both paucal and nonpaucal cardinals.

(37) a. četyre   dolgix   čása/časá   Russian
        four.ACC/NOM long.PL.GEN hour.SG.GEN/PAUC
        'four long hours'
    b. pjat'   dolgix   časov
        five.ACC/NOM long.PL.GEN hour.PL.GEN
        'five long hours'

With feminine nouns an additional option becomes available: with paucal cardinals (38), but not with nonpaucal cardinals (39), the attributive adjective may also be marked *nominative* plural.

(38) a. četyre   dolgix   minuty
     four.ACC/NOM long.PL.GEN   minute.PAUC/SG.GEN/PL.NOM
     'four long minutes'

   b. četyre   *dolgie*   minuty
     four.ACC/NOM long.PL.NOM   minute.PAUC/SG.GEN/PL.NOM
     'four long minutes'

(39) a. pjat'   dolgix   minut
     five.ACC/NOM long.PL.GEN   minute.PL.GEN
     'five long minutes'

   b. *pjat'   *dolgie*   minut
     five.ACC/NOM long.PL.NOM   minute.PL.GEN

The reason for this asymmetry is clear from the glosses: for the majority of feminine nouns the nominative-plural form and the genitive-singular form are phonologically identical. When this is not the case, as with masculine nouns (40), the nominative form becomes unavailable.[15]

(40) četyre   dolgix/*dolgie   čása/časá
     four.NOM/ACC long.PL.GEN/PL.NOM   hour.SG.GEN/PAUC
     'four long hours'

Both case-marking options for attributive adjectives can be easily derived now. The genitive-plural marking is the impoverishment of number: the more marked value (paucal) surfaces as a less marked value (plural), with case marking remaining the same. The resulting genitive-plural form can be further impoverished, yielding the nominative-plural form of the adjective in examples like (38b): the more marked case value is replaced by a less marked case value. The question is, however, why genitive plural is not impoverished to nominative plural in other environments. Given the fact that here we obtain the nominative-plural form from the genitive-plural form (rather than directly from the genitive paucal), the syncretism between genitive singular and nominative plural does not appear to be relevant (41).

(41) a. ošibka molodyx/*molodye   devušek
     error   young.PL.GEN/PL.NOM girl.F.PL.GEN
     'a/the mistake of (the) young girls'

   b. pro   dal'nie/*dal'nix   dorogi
     about faraway.PL.NOM/PL.GEN road.F.PL.LOC/NOM
     'about faraway roads'

Deriving the nominative-plural form from the genitive-paucal form requires the manipulation of two features simultaneously, though this time their relevant markedness goes in the right direction: nominative is less marked than genitive and plural is less marked than paucal.

In light of the above, we assume with Mel'čuk (1985), Franks (1994, 1995), and Rappaport (2002, 2003a,b) that paucal is a case rather than a number. As discussed in chapter 3, we hypothesize that the lexical NP that a cardinal combines with is necessarily atomic, that is, semantically singular. We now need to determine whether the noun in the paucal case is singular syntactically as well (like it is with the cardinal 'one') or syntactically plural (as it is with all other cardinals).

6.3.6   Paucal Case, Plural Number

The first empirical argument in favor of the idea that paucal NPs are syntactically plural is the simple fact that in oblique cases, where the case marking becomes homogeneous, the entire lexical NP is marked plural:

(42) a. okolo trëx   *sinix*   *čašek*
         around three.GEN blue.PL.GEN cup.PL.GEN
         'around three cups'
     b. četyr'mja *levymi*   *rukami*
         four.INS left.PL.INS hand.PL.INS
         'with four left hands'

The hypothesis that the plural marking in cardinal-containing NPs is a result of agreement (chapter 4) predicts that any NP denoting a plurality, including NPs headed by a paucal cardinal, will be morphologically plural, and the plural marking on the attributive APs supports this conclusion. As a result, (37a) is represented as in (43).

(43) četyre   dolgix   čása/časá
     four.NOM/ACC long.PL.PAUC hour.SG.GEN/PAUC
     'four long hours'

The realization of the paucal case in Russian is gender-dependent: an unaccented [a] for masculine and neuter nouns and an unaccented [y] (surfacing as [i] after consonants that are underlyingly palatalized) for others, including first-declension masculine nouns. While Russian plurals usually show no gender distinctions, the paucal quantifier 'both' and the paucal cardinal 'two' do not violate this generalization as long as it is assumed (which we do) that the lexical NP is semantically singular.

Given the syntactic plurality of the lexical NP, the two case-marking options for attributive adjectives have to be derived from the featural specification of the internally complex paucal case. One such solution is proposed by Franks (1995, 105).

The behavior of feminine adjectival surnames derived from possessive adjectives (Franks 1995, 52), like the one in (28a) above, can be taken as evidence for either singular or plural marking on the lexical NP in the context of a paucal cardinal. The feminine paucal ending [y] in (28b) is identical to the nominative-plural exponent of both nouns and adjectives, as well as to the genitive-singular exponent of feminine nouns. As noted by Garde (1998, 207, 230), the declensional pattern of surnames derived with the suffixes -in- and -ov- differs from that of possessive adjectives in the locative singular: compare (44a), the locative form of the surname, to (44b), the locative form of the possessive adjective.

(44) a. ob   Ivanove
         about Ivanov.SG.LOC
         'about Ivanov'
     b. ob   Ivanovom   brate
         about Ivan.POSS.SG.LOC brother.SG.LOC
         'about Ivan's brother'

In other words, it could be argued that such surnames, which already have to be postulated to have a special declensional pattern, also have the dedicated paucal form in the singular (Franks 1995). Conversely, the fact that the adjectival genitive-singular form is ungrammatical with cardinals (28b) could be taken as evidence that the lexical NP in the context of a paucal cardinal is marked plural and takes the paucal plural ending -y-, like other feminine nouns.[16] We adopt the latter hypothesis, which is more economical.

To summarize, plural marking on attributive adjectives, on the paucal cardinal itself, and on the entire lexical NP in oblique cases strongly suggests that despite the fact that the paucal form is homophonous with the genitive singular, it is nonetheless marked plural. Conversely, no evidence apart from this syncretism can be adduced in favor of singular marking.

6.3.7 Evidence from Fractional Cardinals

Another piece of evidence in favor of treating paucal as case rather than number comes from the cardinal *četvert'* 'quarter' and the morphologically bound cardinal *pol-* 'half'. Semantically both cardinals require a singular NP, so the surface genitive singular that they combine with is unrevealing. However, as the five relevant nouns confirm, *četvert'* and *pol-* assign paucal case to their sister (see section 6.4.2; although, as discussed in that section, with *četvert'* genitive singular is preferable).

Given the fact that semantically singular nouns can surface in the paucal form, it can still be hypothesized that paucal is a number (rather than case), which, just like plural, is assigned as a result of agreement. The problem with this view, however, is that paucal has been shown to surface as plural on adjectives with the cardinals *poltorá/poltorý* 'one and a half.M/F', *dva/dve* 'two.M/F', *tri* 'three', and *četýre* 'four' and the quantifier *óba/óbe* 'both.M/F' (see (31a)). This is not the case with *četvert'* 'quarter' and *pol-* 'half', where the adjective surfaces as genitive singular, as (45) shows.

(45) a.   pol-svetov-ogo   goda
          half.ACC/NOM-light-ADJ.M.SG.GEN year.PAUC/SG.GEN
          'half a light-year'
     b.   četvert'   morsk-oj   mili
          quarter.ACC/NOM sea-ADJ.F.SG.GEN mile.PAUC/SG.GEN
          'a quarter of a nautical mile'

In other words, the number marking on the AP reflects the number of the entire numeral NP: when it denotes a semantic plurality, the AP is plural, and when it does not, the AP is singular. Since the noun itself takes the paucal form in both cases, the paucal form cannot be taken to correspond to a particular number specification. It therefore has to be treated as a case.

6.3.8 Summary
In this section, we have examined the behavior of the lower cardinals in Russian and have argued that they have some properties of nouns as well as some properties of adjectives, providing further evidence against a binary adjective–noun distinction for cardinals. We have also examined the special paucal form of nouns appearing with the paucal cardinals 'two', 'three', 'one and a half', and 'four' and the paucal quantifier 'both'. A critical evaluation of the available evidence strongly suggests that in Russian, at least, paucal is a case rather than a number.

6.4 Lexical Fractions

Having examined the behavior of both nominal and adjectival simplex cardinals, we next consider how fractional cardinals behave with respect to the relevant diagnostics. In this section, we will show that fractions in Russian do not form a homogeneous class: while some fraction names are fully nominal, others behave like simplex cardinals with respect to agreement and case assignment and should in fact be analyzed as cardinals. Our conclusion is that fractions in Russian, just like cardinals, fall along a nominal–adjectival continuum and, in some cases, behave like defective nouns (see table 6.1). For more discussion of the syntax of fractions, see chapter 10.

6.4.1 Derived versus Lexical Fractions

Like any familiar language, Russian has a productive way of forming fraction names. Most fraction names in Russian are homophonous with ordinals, as shown in (46a–b). The same is true for English (46c–d) and French.

(46) a. desjataja   stranica
       tenth.NOM page.NOM
       'the/a tenth page'
    b. odna   desjataja   torta
       one.F.NOM tenth.NOM cake.SG.GEN
       'one-tenth of a cake'
    c. the third house
    d. one-third of the cake

Despite their shape (such fractions decline precisely like the regular adjectives from which they are null-derived) these fractions are clearly nominal, as shown by (47). They appear in argument positions (47a–b). Their form is the same with masculine nouns like *tort* 'cake' and feminine ones like *picca* 'pizza' (47a), and with inanimate versus animate nouns (47a–b); this indicates that, unlike adjectives, these fractions do not agree in gender or animacy with the following NPs (compare (47) to (26), with adjectival lower cardinals). Like other nouns, fractions show the heterogeneous pattern of case assignment: the genitive case that they assign persists in oblique cases (47c).[17] Given that they do not show agreement (unlike adjectives) and are fully specified for φ-features (unlike cardinals), we conclude that derived fractions in Russian are fully nominal.

(47) a. On s'el   odnu   desjatuju torta/piccy.
       he eat.M.PAST one.F.ACC tenth.ACC cake.SG.GEN/pizza.SG.GEN
       'He ate one-tenth of the cake/pizza.'
    b. Bylo vidno tol'ko odnu   desjatuju slona.
       was seen only one.F.ACC tenth.ACC elephant.SG.GEN
       'It was possible to see only one-tenth of the elephant.'
    c. s   odnoj   desjatoj torta/*tortom
       with one.F.INS tenth.INS cake.SG.GEN/INS
       'with one-tenth of the cake'

Apart from these derived fractions, Russian also has several lexical fractions, some of which may also function as cardinals. Lexical fractions do not behave as a homogeneous class (Mel'čuk 1985). We will follow Mel'čuk (1985, 322–325) in arguing that whereas *tret'* 'a third' and *polovina* 'a half' are regular nouns, *četvert'* 'quarter' and *pol-* 'half' are cardinals, albeit not of the same kind (we argue that *četvert'* is ambiguous between nominal and cardinal readings). Applying the same diagnostics to fractional cardinals that we applied to simplex cardinals, we will also show that syntactic differences between *pol-* and *četvert'* parallel those between most simplex cardinals and cardinals like 'thousand'. (For the semantics of fractions see chapter 10.)

6.4.2   Case Assignment with Fractional Cardinals

Just like the second-declension cardinals *sorok* 'forty', *devjanosto* 'ninety', and *sto* 'hundred', *pol-* 'one-half' and *poltora* 'one and a half' only have two morphological cases:[18] the direct one, as in (48a–b) and (49a–b), and the oblique one, as in (48c) and (49c) (Garde 1998, 242).[19] The cardinal *poltora* 'one and a half' is derived from *pol-vtora* 'half of the second' (Vasmer 1986; see Hurford 1975 for similar derivations), which explains why both its parts decline independently. Even though it has only two case forms, in the direct cases *poltora* shows gender agreement with the lexical NP (due to the fact that the ordinal 'second' is an adjective), as shown in (49).[20]

(48) a. za pol-minuty
         for half.ACC/NOM-minute.F.PAUC/SG.GEN
         'in half a minute'

     b. za pol-čásá   do   rassveta
         for half.ACC/NOM-hour.PAUC until dawn.GEN
         'half an hour before dawn'

     c. v polu-čase   ezdy   otsjuda
         in half.OBL-hour.LOC traveling.GEN from-here
         'half an hour's traveling time from here'

(49) a. za pol-tory   minuty
         for half.ACC/NOM-second.F.ACC/NOM minute.F.PAUC/SG.GEN
         'in one and a half minutes'

     b. za pol-tora   čásá
         for half.ACC/NOM-second.M.ACC/NOM hour.M.PAUC
         'in one and a half hours'

     c. v polu-tora   minutax/časax
         in half.OBL-second.OBL minute.F.LOC.PL/hour.M.LOC.PL
         'in one and a half minutes/hours'

The fractional cardinals *pol-* and *poltora* behave differently from simplex cardinals in terms of case on the lexical NP. As shown in (48a) and (49a), the lexical NP appears in the genitive singular rather than genitive plural. As discussed in section 6.3.2, for most nouns, the genitive-singular form is syncretic with the paucal form, but for several nouns, including *čas* 'hour' and *šar* 'balloon', the two forms are distinct. As observed by Mel'čuk (1985, 323–324), only the lexical noun *čas* 'hour' systematically appears in the paucal form with the cardinals *pol-* 'half' and *poltora* 'one and a half' (50a–b); the remaining four nouns that have the paucal form preferentially appear in the genitive singular with *pol-* (50c), but occur in the paucal with *poltora* (50d).

(50) a. za pol-časá / *pol-čása.
for half.ACC/NOM-hour.PAUC half.ACC/NOM-hour.SG.GEN
'in half an hour'

b. za pol-tora časá/*čása
for half.ACC/NOM-second.M.ACC/NOM hour.PAUC/SG.GEN
'in one and a half hours'

c. *pol-rjadá / pol-rjáda
half.ACC/NOM-row.PAUC half.ACC/NOM-row.SG.GEN
'half a row'

d. pol-tora rjadá/?rjáda
half.ACC/NOM-second.M.ACC/NOM row.PAUC/SG.GEN
'one and a half rows'

The truly nominal fractions *polovina* 'half' and *tret'* 'third' require the genitive singular on all lexical NPs, including *čas*.

(51) polovina / tret' čása/*časá
half.NOM third.ACC/NOM hour.SG.GEN/PAUC
'half an hour/a third of an hour'

Another difference found in the case-assignment patterns of different fractional cardinals is whether they show the homogeneous or heterogeneous pattern of oblique-case assignment (see section 6.2.3). As shown in (48c) and (49c), *pol-* and *poltora* exhibit the homogeneous pattern, just like most simplex cardinals. This pattern is obligatory, as (52a) establishes. But as shown in (52b), *tret'* and *polovina* require the heterogeneous pattern of case assignment, like regular nouns.[21]

(52) a. polu-*časá/*čása/časom ranee
half.OBL-hour.PAUC/SG.GEN/INS earlier
'half an hour earlier'

b. tret'ju/polovinoj *časá/čása/*časom ranee
third/half.INS hour.PAUC/SG.GEN/INS earlier
'a third of an hour earlier/half an hour earlier'

To sum up, we see a clear difference between, on the one hand, *pol-* and *poltora*, which require the paucal form for *čas* but allow (or require) the genitive singular for other nouns, and, on the other hand *tret'* and *polovina*, which require the genuine genitive singular for all lexical NPs. Furthermore, we see that the surface genitive case appearing with *pol-* is underlyingly paucal, as shown by the fact that it follows the homogeneous case pattern in the oblique cases. If *pol-* assigned a true genitive case, we would have expected it to follow the heterogeneous case pattern, as do genitive-case-assigning regular nouns. In fact, *tret'* and *polovina* assign regular genitive case, as evidenced by the heterogeneous pattern of case assignment.

6.4.3 Gender Agreement and Adjectival Modification

Like simplex cardinals, *pol-* triggers plural agreement, rather than gender agreement, on both determiners and higher adjectives; also like simplex cardinals, it allows plural premodifiers (cf. (18) above). This is all illustrated in (53a) (showing plural agreement on adjectives; the same facts hold for determiners). In contrast, *tret'* 'a third' and *polovina* 'a half' do not allow plural agreement or a plural premodifier and obligatorily trigger gender agreement, as shown in (53b).

(53) a. *celaja/*celyj/*celoe/celyx   *pol-časa   /
         whole.F.NOM/M.NOM/N.NOM/PL.GEN   half.ACC/NOM-hour.SG.GEN
         pol-časá
         half.ACC/NOM-hour.PAUC
         'a whole half an hour'

     b. celaja/*celyj/*celyx   tret'   / polovina
        whole.F.NOM/M.NOM/PL.GEN third.ACC/NOM half.NOM
        čása/*časá
        hour.SG.GEN/PAUC
        'a whole third/half of an hour'

Četvert' 'quarter', on the one hand, shows the same behavior as *million* 'million' and *tysjača* 'thousand' in that it is compatible with both a genitive-plural premodifier and a premodifier agreeing with it in number and gender, as shown in (54a). With the lexical NP *čas*, the form of the premodifier is in direct correlation with the case on *čas*, as shown in (54b–c).

(54) a. celaja/celyx   četvert'   minuty
        whole.F.NOM/PL.GEN quarter.ACC/NOM hour.SG.GEN/PAUC
        'a whole quarter of a minute'

     b. celaja/*celyx   četvert'   čása
        whole.F.NOM/PL.GEN quarter.ACC/NOM hour.SG.GEN
        'a whole quarter of an hour'

c. *celaja/celyx   četvert'   čásá
   whole.NOM.F/PL.GEN quarter.ACC/NOM hour.PAUC
   'a whole quarter of an hour'

This suggests that, like *million* and *tysjača*, *četvert'* can be either a noun or a cardinal. When *četvert'* is a cardinal, it takes a paucal complement and triggers plural agreement (54c), just like *pol-* (53a). When it is a noun, it takes a genitive-singular complement and triggers gender agreement (54b), just like the nouns *tret'* and *polovina* (53b). Likewise, if *četvert'* forms part of a complex cardinal ('one-quarter', 'three-quarters'), as in (55a), it also requires the genitive-singular form, just like other fractions (55b). This makes *četvert'* even more comparable to *million* and other higher cardinals: their only distinctions from nouns are the ability to trigger plural agreement (54) and the ability to combine with a premodifier in the genitive plural.

(55) a. tri   četverti   čása/*čásá
        three.NOM quarter.SG.GEN hour.SG.GEN/PAUC
        'three-quarters of an hour'
     b. dve   treti   čása/*čásá
        two.F.NOM third.SG.GEN hour.SG.GEN/PAUC
        'two-thirds of an hour'

We thus have evidence that fractional cardinals, like simplex cardinals, fall into several categories. The fractions *pol-* and *poltora* behave like the simplex cardinals 'five' through 'hundred' in their homogeneous case-assignment patterns and the fact that they obligatory trigger plural agreement rather than gender agreement on the modifier. In contrast, *četvert'*, like *million*, exhibits the heterogeneous pattern of case assignment and optionally triggers either gender agreement or plural agreement. Finally, *tret'* and *polovina* behave in every respect like true nouns, as do measure nouns like *djužina* 'dozen' and *desjatok* 'a group of ten'.

6.5 Predicate Agreement with Russian Cardinal-Containing NPs

As noted by Franks (1994, 1995), Neidle (1988), and Pesetsky (1982), among many others, cardinal-containing NPs in Russian can both trigger plural agreement on the predicate (56a) and occur with default (neuter-singular) marking on the predicate (56b). Despite the fact that most Russian cardinals, including *pjat'* 'five', belong to the third declension (which contains feminine nouns plus eleven exceptions), gender agreement with them is impossible (56c).

(56) a. K nam prišli   pjat' krasivyx   devušek.
         to us   arrived.PL five beautiful.PL.GEN girl.PL.GEN
         'Five beautiful girls arrived at our place.'

     b. K nam prišlo   pjat' krasivyx   devušek.
         to us   arrived.N.SG five beautiful.PL.GEN girl.PL.GEN
         'Five beautiful girls arrived at our place.'

     c. *K nam prišla/prišël   pjat' krasivyx   devušek.
         to us   arrived.F.SG/M.SG five beautiful.PL.GEN girl.PL.GEN

There are interpretational differences between agreeing and nonagreeing cardinal-containing-NP subjects. Pereltsvaig (2006b) shows that the nonagreeing type cannot be definite or specific (57), outscope other quantifiers (58), or (with certain caveats, on which see Matushansky and Ruys 2015a,b) control PRO and bind freestanding reflexives.

(57) a. Pjat' knig   na stole byli/*bylo   moi.
         five books on table were.PL/was.N.SG mine.PL
         'The five books on the table were mine.'

     b. Kakie-to tri   knigi prodajutsja/*prodajëtsja deševle.
         some   three books sell.3PL.REFL/3SG.REFL   cheaper
         'Some three books are being sold cheaper.'

(58) a. Každyj raz pjat' xirurgov operirovali Bonda.   ∀ > 5, 5 > ∀
         every   time five surgeons operated.PL Bond
         'Every time five surgeons operated on Bond.'

     b. Každyj raz pjat' xirurgov operirovalo Bonda.   ∀ > 5, *5 > ∀
         every   time five surgeons operated.N.SG Bond
         'Every time five surgeons operated on Bond.'

Matushansky and Ruys (2015a,b) show that, on the other hand, default agreement is obligatory with indefinite measure phrases (59) and with indefinite event-oriented readings of cardinal-containing NPs (60); Krifka (1990b) argues that these readings involve degrees.[22]

(59) a. Prošlo/*prošli pjat' let.
         went.N.SG/PL   five years
         'Five years passed.'

     b. Na ètot pirog nužno/*nužny   tri   stakana   muki.
         on this pie   necessary.N.SG/PL three glass.PAUC flour.GEN
         'This pie requires three cups of flour.'

(60) a. Četyre tysjači korablej   Plural: individual ships only
    four thousand.SG.GEN ship.PL.GEN
    prošli čerez šljuz.
    passed.PL through lock
    'Four thousand ships passed through the lock.'

    b. Četyre tysjači korablej   Default: no commitment
    four thousand.SG.GEN ship.PL.GEN
    prošlo čerez šljuz.
    passed.N.SG through lock
    'Four thousand ships passed through the lock.'

To capture these interpretational differences, it has been variously suggested that nonagreeing cardinal-containing noun phrases are QPs or NPs rather than DPs (Pesetsky 1982; Franks 1994; and Pereltsvaig 2006b), that they do not raise to a high enough position (Pesetsky 1982; Franks 1994; and Stepanov 2001), and that they uniformly denote degrees (Matushansky and Ruys 2015a,b). None of these views attribute agreement failure to the cardinals themselves. That this is correct is shown by the fact that agreement failure is not specific to cardinal-containing NPs but also occurs with vague measures, as illustrated in (61) for the vague measure *rjad* 'a number of', literally 'series'.

(61) Bylo namečeno rjad konkretnyx voprosov.
    was.N.SG sketch.PTCP.N.SG series concrete.PL.GEN question.PL.GEN
    'There was sketched a number of concrete questions.'
    (Graudina, Ickovič, and Katlinskaja 1976)

To sum up, while cardinal-containing NPs can fail to trigger number agreement in Russian, this failure is not a unique syntactic property of cardinals. The associated interpretational effects suggest a link to a measure denotation (Matushansky and Ruys 2015a,b).

6.6 Special Count Forms in Russian and Bulgarian

As discussed in section 6.2, the cardinals 'five' and higher in Russian share some properties with regular nouns but also differ from them on such diagnostics as gender agreement and homogeneity versus heterogeneity of case marking under oblique-case assignment. In this section, we discuss one more difference between higher cardinals and regular nouns, namely, the existence of special count forms for some lexical NPs, which only surface with a cardinal. The relevant data come from Russian as well as Bulgarian.

6.6.1 Count Forms in Russian

Another important difference between cardinals and nouns that assign genitive case is that the form that follows the cardinal in the direct cases is not always identical to the genitive. In addition to the dedicated paucal form that a handful of nouns have (see section 6.3; cf. Zaliznjak 1967; Franks 1994, 1995; Yadroff 1999; Rappaport 2002, 2003a,b; Rakhlin 2003; Bailyn and Nevins 2008), Mel'čuk (1985, 430–437) observes that Russian nouns denoting units of measure have a special adnumerative form that they may appear in when combining with a cardinal, as in (62b) and (63b), but not otherwise, as shown in (62a) and (63a). When the regular genitive plural form surfaces with a cardinal for these nouns, as in the use of *kilogrammov* in (62b), the meaning is no longer that of a measure noun but of entities (e.g., kilogram-sized packages).

(62) a. Vsex   ètix   kilogrammov   / *kilogramm
         all.PL.GEN this.PL.GEN kilogram.PL.GEN   kilogram.ADN/NOM.SG
         budet   malo.
         be.FUT.3SG too.little
         'All these kilograms will not be enough.'
     b. Nam   nado   dvadcat'   kilogrammov   /
         we.DAT necessary twenty.ACC/NOM kilogram.PL.GEN
         kilogramm.
         kilogram.ADN/NOM.SG
         'We need twenty kilograms.'

(63) a. Vsex   ètix   rentgenov   / *rentgen
         all.PL.GEN this.PL.GEN roentgen.PL.GEN   roentgen.ADN/NOM.SG
         budet   malo.
         be.FUT.3SG too.little
         'All these roentgens will not be enough.'
     b. Nam   nado   dvadcat'   *rentgenov   /
         we.DAT necessary twenty.ACC/NOM roentgen.PL.GEN
         rentgen.
         roentgen.ADN/NOM.SG
         'We need twenty roentgens.'

Importantly, this adnumerative form is used even if the entire cardinal-containing NP is marked genitive (64), showing that the homogeneous and heterogeneous patterns of case assignment in numeral NPs cannot, irrespective of the structural–inherent divide, be explained by the assumption that the externally assigned case overrides the internally assigned case.

(64) bolee semi   *rentgenov   / rentgen
     more seven.GEN   roentgen.PL.GEN roentgen.ADN/NOM.SG
     'more than seven roentgens'

As shown above, genitive case is part of the oblique paradigm triggering the homogeneous case-assignment pattern with Russian cardinals (see section 6.2.3). The simple overriding pattern would require the lexical NP in (64) to be marked with the usual genitive case rather than the structural genitive assigned by the cardinal, and the adnumerative form would therefore be unexpected. Conversely, if the surface case is a realization of all the relevant case features assigned to the noun, the features assigned only by cardinals would remain on the noun even in the context of the external genitive, leading to the use of the adnumerative form for the lexical-semantic class of the units of measure.

6.6.2 Suppletion with Count Forms in Russian

Like in many other languages, in Russian a handful of nouns have a suppletive plural form: *čelovek* 'person', *god* 'year', *cvetok* 'flower', and *rebënok* 'child'. When combined with a cardinal lower than 'five', they all use the singular rather than the plural form, as the examples in (65) illustrate.[23]

(65) a. odin   čelovek/god/cvetok/rebënok
        one.M.NOM   person/year/flower/child
        'a person/year/flower/child'
     b. dva/tri/četyre   čeloveka   / goda   / cvetka   /
        two/three/four.NOM man.PAUC/SG.GEN   year.PAUC flower.PAUC
        rebënka
        child.PAUC
        'two/three/four men/years/flowers/children'

With the cardinals 'five' and higher, the behavior of the four nouns diverges: while *čelovek* 'person' and *cvetok* 'flower' appear in the same form as the singular (66), *god* 'year' and *rebënok* 'child' become suppletive (67).

(66) a. Ona oprosila   pjat'   čelovek/*ljudej.   Nonsuppletive
        she question.PAST five.ACC person.PL.GEN/people.GEN
        'She questioned five people.'
     b. Ona sorvala   pjat'   cvetkov/*cvetov.
        she pluck.PAST five.ACC flower.PL.GEN/flowers.GEN
        'She plucked five flowers.'

(67) a. Ona oprosila   pjat'   *rebënkov/detej.   Suppletive
        she question.PAST five.ACC   child.PL.GEN/children.GEN
        'She questioned five children.'

     b. Ona ždala   pjat'   *godov/let.
        she wait.PAST   five.ACC   year.PL.GEN/years.GEN
        'She waited for five years.'

Finally, *god* 'year' is unique in that its true plural form is not suppletive, as
shown in (68d). In contrast, *čelovek* 'person', *rebënok* 'child', and *cvetok*
'flower' have only a suppletive plural form, as shown in (68a–c). (The starred
form in (68a–c) is what would be the nonsuppletive plural of the corresponding
noun, formed in accordance with the rules of Russian morphology; these forms
do not actually exist in Russian.)

(68) a. Po ulice šli   ljudi/*čeloveki.   Suppletive
        on street went.PL people.NOM/person.PL.NOM
        'There were people walking down the street.'

     b. V sadu   rosli   cvety/*cvetki.   Suppletive
        in garden grew.PL flowers.NOM/flower.PL.NOM
        'There were flowers growing in the garden.'

     c. Po ulice šli   deti/*rebënki.   Suppletive
        on street went.PL children.NOM/child.PL.NOM
        'There were children walking down the street.'

     d. Prošli   dolgie *leta/gody.   Nonsuppletive
        passed.PL long   years.NOM/year.PL.NOM
        'Long years passed.'

The behavior of suppletive plural nouns fits precisely into the case-marking
pattern exhibited in the oblique cases by nouns denoting units of measure. Just
like measure nouns, a suppletive noun takes the suppletive plural form both
when it is a true plural (68a–c) and when it occurs with a cardinal in the oblique
cases (excepting genitive), as in (69a–c). The exception is again the noun *god*
'year', which occurs in the nonsuppletive plural both when it is a true plural
(68d) and with a cardinal in an oblique case (69d).

(69) a. o   trëx/pjati   ljudjax/*čelovekax   Suppletive
        about three.LOC/five.LOC people.LOC/person.LOC.PL
        'about three/five people'

     b. o   trëx/pjati   cvetax/*cvetkax   Suppletive
        about three.LOC/five.LOC flowers.LOC/flower.PL.LOC
        'about three/five flowers'

     c. o   trëx/pjati   detjax/*rebënkax   Suppletive
        about three.LOC/five.LOC children.LOC/child.LOC.PL
        'about three/five children'

     d. o   trëx/pjati   *letax/godax   Nonsuppletive
       about three.LOC/five.LOC years.LOC/year.PL.LOC
       'about three/five years'

Just like nouns denoting units of measure (Mel'čuk 1985, 430), in the genitive case *čelovek* 'person' (70a) and *cvetok* 'flower' (70b) retain their singular stem (albeit with the genitive-plural ending), while *god* 'year' (70c) and *rebënok* 'child' (70d) follow the plural pattern.

(70) a. bez   trëx/pjati   *ljudej/čelovek
         without three.GEN/five.GEN people.GEN/person.PL.GEN
         'without three/five people'
      b. bez   trëx/pjati   *cvetov/cvetkov
         without three.GEN/five.GEN flowers.GEN/flower.PL.GEN
         'without three/five flowers'
      c. bez   trëx/pjati   detej/*rebënkov
         without three.GEN/five.GEN children.GEN/child.PL.GEN
         'without three/five children'
      d. bez   trëx/pjati   let/*godov
         without three.GEN/five.GEN years.GEN/year.PL.GEN
         'without three/five years'

To summarize, suppletion patterns with *god* 'year' and *rebënok* 'child' support the division of cardinals into two major classes, nominal and adjectival (see chapter 3 as well as sections 6.2 and 6.3 above). On the other hand, the suppletion patterns of *čelovek* 'person' and *cvetok* 'flower', as well as the count forms of nouns denoting units of measure, support the hypothesis that the plural marking appearing in the context of a cardinal differs from the genuine semantic plural.

The existence of count forms can be taken as evidence for either the special number status of the lexical NP combining with a cardinal (see chapter 4) or the special nature of the case that they assign. Bulgarian provides some evidence that the former might be the case.

6.6.3 Count Forms in Bulgarian

While Bulgarian has no morphological case marking, it has special count forms (a.k.a. numerical forms or adnumerative forms) for masculine nouns with all cardinals except 'one'. (While Scatton (1984, 139) restricts numerical forms to [−human] masculine nouns, Leafgren (2011, 47) notes that they are optionally used with [+human] masculine nouns as well; see also Mel'čuk 1985, 437. For [+human] examples, see (73b, d).) As shown in table 6.7, the count form of masculine nouns is distinct from both singular and plural forms.

However, when there is a modifying adjective after the numeral, a plural form
is preferentially used, as in (71).[24] The effect of adjectival modification parallels the one in Russian, illustrated in (31) above, where the genitive singular
is used in place of the paucal in the presence of modification. The facts of both
Russian and Bulgarian suggest that the special count form can only occur when
the lexical NP consists only of the noun; we leave the question of why this
should be so for future research.

In the case of feminine and neuter nouns, the form used with cardinals
higher than 'one' is the same as the plural, as shown in the last column of
table 6.7.

(71) a. dva jezika   Bulgarian
        two language.ADN
        'two languages'
     b. dva săsedni   slavjanski jezici
        two neighboring Slavic   language.PL
        'two neighboring Slavic languages'

Our hypothesis that the lexical NP appearing with a cardinal is semantically
singular even if morphologically plural (chapter 2) is compatible with these
data. Although diachronically derived from the dual, synchronically, as is easy
to see, the count form morphologically contains the nominative-singular form,
lacking certain alterations that would be expected if it were built from the bare
stem: the count-form ending *-a* does not bear stress even with postaccenting
stems (72a), and the zero-alternating stem-final vowels are anomalously
retained (72b).

(72) a. kólko *gráda* 'how many towns' (*grád* 'town': cf. *gradắt* 'the town',
        gradové 'towns')
     b. dva *ógʌnja* 'two fires' (*ógʌn* 'fire': cf. *ógnjat* 'the fire', *ogn'óve*
        'fires')

Table 6.7
Bulgarian nominal forms

	M			F
	'city'	'owner'	'language'	'book'
SG	grad	stopánin	jezik	kniga
PL	gradové	stopáni	jezici	knigi
Form used with cardinals, e.g., *dva* 'two', *pet* 'five'	gráda	stopánina/stopáni	jezika	knigi

The assumption that the count form is constructed on the basis of the singular form explains the appearance of the root vowel л in (72b) as well as the fact that the stress pattern in the count form is the same as in the singular in (72a). The explanation for both facts follows from the assumption that in such examples there is an abstract stem-final vowel (*yer*) that gives rise to vowel–zero alternations and triggers stress retraction (see Scatton 1975).

A further interesting fact about Bulgarian has to do with the suffix *-ma*, which is attached to cardinals 'two' through 'six' when the lexical NP is masculine and human (Scatton 1984). Hurford (2003) analyzes *-ma* as a bound numeral classifier, which is restricted to the natural semantic class of male humans. While Hurford discusses only the case of *-ma* occurring with a plural NP, as in (73a), according to Roumyana Pancheva (pers. comm.) *-ma* is also marginally possible with the count form of the NP (73b); furthermore, while *-ma* cannot be omitted with a plural form (73c) that is masculine and human, it can (marginally) be omitted with a count form (73d). These judgments were confirmed by a Google search. The behavior of *-ma* is similar, but not entirely identical, to the behavior of *broj* ('count'), which also functions as a classifier: as shown in (73e), the lexical NP following *broj* is obligatorily plural (note that *broj* itself occurs in the count form, like any masculine noun that directly follows a cardinal).

(73) a. dva-ma učiteli
         two-CL teacher.PL
         'two teachers'

    b. ?dva-ma učitelja
        two-CL teacher.ADN
        'two teachers'

    c. *dva učiteli
        two teacher.PL

    d. ??dva učitelja
        two teacher.ADN
        'two teachers'

    e. dva broja   bileti/*bileta
        two count.ADN ticket.PL/ADN
        'two tickets'

On our analysis, the morphologically plural form used with masculine nouns in Bulgarian is semantically plural, which is why it does not occur with numerals (which require a semantically singular sister NP), but does occur with *broj* and with *-ma*, which apply to a semantically plural NP. More puzzling for our analysis is why *-ma* can co-occur with the count form (73b) if the count form

is semantically singular. However, note that neither (73b) (count form with
-*ma*) nor (73d) (count form without -*ma*) is entirely well-formed.

In the case of feminine and neuter nouns, the count form is unavailable, and
cardinals higher than 'one' occur with a plural lexical NP, which we would
analyze as a result of agreement (see chapter 4). Plural feminine and neuter
nouns used with cardinals are not used with -*ma*, consistent with the view that
their plural form is not a true semantic plural (and hence does not require a
classifier).

Another fact about the count form is that with higher cardinals like 'hundred',
'thousand', and 'million', the count form becomes optional for masculine
nouns, as shown in (74). The plural is ungrammatical or strongly dispreferred
for lower cardinals, though for some speakers it becomes more acceptable the
higher the cardinal (74a–b) (Roumyana Pancheva, pers. comm.). For the cardinals 'hundred' and above, while the count form is still preferred, the plural
form is also possible; this holds both with simplex numerals (74c, e, g) and
complex numerals (74d, f) (Roumyana Pancheva, pers. comm.).

(74) a. dva bileta   / *bileti
        two ticket.ADN   ticket.PL
        'two tickets'
     b. pet bileta   / ?/??bileti
        five ticket.ADN   ticket.PL
        'five tickets'
     c. sto   bileta   / ?bileti
        hundred ticket.ADN   ticket.PL
        'a hundred tickets'
     d. tri-sta   bileta   / ?bileti
        three-hundred.ADN ticket.ADN   ticket.PL
        'three hundred tickets'
     e. hiljada   bileta   / ?bileti
        thousand ticket.ADN   ticket.PL
        'a thousand tickets'
     f. dve hiljadi   bileta   / ?bileti
        two thousand.PL ticket.ADN   ticket.PL
        'two thousand tickets'
     g. million bileta   / ?bileti
        million ticket.ADN   ticket.PL
        'a million tickets'

While the count form is impossible with nominal numerals like 'dozen'
(75a), the fact that it is impossible with measure nouns like 'kilogram' (75b)
explains this restriction if we assume that 'dozen' is also a measure noun rather

than a cardinal. Hypothesizing that the higher cardinals are ambiguous between true cardinals and measure nouns (i.e., their nominal counterparts) explains therefore why the plural form becomes possible.

(75) a. duzina bileti/*bileta
          dozen ticket.PL/ADN
          'a dozen tickets'
     b. kilogram domati/*domata
          kilogram tomato.PL/ADN
          'a kilogram of tomatoes'

As shown in (74g), the count form is optional with higher cardinals such as 'million'. The availability of the count form is affected by the presence of a determiner, as shown in (76). If the cardinal is preceded by a determiner, either singular agreement (76a) or plural agreement (76b) can be triggered (compare to the Russian facts in (15), in section 6.2.5). The plural form is preferred to the count form in the presence of a determiner (76a–b), and the count form is completely ungrammatical if the determiner is singular (76a). This suggests that when 'million' triggers singular agreement on the determiner, it behaves like a noun rather than a cardinal (and hence takes a plural lexical NP as its complement). As expected, lower cardinals are compatible only with a plural determiner (76c–d), with the count form being the preferred or the only option (according to Roumyana Pancheva, pers. comm., the presence of *tezi* 'these' improves acceptability of the plural form for cardinals higher than 'two').

(76) a. tozi million bileti/*bileta
          this million ticket.PL/ADN
          'this million tickets'
     b. tezi   million bileti/??bileta
          this.PL million ticket.PL/ADN
          'these million tickets'
     c. tezi   pet bileta/?bileti
          this.PL five ticket.ADN/PL
          'these five tickets'
     d. tezi   dva bileta/*bileti
          this.PL two ticket.ADN/PL
          'these two tickets'

The facts of Bulgarian, taken together with the facts of Russian discussed in section 6.2.5, provide evidence that the lexical NP combining with a cardinal cannot be analyzed as a straightforward plural or singular form. This supports our analysis, on which the lexical NP is semantically singular and its surface form reflects agreement with the cardinal and can therefore be distinct from the regular plural form (see chapter 4).

6.7   The Declension of Complex Cardinals in Russian

So far in this chapter, we have focused on the lexical categorization of Russian cardinals. We now move to a consideration of complex cardinals in Russian, in light of our proposal that complex cardinals crosslinguistically are composed in the syntax. In chapter 3, we showed that case-assignment patterns within Russian complex cardinals provide evidence for the cascading structure that we proposed, with the complex cardinal not forming a constituent to the exclusion of the lexical NP. Additional strong evidence for the syntactic construction of complex cardinals in Russian comes from the fact that each subconstituent of a complex cardinal declines independently, as shown in (77)–(78), which is as expected if complex cardinals involving multiplication are constructed on the same principles as the combination of a simplex cardinal and a lexical NP (see chapter 3) and if addition is based on conjunction (see chapter 5).

(77) a. dve-sti   šest'   slonov   Paucal multiplier
         two.F.NOM-hundred.PAUC six.NOM elephant.PL.GEN
         'two hundred and six elephants'
     b. dvumja-stami   šest'ju slonami
         two.INS-hundred.PL.INS six.INS elephant.PL.INS
         '(with/by means of) two hundred and six elephants'

(78) a. pjat-sot   šest'   slonov   Higher multiplier
         five.NOM-hundred.PL.GEN six.NOM elephant.PL.GEN
         'five hundred and six elephants'
     b. pjatju-stami   šest'ju slonami
         five.INS-hundred.PL.INS six.INS elephant.PL.INS
         '(with/by means of) five hundred and six elephants'

As is easy to see, the multiplicand (in this case, 'hundred') behaves exactly like a lexical NP: it appears in the case determined by the multiplier in nominative/accusative environments and in the corresponding oblique case in oblique environments. Addends (cardinals that are added together, like the simplex 'six' and the complex 'two/five hundred' above) predictably differ from multiplicands in this respect, always appearing in the case assigned to the NP as a whole.

However, alongside the examples in (77b) and (78b), which characterize modern literary Russian, in modern colloquial Russian in oblique cardinal-containing NPs all constituents of a complex cardinal but the rightmost one (be they addends or multipliers) may fail to appear in the relevant oblique case, surfacing instead in what looks like the genitive-case form. This is illustrated by the sentences in (79), all culled from the web: in (79a–b), the second cardinal inside 'fifty' appears in the instrumental case, but the first does not; in (79c), neither one does.

(79) a. Tut paxnet pjati-desjat'ju   ottenkami   serogo.
      here smells five.GEN-ten.INS shade.PL.INS gray.GEN
      'It smells of the fifty shades of gray here.'

     b. Oderžal   pobedu pjati-desjat'ju   dvumja procentami
       achieved.SG.M victory five.GEN-ten.INS two.INS percent.PL.INS
       golosov.
       voice.PL.GEN
       'I/you/he obtained victory with 52 percent of the vote.'

     c. V igre zadejstvovany dve kolody s   pjati-desjati   dvumja
       in game activated   two packs with five.GEN-ten.GEN two.INS
       kartami.
       card.PL.INS
       'Two packs of fifty-two cards are used in the game.'

It is also possible for nonrightmost cardinals to appear in the nominative (citation) form, as in (80) (see also Es'kova 2011, 278–279). For the cardinals *devjanosto* 'ninety' and *sto* 'hundred' (81), which only distinguish the direct and the oblique cases, the genitive-case form cannot be distinguished from the oblique-case form, but the nominative case can, and the nominative and the oblique forms are used interchangeably. In simplex cardinals, only the oblique option is available.

(80) a. v šest'-sot   tridcati   pjati   naselennyx
      in six.NOM-hundred.PL.GEN thirty.LOC five.LOC inhabited
      punktax
      point.PL.LOC
      'in six hundred thirty-five settlements'

     b. bez   tysjača   tri-sta   pjat'-desjat
       without thousand.NOM three.NOM-hundred.PAUC five.NOM-ten.PL.GEN
       šesti   èkzempljarov
       six.GEN copy.PL.GEN
       'without one thousand three hundred fifty-six copies'

(81) a. dvumja-stami   devjanosto   / devjanosta pjat'ju
    two.INS-hundred.INS ninety.NOM/ACC ninety.OBL five.INS
    studentami
    student.PL.INS
    '(with/by means of) two hundred ninety-five students'

   b. so   sto   / sta   dvadcat'ju pjat'ju
    with hundred.NOM/ACC hundred.OBL twenty.INS five.INS
    zolotymi
    gold.piece.PL.INS
    'with a hundred and twenty-five gold pieces'

On the surface, the failure of case agreement inside the complex cardinal 'fifty' in oblique cases (79) suggests that complex cardinals in colloquial Russian are morphological compounds, constructed in the lexicon rather than in the syntax. Below, we provide evidence against this possibility.

The combining form used in compound derivations involving cardinals may coincide with the genitive form, as (82a) and (83a) show, or it may take the linking vowel generally associated with nominal compounds, as in (83b–d), where the combining forms are distinct from the corresponding genitive forms in (82b–d). The linking vowel in nominal compounds is either -o- or -e- or -i-: -i- is used for the third declension, -e- for stems ending in a palatalized consonant in other declensions, as in (83d), and -o- otherwise. For those cardinals lower than 'thousand' that belong to the third declension, like 'five' in (83e), the combining form is again, like with 'four' (83a), surface-identical to the genitive-case form, this time because the linking vowel in the third declension is surface-identical with the genitive suffix.[25]

(82) a. četyrëx nog   Genitive forms
    four.GEN leg.PL.GEN
    '(of) four legs'

   b. milliona   nog
    million.GEN leg.PL.GEN
    '(of) a million legs'

   c. sta   nog
    hundred.OBL leg.PL.GEN
    '(of) one hundred legs'

   d. tysjači   nog
    thousand.GEN leg.PL.GEN
    '(of) a thousand legs'

   e. pjati   nog
    five.GEN leg.PL.GEN
    '(of) five legs'

(83) a. četyrëx-nog-ij   Compound forms
        four.GEN-leg-ADJ
        'four-legged'
   b. million-o-nog-ij
        million-LNK-leg-ADJ
        'million-legged'
   c. sto-nog-ij
        hundred.NOM/LNK-leg-ADJ
        'hundred-legged'
   d. tysjač-e-nog-ij
        thousand-LNK-leg-ADJ
        'thousand-legged'
   e. pjati-nog-ij
        five.GEN-leg-ADJ
        'five-legged'

Traditional Russian grammars specifically distinguish between the suffix
-i- appearing in cases like (79), which they unhesitatingly treat as a case ending
(Vinogradov 1952, 366; Rozental, Džandžakova, and Kabanova 1998, 232–
233; note that the genitive form of the noun is also used in possessive compounds, as in (84)), and the linking vowel that is used in nominal and adjectival
compounds. The linking vowel is determined by the declension and by the
palatalization of the final segment of the stem, as noted above.

(84) [Mar'j-i   Ivanovn]-in   dom
        Maria-GEN Ivanovna-POSS house.NOM
        'the house of Maria Ivanovna'
        (Rappaport, to appear)

Nevertheless, the varying forms appearing in complex cardinals that fail to
undergo oblique-case spreading, as in (79) and (80), show that this phenomenon cannot be treated as evidence for constructing complex cardinals in the
lexicon. Specifically, while the form of nonagreeing lower cardinals can be
identical to the genitive-case form (79) and, in some cases, to the "linked"
combining form used in compounds (83), the bare (nominative) form is also
available (80), which is excluded from compounds. Indeed, the cardinal 'thousand', when failing to agree, can *only* appear in the nominative form (80b),
whereas in compounds only the combining form (85) is possible.[26]

(85) dvux-*tysjač-e*-pjati-sot-let-ie
        two.GEN-thousand-LNK-five.GEN-hundred.PL.GEN-year-NMN
        'a two thousand five hundredth anniversary'
        (Rozental, Džandžakova, and Kabanova 1998, 233)

Further evidence against treating the case-agreement failure in complex cardinals as a sign of their composition in the lexicon comes from examples like (86)–(87), where only some of the composing cardinals fail to agree in case. If complex cardinals were composed in the lexicon, we would expect the entire complex cardinal to behave as a single noun, with only the rightmost element reflecting case assignment.

(86) s   *vosem'-sot*   dvadcat'ju četyr'mja učastnikami
      with eight.NOM-hundred.PL.GEN twenty.INS four.INS   participant.PL.INS
      'with eight hundred and twenty-four participants'
      (Es'kova 2011, 279)

(87) s   pjat'ju tysjačami   *pjat'-sot*
      with five.INS thousand.PL.INS five.NOM-hundred.PL.GEN
      sem'-desjat   četyr'mja bojcami
      seven.NOM-ten.PL.GEN four.INS   fighter.PL.INS
      'with five thousand five hundred and seventy-four fighters'
      (Vinogradov 1952, 371)

Whatever mechanism is responsible for the failures of the oblique-case marking in complex cardinals in colloquial Russian, it cannot be taken as evidence for treating such a complex cardinal as a constituent to the exclusion of the lexical NP.

6.8 Cardinals in Polish

So far, we have focused on the properties of Russian (and, to a lesser extent, Bulgarian) cardinals. In this section, we turn to Polish cardinals, which—while in many ways similar to Russian cardinals—also exhibit a number of unique properties that have been addressed in the literature. The goal of this section is to examine the behavior of Polish cardinals in light of our proposal, with special reference to the so-called *Accusative Hypothesis* (see section 6.8.1).

In Polish, as in Russian, the lower cardinals 'one' through 'four' exhibit adjectival properties while the cardinals 'five' and higher exhibit (some) properties of nouns, as discussed by Miechowicz-Mathiasen (2011) and Klockmann (2012, 2013, 2017).[27] The adjectival properties exhibited by the lower cardinals include case agreement with the lexical noun and agreement in animacy and gender (at least for some case configurations), as shown in (88a). In contrast, 'five' and higher, when appearing in nominative/accusative case, require the lexical NP to be in the genitive plural, exactly as in Russian, and trigger neuter agreement on the predicate (88b). 'Thousand' and 'million', though, allow both neuter agreement and gender agreement, as shown in (88c).[28]

(88) a. Dwie   dziewczyny / dwa   koty   Polish
two.F.NOM girl.PL.NOM   two.NV.NOM cat.M.PL.NOM
przyszły.
came.NV.3PL
'Two girls/cats came.'

    b. Pięć   dziewczyn/kotów   przyszło.
five.ACC/NOM.NV girl.PL.GEN/cat.M.PL.GEN came.N.3SG
'Five girls/cats came.'

    c. Tysiąc   listów   przyszło/przyszedł do
thousand.M.NOM letter.PL.NV.GEN came.N.3SG/M.3SG to
Piotra.
Peter.GEN
'A thousand letters came to Peter.'

According to Rutkowski (2007, 216–220; also Wągiel 2015), all Polish cardinals have fully nominal counterparts: for example, for 'five', *pieć* is the numeral form and *piątkę* is the nominal form. In direct cases, both forms of 'five' require the lexical NP to be in the genitive plural (89a–b); in oblique cases, the numeral form shows homogeneous case assignment (89c) while the fully nominal form shows heterogeneous case assignment (89d), reminiscent of what happens with 'five' versus 'dozen' in Russian.

(89) a. Widzę pięć   lingwistek.
    I.see   five.ACC linguist.F.PL.GEN
    'I can see five female linguists.'

    b. Widzę piątkę   lingwistek
    I.see   five.ACC linguist.F.PL.GEN
    'I can see five female linguists.'

    c. Pracuję z   pięcioma lingwistkami/*lingwistek.
    I.work with five.INS   linguist.F.PL.INS/GEN
    'I work with five female linguists.'

    d. Pracuję z   piątką lingwistek/*lingwistkami.
    I.work with five.INS linguist.F.PL.GEN/INS
    'I work with five female linguists.'

In the case of the paucal cardinals 'two' through 'four', while the numeral form shows gender agreement with the lexical NP (90a) (more on this in the next section), the fully nominal form behaves just like the nominal forms of higher cardinals, requiring genitive plural on the lexical NP (90b).

(90) a. Dwaj   lingwiści   napisali   książkę.
    two.V.NOM linguists.V.NOM wrote.V.PL book.ACC
    'Two linguists wrote a book.'

b. Dwójka   lingwistów   napisała książkę.
two.F.NOM linguists.PL.GEN wrote.F.SG book.ACC
'A group of two linguists wrote a book.'
(Wągiel 2015)

Thus far, with respect to NP-internal syntax, Polish cardinals behave just like Russian ones, hence the same analysis applies. However, Polish differs from Russian with respect to verbal agreement: as shown in (88) and (90), the number agreement on the verb is obligatorily singular or plural depending on the type of cardinal, whereas in Russian both types of agreement are nearly always available (see section 6.5). Polish also differs from Russian with regard to case marking on cardinal-containing NPs in subject position, as discussed next.

6.8.1 The Accusative Hypothesis

Crucial for the understanding of Polish cardinals is the subgender distinction between virile and nonvirile, where virile is defined as masculine and [+human]. The virile–nonvirile distinction is only operative in the plural and is manifested in the agreement of verbs, adjectives, demonstratives/quantifiers, and cardinals and in the realization of accusative case. For nonvirile-gender adjectives and demonstratives, the accusative-plural form is the same as the nominative-plural form, but for virile-gender adjectives and demonstratives, the accusative-plural form is the same as the genitive-plural form (see Brooks 1975, 265).

The virile–nonvirile distinction is also reflected in the behavior of cardinal-containing NPs. For the cardinals 'five' and up, the form appearing with a virile NP is different from that appearing with all other nouns, as illustrated in (91a) for 'five' (compare to (88b)). With a virile NP, the same form (e.g., for 'five', *pięciu*) is used for all cases other than instrumental, indicating that this is the default oblique form; see table 6.8. 'Two' through 'four' are claimed to have two different virile forms in subject position: a nominative form, where the cardinal agrees with the lexical NP in case and co-occurs with plural agreement on the verb (91b), and a syncretic genitive/accusative form, where it behaves exactly like 'five' and up (91c).[29] In the feminine and neuter only one form is available; the cardinal agrees in gender with the lexical NP and the predicate is obligatorily plural (92).

(91) a. Pięciu   chłopców   przyszło.
       five.V.OBL boy.PL.GEN came.N.3SG
       'Five boys came.'

Table 6.8
Case paradigm of *pięć* 'five' in Polish, assuming the Accusative Hypothesis (Miechowicz-Mathiasen 2011, table 9)

Case	'these five men' (v)	'these five women' (NV)	'these five houses' (NV)
nom	~~*ci pięciu mężczyzn-gen~~	*te pięć kobiet-gen*	*te pięć domów-gen*
ACC	*tych pięciu mężczyzn-GEN*	*te pięć kobiet-GEN*	*te pięć domów-GEN*
GEN	*tych pięciu mężczyzn*	tych pięciu kobiet	tych pięciu domów
DAT	tym pięciu mężczyznom	tym pięciu kobietom	tym pięciu domom
INS	tymi pięcioma mężczyznami	tymi pięcioma kobietami	tymi pięcioma domami
LOC	tych pięciu mężczyznach	tych pięciu kobietach	tych pięciu domach

  b. Dwaj chłopcy przyszli.
   two.V.NOM boy.PL.NOM came.V.3PL
   'Two boys came.'
  c. Dwóch chłopców przyszło.
   two.V.GEN/ACC boy.PL.GEN came.N.3SG
   'Two boys came.'

(92) a. Dwie dziewczyny pływały.
   two.F.NOM girl.PL.NOM swam.NV.PL
   'Two girls swam.'
  b. Dwa krzesła/stoły rozbiły się.
   two.NV.NOM chair.N.PL.NOM/table.M.PL.NOM break.PAST.NV.PL REFL
   'Two chairs/tables broke.'
  (Klockmann and Šarić 2015)

When a cardinal-containing NP whose cardinal is 'five' or higher combines with a determiner, we observe different patterns with virile and nonvirile NPs. For a nonvirile NP, the determiner can be either genitive (*tych*) or the syncretic nominative/accusative form (*te*), as shown in (93a). For virile NPs, the determiner must be in the syncretic genitive/accusative form (*tych*), with the nominative form (*ci*) disallowed, as shown in (93b). These facts are summarized in table 6.8.

(93) a. Tych / te pięć
   this.PL.NV.GEN this.PL.NV.ACC/NOM five.NV.ACC/NOM
   kobiet/okien/kotów stało.
   woman/window/cat.PL.GEN stood.N.3SG
   'These five women/windows/cats stood.'
  b. Tych / *ci pięciu mężczyzn stało.
   this.PL.V.GEN this.PL.V.ACC/NOM five.V.OBL man.PL.GEN stood.N.3SG
   'These five men stood.'

These empirical facts give rise to the so-called Accusative Hypothesis, according to which the Polish cardinals 'five' and up do not have a nominative form and appear in subject position in the accusative (Schenker 1971; Franks 1995, 2002; Przepiorkowski 1997; Rutkowski and Szczegot 2001; and Rutkowski 2002, 2007, among others). Noting that this hypothesis is simply a restatement of the facts, Miechowicz-Mathiasen (2011) proposes instead a null light preposition that assigns accusative case. Her proposal is, however, no less problematic: the semantics of the putative preposition is not stated and it is not clear what prevents cardinal-containing NPs from appearing without it in subject position, as they must presumably do in all other syntactic environments.

We therefore deny that cardinal-containing NPs in subject position are accusative. Instead, we propose that the observed pattern is due to the more general algorithm of default case realization, which would also account for the more general (language-specific) syncretism in the realm of the accusative case, which, in Polish as in Russian, involves syncretism with genitive and nominative. Taking into consideration the fact that in the singular the syncretism with the genitive is governed by the feature [αanimate] rather than [αhuman], we propose that the set of vocabulary-insertion rules for case morphology for nouns in Polish is terminated by statements with the effect of those in (94), with no other lexical entries provided for the realization of the accusative case in the plural.

(94) [+masculine, +human, +plural] $\rightarrow$ genitive plural
    [+masculine, +animate, +singular, +accusative] $\rightarrow$ genitive
    otherwise $\rightarrow$ nominative

In other words, when more specific vocabulary-insertion rules do not apply (as would be the case for accusative, as well as the absence of case specification), the default realization is determined by the presence of the feature complexes [+masculine, +human] (i.e., virile) and [+masculine, +animate], respectively, for plural NPs and singular NPs. We contend that what happens with cardinal-containing NPs is that in subject position they are not assigned any case, either nominative or accusative, and that is why these elsewhere conditions apply. To answer the question of why cardinal-containing NPs are not assigned case, we invoke our previous assumption that Polish cardinals higher than 'four' are defective. We propose, similarly to Klockmann (2012, 2013), that it is this deficiency that prevents NPs headed by these cardinals from entering into an agreement relation with Tense and hence from being assigned case.

Independent evidence for the deficiency of higher cardinals comes from the fact that they do not trigger agreement on the verb (see (88b), (91a)). To explain the facts, Klockmann (2012, 2013) proposes that cardinals are ϕ-defective in that they lack the gender feature, which precludes agreement with T. The problem with this proposal is that cardinal-containing NPs headed by higher cardinals must be formally specified as either virile or not, as shown by the fact that in subject position such cardinal-containing NPs bear the surface genitive case when virile and the surface nominative case when non-virile. For NP-internal concord to happen, the cluster of features corresponding to the virile should spread to the NP as a whole, including the cardinal, whose surface form in the subject position is also affected by whether the noun is virile (93). We must therefore conclude that it cannot be the case that the higher cardinals lack gender features altogether.

Further evidence against the connection between case and the presence of the gender features comes from the lower cardinals 'two' through 'four'. As discussed above, the lower cardinals agree in gender with the lexical NP and trigger agreement on the verb. However, we contend that they exhibit exactly the same case-assignment pattern as do the higher cardinals: the default case realized as genitive for virile NPs (91c) and as nominative for nonvirile NPs (92). As for the form in (91b), there are reasons to believe that it should not be treated as a cardinal. On the one hand, as noted by Swan (2002, 190), this form can only be used for all-male groups, which makes it different from all other instances of the virile, which are compatible with a female-male mixture. On the other hand, as we will see below, this form cannot appear in complex cardinals.[30]

Once the confusing factor introduced by *dwaj* 'two' is removed, we see that the lower cardinals 'two' through 'four' form an intermediate group between the adjectival 'one' and the more nominal higher cardinals lower than 'thousand', as summarized in table 6.9. The connection between the surface (i.e., morphological) case on the cardinal and the ability of the corresponding cardinal-containing NP to trigger agreement on the predicate is unmistakable. The feature (or rather, feature complex) responsible for the virile gender is not

Table 6.9
The Polish cardinal "squish"

	'one'	'two'–'four'	'five' →
ϕ-agreement	Yes	Yes	No
Predicate agreement	Yes	If nonvirile	No
Case assignment to lexical NP	No	No	Yes (visible in direct cases, nonvirile)

involved: it is irrelevant for cardinals higher than 'five'. The cardinal itself is not the deciding factor: the behavior of the lower cardinals 'two' through 'four' depends on the featural specification of the lexical NP they combine with. It is impossible to say whether the form used with virile cardinals (91c) itself shows agreement (the cardinals in question do not distinguish for gender in the genitive, locative, and dative cases), yet the lack of predicate agreement is compatible with the more general inability to participate in agreement.

The correlation between case marking and agreement cannot be reasonably attributed to different syntactic configurations: it seems highly unlikely that the virile cardinals 'two' through 'four' and their nonvirile counterparts should have such a radically different syntax, especially since the difference in surface case marking between virile and nonvirile cardinal-containing NPs is replicated with cardinals higher than 'five', which do not trigger agreement. It seems therefore that the proper generalization has to do indeed with the surface case: when the subject surfaces in the nominative case, it triggers an agreement on the predicate; otherwise it does not. Leaving for future research a formal account of this phenomenon, we turn now to additive complex cardinals, which cast further light on the syntax of cardinals in Polish.

6.8.2 Complex Cardinals Involving Addition

When the virile cardinal forms of 'two', 'three', and 'four' are part of a complex cardinal involving addition, they must appear in the syncretic genitive/accusative form, as shown in (95a–b); this restriction does not hold for their nonvirile counterparts (95c). Furthermore, the cardinal 'one', which on its own behaves like an adjective, exhibiting gender agreement (as well as number agreement) with the lexical NP (96a–c), occurs in an uninflected form in coordinated cardinals (96d).

(95) a. *Dwadzieścia/dwudziestu dwaj/trzej/czterej　　chłopcy
　　　　twenty.NV/V　　　　　　　two/three/four.V.NOM boy.PL.NOM
　　　　przyszli.
　　　　came.V.3PL
　　b. Dwudziestu dwóch/trzech/czterech　　chłopców　　przyszło.
　　　　twenty.V　　two/three/four.V.ACC/GEN boy.PL.GEN came.N.3SG
　　　　'Twenty-two/three/four boys came.'
　　c. Dwadzieścia trzy　　　　　koty　　　bawiły　　się.
　　　　twenty.NV　　three.NV.NOM cat.PL.NOM play.NV.PL REFL
　　　　'Twenty-three cats were playing.'
　　　　(Swan 2002, 199)

(96) a. Jeden   chłopiec przyszedł.
     one.M.NOM boy.NOM came.M.SG
     'One boy came.'

   b. Jedna   dziewczyna przyszła.
     one.F.NOM girl.NOM   came.F.SG
     'One girl came.'

   c Jedno   dziecko   przyszło.
     one.N.NOM child.NOM came.N.SG
     'One child came.'

   d. Dwudziestu jeden chłopców/dziewczyn/dzieci przyszło.
     twenty.NV one   boy/girl/child.PL.GEN   came.N.3SG
     'Twenty-one boys/girls/children came.'

This is different from Russian, where 'one' retains its ability to agree for gender and number in coordination:

(97) a. odin   kot   / odna   koška   Russian
       one.M.SG.NOM cat.M.NOM one.F.SG.NOM cat.F.NOM
       'one male cat/one female cat'

   b. dvadcat'   odin   kot
     twenty.NOM one.M.SG.NOM cat.M.NOM
     'twenty-one male cats'

   c. dvadcat'   odna   koška
     twenty.NOM one.F.SG.NOM cat.F.NOM
     'twenty-one female cats'

While the Polish loss of declinability in a complex cardinal (96d) is in itself a puzzle, it also seems incomprehensible why 'one' should lose the ability to decline while the cardinals 'two', 'three', and 'four', which would seem to be adjectival to almost the same extent, do not. To understand this, we appeal once again to the different featural specification of different cardinals in Polish. Crucially, as discussed above, the accusative syncretism for masculine nouns in Polish is regulated by different features in the singular and plural: in the singular it is the feature [+animate] that yields the surface genitive, while in the plural the relevant feature is [+human]. While the former entails the latter semantically, it seems not unreasonable to assume that cardinals higher than 'one' are not formally specified for the feature [αanimate], which is irrelevant for plurals. This full specification for φ-features renders the cardinal *jeden* 'one' fully adjectival and the NP containing it fully nominal.

Recall from chapter 5 that, on our view, additive complex cardinals involve NP-level coordination. The coordination of a cardinal-containing NP with a regular NP does not give rise to any problems, as shown by examples like (98). However, in (98) it is two type e expressions that are coordinated, and the masculine human feature specification of the resulting expression can be determined from their reference.

(98) Róża i   pięć chłopców   Polish
     Rosa and five boy.PL.GEN
     'Rosa and five boys'

In an additive complex cardinal, on the other hand, as discussed in chapter 5, the semantic type of the coordinated NPs is $\langle e, t \rangle$, which means that the featural specification of the result can only be established on formal grounds. We propose that what distinguishes Russian and Polish cardinal-containing NPs is the fact that in Polish different features are involved in the formal specification of the singular and plural. The problem does not arise for the plural adjectival cardinals 'two', 'three', and 'four' because they are not specified for the feature [αanimate].

6.8.3 'Thousand' and 'Million'

As we saw above, with both virile and nonvirile NPs that follow a cardinal 'five' or higher, the verb is obligatorily singular, in both definite and indefinite environments. However, as noted by Miechowicz-Mathiasen (2011), the semilexical cardinal 'thousand' (and potentially other semilexical cardinals, like 'million', 'billion', etc.) behaves subtly differently. In indefinite contexts (99a), both gender agreement ('thousand' is masculine) and the default neuter-singular agreement are possible (however, we note that the agreeing masculine form is dispreferred by some speakers). In the presence of an agreeing definite determiner (99b), gender agreement is obligatory.

(99) a. Tysiąc   listów   przyszło/%przyszedł do Piotra.
        thousand.M.NOM letter.NV.PL.GEN came.N.3SG/M.3SG   to Peter.GEN
        'A thousand letters came to Peter.'
     b. Ten   tysiąc   listów   *przyszło/przyszedł
        this.M.NOM thousand.M.NOM letter.NV. PL.GEN came.N.3SG/M.3SG
        do Piotra.
        to Peter.GEN
        'These one thousand letters came to Peter.'

This description of the facts is, however, incomplete. As with other cardinals, the demonstrative can also take the plural form, in which case it can surface as either genitive (100a) or nominative (100b); in both of these cases, agreement on the verb is obligatorily neuter singular.

(100) a. Tych   tysiąc   listów   przyszło/*przyszedł
        this.PL.GEN thousand.M.NOM letter.NV.PL.GEN came.N.3SG/M.3SG
        do Piotra.
        to Peter.GEN
        'These one thousand letters came to Peter.'

     b. Te   tysiąc   listów   przyszło/*przyszedł
        this.PL.NOM thousand.M.NOM letter.NV.PL.GEN came.N.3SG/M.3SG
        do Piotra.
        to Peter.GEN
        'These one thousand letters came to Peter.'

We propose that the semilexical cardinal 'thousand' is optionally φ-defective. It can behave like a regular (measure) noun, in which case it triggers masculine agreement on the determiner as well as on the verb (hence obligatory gender agreement on the verb when the determiner agrees for gender, as in (99b)). It can also behave like the other cardinals, in which case it lacks gender features and occurs with a neuter singular verb and a plural determiner (for the specifics of adjective and determiner agreement, see the next section).

Supporting evidence for this proposal comes from similar behavior of measure nouns like 'kilogram'. As shown in (101), indefinite measure NPs allow both gender agreement and the default neuter-singular agreement on the verb (101a), while the presence of a gender-agreeing determiner forces gender agreement on the verb as well (101b), exactly as for 'thousand' (compare (101a–b) to (99)). This indicates that measure nouns, like 'thousand', can optionally be φ-defective. Unlike 'thousand', 'kilogram' cannot combine with a plural determiner (101c), which is expected given that there is no plural number feature on 'kilogram'.[31] In other words, we get further support for treating some cardinals as defective (measure) nouns.

(101) a. Kilogram   listów   przyszło/przyszedł do Piotra.
        kilo.M.NOM letter.NV.PL.GEN came.N.3SG/M.3SG to Peter.GEN
        'A kilogram of letters came to Peter.'

     b. Ten   kilogram   listów   *przyszło/przyszedł
        this.M.NOM kilo.M.NOM letter.NV.PL.GEN came.N.3SG/M.3SG
        do Piotra.
        to Peter.GEN
        'This kilogram of letters came to Peter.'

     c. *Tych/te   kilogram   listów
        this.PL.GEN/PL.NOM kilo.M.NOM letter.NV.PL.GEN
        przyszło/przyszedł do Piotra.
        came.N.3SG/M.3SG to Peter.GEN

Inside complex cardinals involving addition, 'thousand' behaves just like a lexical NP in terms of its case and number. Thus, the behavior of complex cardinals involving addition in Polish is consistent with the cascading-NP structure, which leads Miechowicz-Mathiasen (2012) to adopt our approach for them.

6.8.4 Adjective Agreement with Polish Cardinals

For nonvirile nouns in Polish, both the attributive adjective that combines with a cardinal-containing NP and the predicative adjective that follows a copular verb can bear either genitive or accusative case, as shown in (102) below. Genitive and accusative case are also possible in a depictive construction (103a), but not with a nominal predicate, which allows only the instrumental (103b). While following Przepiórkowski and Patejuk (2012) in glossing the corresponding forms as genitive and accusative, we note that the accusative form for plural nonvirile adjectives is systematically syncretic with the unmarked (nominative) form (Brooks 1975, 265).

(102) a. Nastepne kilkadziesiat   metrów   było   czyste.
          next.ACC several.ten.PL.ACC meter.PL.GEN was.N.3SG clean.ACC
          'The next few tens of meters were clean.'

      b. Kolejnych jedenascie zarzutów   było   podobnych.
          further.GEN eleven.ACC charge.PL.GEN was.N.3SG similar.GEN
          'A further eleven charges were similar.'

      c. Kolejne   piecdziesiat aut   zostało   uszkodzonych.
          further.ACC fifty.ACC   car.PL.GEN   became.N.3SG damaged.GEN
          'A further fifty cars became damaged.'

      d. Minionych dwanascie miesiecy   było   ajgorsze w
          past.GEN   twelve.ACC month.PL.GEN was.N.3SG worst.ACC in
          historii.
          history
          'The past twelve months were the worst in history.'
      (Przepiórkowski and Patejuk 2012, (7), (9)–(11))

(103) a. Pięć osób   przyszło   pijanych/pijane.
          five person.PL.GEN arrived.N.3SG drunk.GEN/ACC
          'Five people arrived drunk.'

      b. Pięć osób   było/zostało
          Five person.PL.GEN was.N.3SG/became.N.3SG
          nauczycielami/*nauczycieli.
          teacher.PL.INS/PL.ACC/PL.GEN
          'Five people were/became teachers.'
      (Barbara Citko, pers. comm.)

In the spirit of the proposal advanced by Przepiórkowski and Patejuk (2012), we suggest that both attributive and predicative adjectives (as well as determiners: see (93a)) can agree either with the lexical NP (which is marked genitive) or with the entire cardinal-containing NP (which is not case-marked in the subject position in our analysis, and surfaces as nominative/accusative). In the Minimalist perspective, this can be formalized as follows: adjectives are known to agree for number and gender, and therefore they should bear two uninterpretable features. Assuming that they do not probe simultaneously, we hypothesize that the optionality in adjectival case marking arises from the order of probing. If the first feature to probe is uninterpretable number, then it finds its interpretable counterpart on the cardinal. If the first feature to probe is gender, then, cardinals being unspecified for gender, it finds its interpretable counterpart on the lexical NP. Either way, the case feature is valued as a free rider on the first agreement relation. Nominal predicates, on the other hand, cannot agree with the subject (103b), independently of whether it contains a cardinal: instrumental case on the nominal predicate is obligatory (Bailyn 2001).

Turning now to cardinal-containing subject NPs that contain a virile lexical NP, only genitive on the adjectives is possible (104), consistent with what we see on determiners in (93b). This outcome is predicted both by the Accusative Hypothesis (since, for virile nouns, accusative is syncretic with genitive) and by our counterproposal in section 6.8.1.

(104) Następnych / *następne kilkadziesiat mężczyzn było
next.GEN next.ACC/NOM several ten.PL.ACC man.PL.GEN was.N.3SG
czystych / *czyste.
clean.GEN clean.ACC/NOM
'The next few tens of men were clean.'
(Barbara Citko, pers. comm.)

A potential problem for this proposal comes from case marking on adjectival predicates in control structures, where both genitive and accusative marking on the adjective inside the control structure are also allowed, with both primary predication (105a) and in depictives (105b). On the assumption that the subject in control structures is PRO, which presumably does not have any internal structure, the availability of genitive marking on the adjective is unexpected (cf. Witkoś 2008).

(105) a. Pięć osób probowało być szczesliwych/szczesliwe.
Five person.PL.GEN tried.N.3SG be.INF happy.GEN/NOM
'Five people tried to be happy.'

b. Pięć osób   probowało przyjść   pijanych /
   five person.PL.GEN tried.N.3SG arrive.INF drunk.GEN
   pijane.
   drunk.ACC/NOM
   'Five people tried to arrive drunk.'
   (Barbara Citko, pers. comm.)

Two solutions can be envisaged to this problem. The first one relies on the movement theory of control (Hornstein 1999, 2001; Nunes 1995). Under this view, the subject of control structures is the same cardinal-containing NP that surfaces in the subject position of the main clause, and therefore the adjective can agree with the lexical NP, as discussed above. The alternative relies on Landau 2015b, which proposes that agreement with the controller proceeds via prior agreement of the infinitival C^0. Assuming that the infinitival C^0 agrees in the same way as all other agreeing elements (i.e., can probe either the entire cardinal-containing NP or just the lexical NP), further agreement inside the control structures will depend on the choice made in the matrix clause.

6.8.5 Summary

To sum up, Polish cardinals behave like Russian cardinals with respect to the adjectival–nominal divide between lower and higher cardinals, as well as the more noun-like behavior of so-called lexical cardinals. Polish differs from Russian in the behavior of cardinals with virile nouns and in the obligatory neuter-singular marking on the predicate for cardinals 'five' through 'hundred'; we have analyzed both phenomena in terms of ϕ-deficiency of higher cardinals.

6.9 Conclusion

In this chapter, we have discussed the behavior of cardinals in several Slavic languages, with a particular focus on Russian and Polish and on the lexical category of cardinals. The declensional patterns of Slavic cardinals, as well as their behavior with respect to case assignment and gender agreement, shows that cardinals are not a separate class of functional elements. Lower cardinals ('one' through 'four') have more in common with adjectives, and the highest cardinals ('thousand', 'million', and to a lesser extent 'five' through 'hundred') have more in common with nouns (cf. Hurford 1975). At the same time, 'two' through 'four' show a mixed pattern of nominal-adjectival behavior, while 'five' through 'hundred' lack many of the characteristics of "regular" nouns.

We have analyzed cardinals in terms of feature (under)specification, and have shown how this analysis can capture a number of apparently peculiar characteristics, such as homogeneous case assignment in Russian and the behavior captured under the Accusative Hypothesis in Polish. Additionally, we have shown that the behavior of complex cardinals in both Russian and Polish supports our analysis that complex cardinals are composed in the syntax. In addition to these main points, this chapter also discussed a number of Slavic-specific properties, such as the status of paucal marking with cardinals, the behavior of virile nouns in Polish, and the special count form in Bulgarian.

7   Cardinals in Other Languages

In this chapter, we take an in-depth look at cardinals from several different languages: three Semitic languages, Modern Hebrew, Standard Arabic, and Lebanese Arabic; Welsh, both Biblical and Modern; and Dagaare, a language of the Gur family. These particular languages are chosen because they have interesting properties with respect to cardinal formation. While the behavior of cardinals in Dagaare and Welsh provides strong evidence for our proposal, the behavior of cardinals in Semitic languages provides possible challenges, which we address below.

It is beyond the scope of this work to address in detail every aspect of cardinals in each language under consideration. In choosing which issues to focus on for each language, we are driven both by what is known about the cardinal system in the language and by what is interesting or unusual about this system when compared with other languages discussed so far in this work. Thus, for Dagaare (section 7.1), where less information is available than in the case of the other languages, we focus on the overall properties of both simplex and complex cardinals. We show that the cardinal system of Dagaare behaves exactly as we predict cardinal systems to behave, carrying the cascading structure on its sleeve, so to speak. In the case of Biblical Welsh (section 7.2), we focus on its two fairly unique features, as described by Hurford (1975): the availability of NP deletion almost anywhere inside a coordinated NP and the existence of a variety of arithmetic operations beyond addition and multiplication. We show that both phenomena are easily explained on our analysis. Turning to Modern Welsh (section 7.3), our focus shifts to the existence of two distinct syntactic patterns for cardinals, which we analyze as corresponding to our cascading structure and to pseudopartitives, respectively.

In the case of Modern Hebrew (section 7.4), our main concern is with complex-cardinal formation. We show that the distribution of bound versus free forms of cardinals inside Hebrew complex cardinals is a potential challenge to our proposal, but not incompatible with it. Finally, in the case of

Arabic (section 7.5), our attention turns to patterns of case assignment and agreement with cardinal-containing NPs.

7.1 Dagaare

The cardinal system of Dagaare, a Gur language spoken in Ghana and Burkina Faso, is discussed by Hiraiwa (2013).

In this language, the cardinal *yeni* 'one' combines directly with a singular NP (1). It further appears from the description by Bodomo (2000) that 'one', like other adjectives in Dagaare, combines with the noun by forming a compound with the nominal stem: compare (2a–b) to (2c).[1]

(1) a. nen   ∅-yeni   Dagaare
       person one
       'one person'
    b. baa ∅-yeni
       dog one
       'one dog'

(2) a. nén-yéni
       HUMAN.COUNT-one
    b. bón-yéni
       NHUMAN.COUNT-one
    c. pòg-vèlàà   là
       woman-beautiful FOC
       'That is a beautiful woman.'
    (Bodomo 2000)

All cardinals higher than 'one' require the plural form of the lexical NP (3).[2] Simplex cardinals higher than 'one' can be divided into two sets, multiplicands and nonmultiplicands. Multiplicands ('ten', 'twenty', 'hundred', etc.) combine with the lexical NP directly (3c–d), while all other simplex cardinals first combine with one of two markers: *ba* if the lexical NP is [+human] and *a* if it is [−human], as shown in (3a–b). While Hiraiwa calls these markers noun-class markers, we term them instead noun-concord markers.

(3) a. noba   ba-yi
       people NC.HUMAN-two
       'two people'
    b. baare   a-yi
       dog.PL NC.NHUMAN-two
       'two dogs'

c. noba   pie
   people ten
   'ten people'
d. baare   pie
   dog.PL ten
   'ten dogs'

Following the description given by Bodomo (2000), we propose that these noun-concord markers are agreement markers for the feature [±human], and that their presence is indicative of the adjectival status of lower cardinals, as opposed to the nominal status of 'ten', 'twenty', and other multiplicands. This brings Dagaare in line with other languages, which, as discussed earlier (see chapter 3), also exhibit an adjectival–nominal cline between lower and higher cardinals (Hurford 1975). While Hiraiwa treats *a* and *ba* as noun-class markers, the noun-class system of Dagaare is in fact considerably more complex, with ten noun classes, of which the [+human] class is only one: see Bodomo and Marfo 2006. The fact that cardinals come with only two types of markers rather than ten suggests that we are dealing with agreement on the cardinal for the [±human] feature, not with a noun-class marker. If Dagaare noun-class markers in fact only surface on nouns, as Bodomo and Marfo seem to suggest, then noun classes in Dagaare are inflectional classes rather than agreement classes, and agreement is only for the [±human] feature. Furthermore, the same agreement markers that appear on lower cardinals appear as well on both personal third-person pronouns, where they reflect agreement with the antecedent NP (4), and nominal demonstratives, where they reflect agreement with the lexical NP that would appear with the demonstrative (5).

(4) a. bà/báná
       NC_{HUMAN}.SG.NOM/PL.NOM
    b. bà
       NC_{HUMAN}.OBL
    c. à/áná
       NC_{NHUMAN}.SG.NOM/PL.NOM
    d. à
       NC_{NHUMAN}.OBL
    (Bodomo 2000, 18)

(5) a. bánàng
       those.ones.HUMAN
    b. ánàng
       those.ones.NHUMAN
    (Bodomo 2000, 18)

We now move on to complex-cardinal formation in Dagaare. Given that Dagaare is head-final in nominal syntax (except for the definite article: Bodomo 1997), it is unsurprising that the lexical NP precedes the cardinal, as in (3): on our analysis, the cardinal is the head of the entire cardinal-containing NP, so we expect it to occur head-finally in a head-final language. Complex cardinals involving multiplication are consistent with our cascading-structure analysis, in that the multiplicand precedes the multiplier and follows the lexical NP, as shown in (6). The fact that the agreement marker on 'two' is [−human] in (6a) as well as (6b) is indicative of the fact that 'two' agrees not with the lexical NP but with 'twenty', which we have already hypothesized to be nominal based on the fact that it does not show agreement.

(6) a. noba   lezaɛ   a-yi
       people twenty NC$_{NHUMAN}$-two
       'forty people'
   b. baare   lezaɛ   a-yi
       dog.PL twenty NC$_{NHUMAN}$-two
       'forty dogs'

Turning to complex cardinals involving addition, we find that the Dagaare facts are compatible with our analysis of addition as coordination plus NP deletion. As shown in (7), the second conjunct ('two') is marked with the [+human] agreement marker, indicative of the presence of the lexical NP 'people' as sister to 'two' in the underlying representation. We thus propose the structure in (8) for (7a), with deletion of the lexical NP in the rightmost conjunct.[3]

(7) a. noba   lezaɛ   a-yi   ne ba-yi
       people twenty NC$_{NHUMAN}$-two and NC$_{HUMAN}$-two
       'forty-two people'
   b. noba   koore   a-naare   ne lezaɛ   a-yi   ne
       people hundred NC$_{NHUMAN}$-four and twenty NC$_{NHUMAN}$-two and
       ba-yi
       NC$_{HUMAN}$-two
       'four hundred forty-two people'

(8)

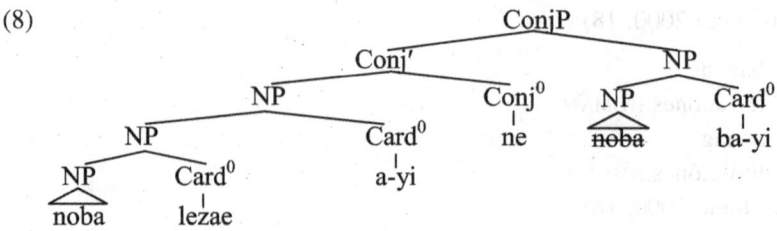

Our analysis is much simpler than that proposed by Hiraiwa (2013), which involves multiple movements and additional projections such as CaseP; what the two analyses have in common, however, is that the lexical NP is underlyingly present in both conjuncts, with the result that a complex cardinal does not form a unit to the exclusion of the lexical NP.

To sum up, the facts of Dagaare provide strong evidence in favor of our proposal for the composition of complex cardinals.

7.2   Biblical Welsh

In this section, we consider Biblical Welsh, a variety of Welsh into which the Bible was translated some four hundred years ago. According to Hurford (1975), Biblical Welsh is distinct from both Modern Welsh (discussed in Hurford 2003; see also section 7.3) and Classical Welsh, and it presents a particularly interesting numeral system. Our data come primarily from Hurford 1975, which discusses at length the numeral system of Biblical Welsh, using a 1620 version of the Welsh Bible.

Biblical Welsh provides clear evidence for our proposal from multiple standpoints, showing that complex cardinals are built using syntactic means attested elsewhere in the language. On the one hand, while we have analyzed addition as involving regular coordination in other languages (see chapter 5), in Biblical Welsh it can be constructed with the aid of a PP. On the other hand, the fact that a complex cardinal in Biblical Welsh may contain several instances of the lexical NP provides evidence for our analysis of addition in terms of some form of NP deletion.

Biblical Welsh complex cardinals are derived compositionally from the simplex cardinals given in (9).

(9) a. un, dau, tri, pedwar, pump, chwech, saith, wyth,   Biblical Welsh
          naw, deg
          'one, two, three, four, five, six, seven, eight, nine, ten'
      b. ugain, cant, mil
          'twenty, hundred, thousand'
      (Hurford 1975, 139)

We will now show that the behavior of complex cardinals in Biblical Welsh is consistent with our proposal.[4]

7.2.1   Complementation
Like the other languages we have discussed so far, Biblical Welsh allows cardinals to occur next to a lexical NP or—in the case of complex cardinals

involving multiplication—next to another cardinal. In most cases, the lexical NP that occurs with a cardinal in Biblical Welsh is singular, as shown in (10).

(10) a. pedwar brenhin
         four.M   king.SG
         'four kings'
     b. chwe diwrnod a   thri   ugain
        six   day.SG and three twenty
        'sixty-six days'
     (Hurford 1975, 192)

According to Hurford, the noun *blwyddyn* 'year' is an exception to this claim because, when used with cardinals, it obligatorily occurs in the plural, *blynedd*: see (11a) below. On our analysis, plural marking on *blynedd* is not a reflection of true semantic plurality: like the morphologically singular NP that is normally a cardinal's sister (as in (10)), *blynedd* in (11a) is semantically singular, and the plural marking is an idiosyncratic property of this particular noun.[5]

The Welsh cardinals *dau* 'two', *tri* 'three', and *pedwar* 'four' have feminine forms, *dwy*, *tair*, and *pedair*, respectively, and exhibit gender agreement with the lexical NP they combine with, as seen in (10a) versus (11a). The word for 'thousand', *mil*, behaves much like a lexical NP: it is feminine, and it triggers feminine agreement on the cardinals 'two' through 'four', as shown in (11b). Like most lexical NPs, *mil* occurs in the singular when, as in (11b), it follows a cardinal; the plural form of *mil* is *miloedd*.

(11) a. pedair blynedd
         four.F   year.PL
         'four years'
     b. pedair mil
        four.F thousand.SG
        'four thousand'
     (Hurford 1975, 152)

Another way in which *mil* behaves like a lexical NP is that it can occur with the word for 'one' when expressing the meaning 'one thousand', as shown in (12a); however, this is optional, as shown in (12b). According to Hurford (1975), this optional appearance of 'one' is also found with *cant* 'hundred', but not with other cardinals.

(12) a. un fil   a   chant
         one thousand and hundred
         'one thousand one hundred'

b. mil   a   saith gant
thousand and seven hundred
'one thousand seven hundred'
(Hurford 1975, 144)

We assume that gender specification and the ability to occur with 'one' are properties of nouns whereas gender agreement is a property of adjectives. Then *mil* 'thousand' is clearly nominal, and the words for 'two' through 'four' are just as clearly adjectival. For the other simplex cardinals, there are no clear criteria that would allow us to classify them one way or the other. In the absence of evidence to the contrary, we assume that all simplex cardinals above 'four' in Biblical Welsh are nominal, given the fact that they do not exhibit gender agreement. At the same time, cardinals like 'ten', 'twenty', and so on do not exhibit all the properties of nominals, unlike *mil* 'thousand'. This is similar to what we saw in Russian (see chapter 6).

We now briefly consider complex cardinals involving multiplication. In Biblical Welsh, multiplication applies with the base 'twenty' (13a–b) and with bases that are multiples of a hundred (13c–e): multiplication can be recursively applied, as in (13f). Word boundaries are often deleted, as shown in (13b,d–f); in some cases, boundary deletion is optional (13a).

(13) a. chwech ugain / chwe-ugain
      six   twenty six-twenty
      'one hundred twenty'

    b. de-ugain
      two-twenty
      'forty'

    c. pum cant
      five hundred
      'five hundred'

    d. deu-cant
      two-hundred
      'two hundred'

    e. ddeng-mil
      ten-thousand
      'ten thousand'

    f. saith-ugain-mil
      seven-twenty-thousand
      'one hundred and forty thousand'

(Hurford 1975, 146, 149, 151)

In short, the behavior of complex cardinals involving multiplication is consistent with the recursive structure proposed in chapter 2. We now move on to complex cardinals involving addition.

7.2.2 Lexical NP Deletion

Biblical Welsh provides evidence for our analysis of additive complex cardinals as containing two (or more) underlying forms of the lexical NP (see chapter 5). For complex cardinals above 'forty', Biblical Welsh uses *a(c)* 'and' to indicate addition (see section 7.2.3 on other arithmetic operations in this language). The lexical NP can be overt with each conjunct, as in (14a–b). It is also possible to delete the NP just in the second conjunct (14c), just in the first conjunct (14d), or even in the middle conjunct of three (14e). Further evidence that the lexical NP is underlyingly present even when it is not there on the surface comes from gender agreement: in (14d), 'four' exhibits gender agreement with the deleted NP, *mlynedd*, which is feminine.

(14) a. saith mlynedd ac   wyth gan   mlynedd
         seven year.PL   and eight hundred year.PL
         'eight hundred and seven years'
     b. naw can   mlynedd a   deng mlynedd a   de-ugain   mlynedd
         nine hundred year.PL   and ten   year.PL   and two-twenty year.PL
         'nine hundred and fifty years'
     c. chwe enaid   a   thri   ugain   [enaid]
         six   person and three.M twenty person
         'sixty-six people'
     d. pedair [mlynedd] a   phedwar ugain   mlynedd
         four.F   year.PL   and four.M   twenty year.PL
         'eighty-four years'
     e. bymtheng mlynedd a   phedwar ugain   [mlynedd] ac   wyth
         fifteen   year.PL and four.M   twenty year.PL   and eight
         gan   mlynedd
         hundred year.PL
         'eight hundred and ninety-five years'
     (Hurford 1975, 175, 176, 177, 192)

Once again, *mil* 'thousand' behaves just like a lexical NP: it can be spelled out in both conjuncts (15a), deleted in the first conjunct (15b), or deleted in the second conjunct (15c). Hurford notes that the surface form of examples like (15c) is potentially ambiguous: it is also compatible with the structure in (15d), which has no deleted *mil* and hence has a different meaning.

(15) a. bedair mil   a   saith-ugein-mil
        four.F thousand and seven-twenty-thousand
        'one hundred forty-four thousand'
   b. cant   [mil]   ac ugain mil
        hundred thousand and twenty thousand
        'one hundred twenty thousand'
   c. naw mil   a   de-ugain   [mil]
        nine thousand and two-twenty   thousand
        'forty-nine thousand'
   d. naw mil   a   de-ugain
        nine thousand and two-twenty
        'nine thousand and forty'
    (Hurford 1975, 176, 177, 191)

The fact that the lexical NP (or the word for 'thousand') can be deleted in either conjunct suggests that in Biblical Welsh, additive-complex-cardinal formation is achieved via NP deletion rather than right-node raising (see chapter 5 for more evidence against right-node raising in complex cardinals).

7.2.3 Other Arithmetic Operations in Biblical Welsh

Biblical Welsh makes use of several other operations besides complementation and overt conjunction for building complex cardinals (see also chapter 5). Two cardinals, 'twelve' and 'fifteen' (16), are formed with no overt conjunction and can be treated as single words, much like *fifteen* in English.

(16) a. deu-ddeg
       two-ten
       'twelve'
   b. pym-theg
       five-ten
       'fifteen'
    (Hurford 1975, 139)

Otherwise, addition is done in two different ways. For complex cardinals below 'forty', addition is done using the preposition *ar* 'on', and the lower cardinal must precede the higher one, as shown in (17). For complex cardinals above 'forty' (i.e., those that involve multiplication inside at least one addend), addition is done with *a(c)* 'and', as was already shown in section 7.2.2; both higher-lower and lower-higher orderings are attested, as in (18a–b) versus (18c).

(17) a. un ar ddeg
   one on ten
   'eleven'
  b. pedwar ar bymtheg
   four on fifteen
   'nineteen'
  c. naw ar hugain
   nine on twenty
   'twenty-nine'
  (Hurford 1975, 163–164)

(18) a. cant a phedwar
   hundred and four
   'one hundred and four'
  b. un fil a chant
   one thousand and hundred
   'one thousand one hundred'
  c. deuddeg ac wyth ugain
   twelve and eight twenty
   'one hundred and seventy-two'
  (Hurford 1975, 142, 144, 146)

Multiple appearances of *ar* (19a) or of *a(c)* (19b) are possible. It is of course possible to combine a complex cardinal that contains *ar* with another cardinal, using *a(c)*, as shown in (19c).

(19) a. tri ar ddeg ar hugain
   three on ten on twenty
   'thirty-three'
  b. mil a saith gant a phymtheg ... a thri ugain
   thousand and seven hundred and fifteen   and three twenty
   'one thousand seven hundred and seventy-five'
  c. tri ar ddeg a thri ugain a deu-cant
   three on ten and three twenty and two-hundred
   'two hundred and seventy-three'
  (Hurford 1975, 157, 160, 164)

Our hypothesis that complex cardinals do not form a constituent to the exclusion of the lexical NP they combine with entails that with the adpositional strategy of encoding addition just as with the coordination strategy, multiple instances of the same lexical NP should be assumed, as spelled out in (20). The syntactic structure can then be hypothesized to be that of simple NP-internal PP adjunction (cf. section 5.6.1).

(20) [ddeng mlynedd [$_{PP}$ ar hugain ~~mlynedd~~]] ac   wyth cant   ~~mlynedd~~
      ten   year.PL   on twenty year.PL   and eight hundred year.PL
      'eight hundred and thirty years'
      (Hurford 1975, 182)

The syntax we propose means that nothing special needs to be assumed about the interpretation of the preposition 'on' in complex cardinals.

Finally, Biblical Welsh also uses subtraction (see also section 5.6.2), with the subtrahend, the part being subtracted, appearing before the minuend, the part being diminished, as shown in (21).[6] Only cardinals 'four' and below can be subtrahends, and only cardinals 'twenty' and above can be minuends. The minuend can be either a simplex cardinal, as in (21a), or a complex one, as in (21b). A complex cardinal that involves subtraction can furthermore be conjoined with another cardinal using addition, as in (21c).

(21) a. onid   pedwar cant
        minus four   hundred
        'ninety-six'
     b. namyn tri   pedwar ugain
        minus three four   twenty
        'seventy-seven'
     c. dwy fil   ac onid   pedwar tri   ugain
        two thousand and minus four   three twenty
        'two thousand fifty-six'
        (Hurford 1975, 140, 158)

Subtraction works just like addition, in that there is an underlying NP next to both the subtrahend and the minuend: the NP next to the subtrahend can be either overt (22a) or deleted (22b). Once again, agreement on 'three' in (22b) provides evidence for the underlying deleted NP. And once again, *mil* 'thousand' works just like a lexical NP (22c), in that it can be spelled out multiple times.

(22) a. gant   [~~mlynedd~~] ac onid   tair   blynedd de-ugain   [~~mlynedd~~]
        hundred   year.PL on minus three.F year.PL   two-twenty   year.PL
        'one hundred and thirty-seven years'
     b. onid   tair   [~~mlynedd~~] de-ugain   [~~mlynedd~~] a   chan   mlynedd
        minus three.F   year.PL   two-twenty year.PL   and hundred year.PL
        'one hundred and thirty-seven years'
     c. onid   pedair mil   pedwar ugain mil   a   phum cant
        minus four.F thousand four.M twenty thousand and five   hundred
        'seventy-six thousand five hundred'
        (Hurford 1975, 145, 176, 183)

Thus, Biblical Welsh uses all of the different strategies for nonmultiplicative complex cardinals that we discussed in chapter 5.

7.2.4 Summary

To sum up, the data from Biblical Welsh provide strong support for our analysis of complex cardinals. First, complex cardinals in Biblical Welsh are highly compositional, with the arithmetical operations read off directly from the syntactic structure. Second, Biblical Welsh provides direct evidence in favor of our analysis of complex cardinals involving addition. Third, the use of prepositions to express addition and subtraction offers further support to our view that complex cardinals are constructed via regular syntactic means, and therefore that they should not be viewed as a separate linguistic system.

Alongside Russian (discussed in chapter 6), Biblical Welsh provides further evidence that cardinals form a scale, from those that are clearly adjectival (e.g., 'two' through 'four') to those that behave much like lexical nouns (e.g., 'thousand'), with others falling in between.

7.3 Modern Welsh

As discussed by Hurford (1975, 1987, 2001, 2003), as a rule cardinals within a single language do not all behave identically: while lower cardinals appear to be more adjectival (for instance, agreeing for gender), higher cardinals are unmistakably more nominal. We have already seen this in Russian and Biblical Welsh. Another very special case of this "cardinal cline" comes from Modern Welsh, where lower cardinals preferentially appear with a singular lexical NP whereas higher cardinals prefer a plural lexical NP with the obligatory preposition *o* 'of'.[7] This is illustrated in (23).[8]

(23) a. pedwar drws   Modern Welsh
          four.M door.SG
          'four doors'
      b. cant   *(o) ddrysiau
          hundred of door.PL
          'a hundred doors'
      (King 2003, 111)

The distinction between the two patterns is generally attributed to the fact that Welsh cardinals can be adjectival or nominal (Greene 1992; Williams 1980; Xalipov 1995). The nominal cardinals *pump* 'five', *chwech* 'six', and *cant* 'hundred' differ morphologically from the adjectival forms *pum* 'five', *chwe* 'six', and *can* 'hundred'. Only the nominal long form of the cardinal can occur in the plural pattern, as in (23b) (see (26) and (27) below for further examples), or appear in isolation, as in (24).

(24) Pa sawl   afal   sydd yna? — Pump.
     wh how.many apple.SG are   there   five
     'How many apples are there?' — 'Five.'
     (Williams 1980, 41)

Danon (2012) uses the Welsh data (as well as the Hebrew data discussed in section 7.4) to argue that cardinals crosslinguistically can be projected either as heads or as specifiers, above and beyond their adjectival or nominal properties. In this section we will argue against this hypothesis (for Welsh; see section 7.4 on Hebrew).

7.3.1   Two Syntactic Patterns for Modern Welsh Cardinals

The special interest of Modern Welsh lies in the distribution of the two syntactic structures available for cardinal-containing NPs: the older singular pattern, abundantly exemplified in the Welsh translation of the Bible (see section 7.2), and the modern pseudopartitive pattern, where the lexical NP is obligatorily plural and introduced by the preposition 'of' (see Greene 1992 and G. Roberts 2000 for a discussion of how the situation developed historically). The distribution of the two patterns in Modern Welsh is not altogether straightforward (King 2003, 111). Although Danon (2012), relying on Mittendorf and Sadler (2005), suggests that the two patterns are interchangeable for lower cardinals, King (2003, 111) asserts that the dividing line between the two patterns is hard to draw.

The two options are illustrated again in (25) with 'nine'. With higher cardinals (26), both options are likewise available, though with most nouns the use of the singular form is generally associated with a more formal register.[9]

(25) a. naw ddyn
        nine man.SG
        'nine men'
     b. naw o ddynion
        nine of man.PL
        'nine men'
        (Hurford 2003)

(26) a. can   drws
        hundred door.SG
        'a hundred doors'

b. cant   o ddrysiau
   hundred of door.PL
   'a hundred doors'
(Hurford 2003)

While Thorne (1993, 149; cited in Hurford 2003), Sadler (2000), and Mittendorf and Sadler (2005) claim that with lower cardinals the choice of the pattern has no semantic effect, P. W. Thomas (1996, 314; cited in Borsley, Tallerman, and Willis 2007) indicates that in some environments (with measure-denoting nouns) the use of the plural pattern gives rise to an individuated interpretation, as shown in (27) through (29); it is furthermore impossible in expressions of time and in dates, where the singular pattern is instead obligatory.

(27) a. pump o geiniogau
        five   of penny.PL
        'five pennies (coins)'
     b. pum ceiniog
        five penny.SG
        'five pence (amount of money)'
     (P. W. Thomas 1996, 314; cited in Borsley, Tallerman, and Willis 2007)

(28) a. tri-deg   o geiniogau
        three-ten of penny.PL
        'thirty pennies'
     b. tri-deg   ceiniog
        three-ten penny.SG
        'thirty pence'
     (P. W. Thomas 1996, 314; cited in Borsley, Tallerman, and Willis 2007)

(29) a. gan   gradd
        hundred degree.SG
        'one hundred degrees'
     b. *gan/gant o raddau
        hundred of degree.PL
     (P. W. Thomas 1996, 314; cited in Borsley, Tallerman, and Willis 2007)

The assumption that 'thousand' and 'million' behave like measure nouns explains why the more modern plural pattern is not available inside complex cardinals involving multiplication (Greene 1992): only the older singular pattern is available for combining the multiplier with the multiplicand. The cardinal 'thousand' in (30b) is just like the measure nouns in (30a) in this respect.[10]

(30) a. deng nniwrnod/minud
         ten   day.SG/minute.SG
         'ten days/minutes'
     b. deng mil/miliwn   o ddynion
         ten   thousand.SG/million.SG of man.PL
         'ten thousand/million people'
     (King 2003, 112)

Several questions arise. Do the singular and plural patterns project in the same syntactic structure? If so—as we have assumed elsewhere for the singular versus plural variation in the lexical NP—how is the preposition 'of' accounted for? If not, do the superficially identical surface forms appearing in the two structures (at least for some cardinals) correspond to the same lexical entry? And if the answer to this question is no, do these different lexical entries differ only in their formal properties or do they also have a different semantics?

7.3.2 Lexical Categorization of Welsh Cardinals

The existence of the two patterns is generally derived from the fact that Welsh cardinals can be adjectival or nominal (Greene 1992; Williams 1980; Xalipov 1995), a difference that is morphologically overt in some cases, such as 'five' and 'hundred' (see (23b), (26), and (27) above). The ability of Welsh cardinals to agree for gender, though, does not directly correlate with the syntactic structure they appear in. Gender is overtly indicated on the cardinal in both of the Welsh patterns (compare (31) and (32)), but only for the cardinals *dau/dwy* 'two.M/F', *tri/tair* 'three.M/F', and *pedwar/pedair* 'four.M/F' (Borsley, Tallerman, and Willis 2007, 163). Unlike regular adjectives and nouns, which undergo a phonological change called soft mutation after the definite article *yr* 'the' in feminine DPs, cardinals do not undergo this gender-conditioned change except in substandard Welsh. The standard Welsh is illustrated in (32). (As shown in (33), the cardinal 'two' does undergo soft mutation in definite DPs, regardless of their gender; see Borsley, Tallerman, and Willis 2007, 156.)

(31) a. tair   o ferched
         three.F of girl.PL
         'three girls'
     b. tri   o ffilmiau
         three.M of film.PL
         'three films'
     (Borsley, Tallerman, and Willis 2007, 170–171)

(32) y   tair/pedair/pum   cath
      the three.F/four.F/five.F cat.F
      'the three/four/five cats'
      (Borsley, Tallerman, and Willis 2007, 156)

(33) a. y ddau aderyn
        the two.M bird.SG
        'the two birds'
     b. dau   aderyn
        two.M bird.SG
        'two birds'
      (Borsley, Tallerman, and Willis 2007, 156)

Hurford (2003) assimilates the more modern plural construction, like the example in (34a), to a pseudopartitive (34b), noting that the obligatory matching between the gender on the cardinal and the gender of the lexical NP need not be due to agreement, since it is also available in partitives (34c).

(34) a. tri   o blant
        three.M of child.PL
        'three children'
     b. potel o win
        bottle of wine
        'a bottle of wine'
     c. tri   o 'r plant
        three.M of the child.PL
        'three of the children'
      (Borsley, Tallerman, and Willis 2007, 175)

Along the same lines, Danon (2012) proposes that in the plural construction the cardinal should be treated as the head (35a) and in the singular construction as a specifier (35b).

(35) a.   DP   *cant o ddrysiau* 'hundred doors' (26b)

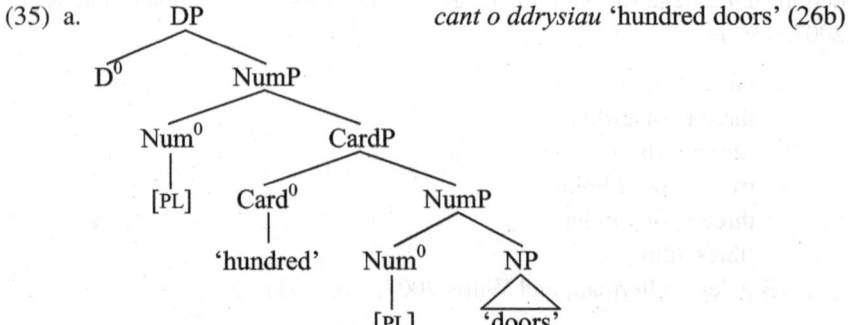

b.

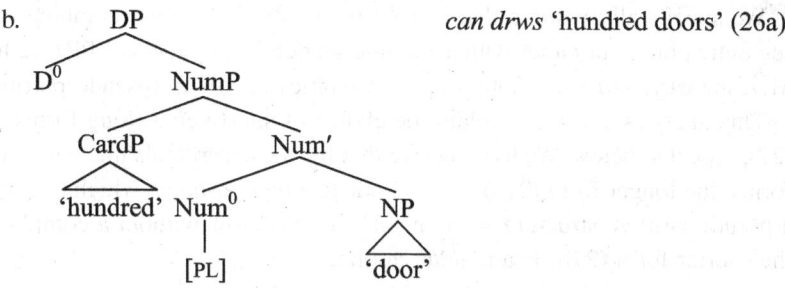

can drws 'hundred doors' (26a)

In chapter 4, we argued against semantic plurality below the cardinal. Furthermore, the structures in (35) require a number of problematic assumptions. First of all, the syntax of Card^0 is unclear: while in (35a) it takes a NumP as a complement, that is, as an internal argument, in (35b) it seems to have no internal arguments, and its mode of composition with the plural lexical NP is presumably different. The question arises whether these differences in syntax correlate with differences in lexical semantics, and if so, whether the same label is suitable for these two types of cardinals.

Secondly, the presence of the second Num^0 in (35a) suggests that the cardinal itself should either be marked plural, if the contribution of Num^0 is limited to overt number marking, or should not itself be plural in the way that the cardinal in (35b) is, if the contribution of Num^0 is semantic. In other words, if the [PL] featural specification of Num^0 contributes semantic plurality, then CardP in (35a) should not be semantically plural.

Thirdly, the more global question of labeling must be addressed: why is the combination of a cardinal with a projection of Num^0 labeled differently in the two structures in (35)? In other words, which property of the cardinal requires it to project in (35a) but not in (35b)?

All these questions are naturally answered if we assume that some cardinals, the more nominal ones, behave like measure nouns and project in the pseudopartitive structure, as in (36), whereas others (such as (26a)) take the lexical NP as a complement, as we have proposed for other languages.

(36)

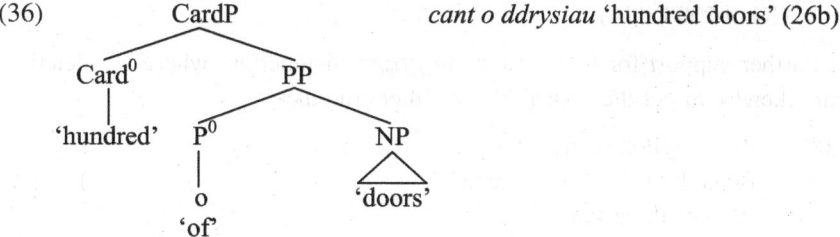

cant o ddrysiau 'hundred doors' (26b)

We specifically assume that in (36) the cardinal does not combine with a semantic plural but rather with an atomic set denoted by the 'of' PP (see Ionin, Matushansky, and Ruys 2006 on the semantics of 'of' in (pseudo)partitives).

This analysis can also explain the choice of short versus long forms, as in (27), repeated below. We hypothesize that for those cardinals that come in two forms, the longer form (27a) corresponds to a regular noun, which can project a pseudopartitive structure with an 'of' PP or appear without a complement; the shorter form (27b) is a regular cardinal.

(27) a. pump o geiniogau
         five   of penny.PL
         'five pennies (coins)'
     b. pum ceiniog
         five penny.SG
         'five pence (amount of money)'
     (P. W. Thomas 1996, 314; cited in Borsley, Tallerman, and Willis 2007)

7.3.3 Complex Cardinals Involving Addition

In Modern Welsh, as in Biblical Welsh, additive complex cardinals provide strong support for our analysis. First of all, they can be formed according to either one of the two patterns: the older singular pattern (37a) or the more modern plural pattern (37b).[11] In the former pattern, the lexical NP is overt in the first rather than the second conjunct, consistent with our analysis of additive-complex-cardinal formation as NP deletion. The position of the lexical NP in (37a) is not amenable to an analysis that treats the complex cardinal as a unit to the exclusion of the lexical NP.

(37) a. pum mlynedd ar   hugain
         five year.PL   and twenty
         'twenty-five years'
     b. tri-deg   tri   o deirw
         three-ten three of bull.PL
         'thirty-three bulls'
     (King 2003, 114)

Further support for this view comes from disjunction, where NP deletion can likewise affect the lexical NP in either conjunct:

(38) a. dwy   gyllell.SG neu dair
         two.F knife   or three.F
         'two or three knives'

b. dwy   neu dair   cyllell.SG
   two.F or   three.F knife
   'two or three knives'
(Williams 1980, 41)

Finally, we address the fact, noted by Mittendorf and Sadler (2005), that the singular pattern appears to be incompatible with coordination underneath the cardinal (39a), which is fine under the plural pattern (39b).

(39) a. *pum bachgen a   merch
         five boy.SG   and girl.SG
     b. pump o fechgyn a   merched
         five   of boy.PL and girl.PL
         'five boys and girls'
(Mittendorf and Sadler 2005)

As discussed in chapter 5, nonintersective coordination of two (semantically singular) lexical NPs underneath a cardinal, as in (39a), can only be derived via the disjunctive 'and', whereas coordination of two semantically plural lexical NPs, as in (39b), can also be derived via the additive interpretation. The facts in (39) can be straightforwardly explained if the Modern Welsh *a* 'and' lacks the disjunctive interpretation. This explanation makes a strong prediction: all types of coordination that can only be derived via disjunctive 'and' should be unavailable in Modern Welsh.

7.3.4 Summary

To sum up, Modern Welsh, like Biblical Welsh, provides strong support for our proposal in the way that it composes additive complex cardinals as well as those involving multiplication. The existence of two different cardinal systems in Modern Welsh is straightforwardly explained on the view, due to Hurford (2003), that nominal cardinals can project a pseudopartitive structure.

7.4 Modern Hebrew

Cardinal-containing NPs in standard Modern Hebrew have two characteristics that set them apart from those in non-Semitic languages: (i) they are frequently analyzed as involving the construct state; and (ii) they exhibit chiastic agreement, or gender polarity. Therefore, before proceeding to our analysis of Hebrew complex cardinals, we give a brief overview of the construct-state construction (section 7.4.1) and of the properties of Hebrew simplex cardinals, including chiastic agreement (section 7.4.2). After these preliminaries, we move on to a discussion of the morphological changes that Hebrew cardinals undergo inside complex cardinals in sections 7.4.3 and 7.4.4. We focus on the formation of complex cardinals involving addition and multiplication in sections 7.4.5 and 7.4.6, respectively. We show that the formation of complex cardinals in Hebrew presents some challenges to our analysis of complex cardinals. In particular, we address the challenges posed by the distribution of the bound and free forms of cardinals. We argue, contra Danon (2012) and Rothstein (2009, 2011a, 2013, 2016, 2017), that Hebrew cardinals in both their bound and free forms are compatible with the cascading structure that we have proposed.

7.4.1 The Construct State in Modern Hebrew

Nouns in Semitic languages come in two forms, the free form and the bound form, appearing in syntactically different environments, the free state and the construct state, as illustrated in (40) for Modern Hebrew. For the construct state, following the terminology of Shlonsky (2004), we term the first noun, which occurs in the bound form, the *head* of the construct state, and the NP that it combines with the *annex*. So in (40b), *dirat* is the head and *ha-more* is the annex. The bound form is not distinct from the free form for all nouns.[12]

(40) a. *Free state*

ha-dira   šel ha-more   Modern Hebrew
the-apartment of the-teacher
'the teacher's apartment/the apartment of the teacher'

   b. *Construct state*

dirat   ha-more
apartment_BOUND the-teacher
'the teacher's apartment'

While the free state (40a) is simply the Semitic counterpart of a prepositional genitive, which also exists in many familiar languages, the construct state (40b) has a number of special characteristics (see, e.g., Ritter 1987; 1988; Borer 1999; 2005, 216; 2011; 2013; and Shlonsky 2004, among many others). The syntax of the construct state involves a head-complement structure (41).

(41)

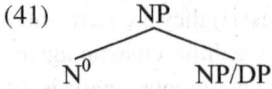

While some researchers (see, e.g., Ritter 1987, 1988; Borer 1996, 1999, 2005) assume that the noun in the bound form starts out in a lower position than the noun in the free form, we believe with Faust (2014) that this is not the case: a movement analysis would be incorrect for quantifiers and cardinals in the construct state, and unlikely to be correct for APs in the construct state.

The head noun in the construct state cannot be preceded by a (prefixal) determiner, and all modifiers appear linearly after the complement. In a definite construct-state NP, the definite article appears in the complement NP, yet it takes scope over the entire construct-state NP (40b).

In Modern Hebrew, the construct state is declining and is not productively used for possessive-genitive formation. As discussed by Zuckermann (2006) and Faust (2014), it is much less common in Modern Hebrew than Biblical Hebrew. In particular, the position of the definite article is shifting, with the prescriptive placement of *ha* 'the' before the complement NP (42a) giving way to the placement of *ha* in front of the entire NP, as in (42b). Note that the form in (42b), unlike the one in (42a), puts *ha* in the position that corresponds to its scope: the entire NP is definite, not just its second component (i.e., (42a–b) both mean 'the arranger of law', not 'an arranger of the law').

(42) a. orex   ha-din
         arranger the-law
         'the lawyer'
    b. %ha-orex din
         arranger law
         'the lawyer'

As shown in (42), the construct-state structure is on its way to being replaced with a morphological compound, with no apparent effect on the meaning. In contrast, as shown by Rothstein (2009 et seq.), in pseudopartitives, use of the free genitive or the construct state does not give rise to the same readings: measure readings are available only to construct-state nominals, an observation that will become relevant later.

7.4.2 Simplex Cardinals: An Overview

Modern Hebrew cardinals form an adjective–noun continuum, differing from each other in their position with respect to the noun, their ability to agree, and the distribution of their bound form. The cardinal 'one' is postnominal and agrees with the lexical NP in both gender (43a–b) and definiteness (43c) (see, e.g., Rosén 1977). These properties make 'one' a clearly adjectival cardinal.

(43) a. sefer   axat
         book.M one.M
         'one book'

    b. delet   exad
       door.F one.F
       'one door'
    c. ha-delet   ha-exad
       the-door.F the-one.F
       'the one door'

All cardinals above 'one' are prenominal. 'Two' through 'ten' have a bound form that is indicative of the construct state,[13] though this form is morphologically distinct from the free form only for masculine cardinals and the cardinal 'two'. For the cardinal 'two', the bound form is obligatory in the presence of an overt lexical NP, as shown in (44a–b). (In the case of NP ellipsis, the free form of 'two' is used.) For 'three' through 'ten', the bound form must be used when the NP is definite, as shown in (44c); compare (44d), where the free form of the cardinal is used inside an indefinite NP. In (44e), the NP is feminine, hence there is no morphological distinction between the bound and free form of the cardinal.

(44) a. šney/*šnayim   sfarim
       two.M.BOUND/two.M book.M.PL
       'two books'

    b. štey/*štayim   dlatot
       two.F.BOUND/two.F door.F.PL
       'two doors'

    c. šlošet   ha-sfarim
       three.M.BOUND the-book.M.PL
       'the three books'

    d. šloša   sfarim
       three.M book.M.PL
       'three books'

    e. šaloš   (ha-)dlatot
       three.F the-door.F.PL
       '(the) three doors'

Simplex cardinals 'ten' and below in Hebrew show *chiastic agreement* (a.k.a. gender polarity; see Hetzron 1967; Halle 1990; Rosenthal 2006, 35–36): the gender suffix on the cardinal is the opposite of what you would expect for its gender. This was illustrated above in (43) with the cardinal 'one'. *Axat* is morphologically feminine, yet used with masculine nouns (43a), while *exad* is unmarked, as expected for masculine forms, yet used with feminine nouns, as in (43b–c). At the same time, the morphologically unmarked form is the syntactic default used in counting (45).

(45) exad, štayim, šaloš, arbaʕ ...   Counting
     one   two   three four
     'one, two, three, four ...'

The use of the morphological feminine with a syntactic masculine and vice versa is found with some lexical nouns as well, as shown in (46). The feminine noun for 'year' takes the masculine plural suffix -*im* (46a), while the masculine noun for 'table' takes the feminine plural suffix -*ot* (46b). And for a handful of nouns, the singular form does not show the syntactic gender: the noun for 'city' is syntactically feminine despite having the zero masculine ending in the singular and the masculine -*im* ending in the plural (46c). Likewise, the noun for 'night' is syntactically masculine despite taking the feminine -*a* and -*ot* endings in the singular and plural (46d).

(46) a. šana, šan-im
        year.F year-PL
        'year', 'years'
     b. šulxan, šulxan-ot
        table.M table-PL
        'table', 'tables'
     c. ʕir-∅, ʕar-im
        city.F city-PL
        'city', 'cities'
     d. layla, leyl-ot
        night.M night-PL
        'night,' 'nights'

Given the imperfect form–gender correlation even with nouns, chiastic agreement in Hebrew cardinals is best analyzed as a purely morphological phenomenon. As suggested by Halle (1990) for Hebrew plurals, we propose that gender morphology can be realized differently with different roots, making it a lexical rather than a syntactic process. We note here that chiastic agreement holds for complex cardinals exactly as for simplex ones: for example, 'three' shows chiastic agreement when simplex as well as when it is inside 'thirteen', 'twenty-three', and so on. Throughout this section, we gloss cardinals that occur with masculine nouns as masculine and those that occur with feminine nouns as feminine: that is, our glosses reflect the gender of the entire cardinal-containing NP, rather than the gender indicated by the morphology of the cardinal.

The fact that lower cardinals agree (chiastically) with the lexical NP in gender, as in (43) and (44), speaks in favor of treating them as adjectival. However, even though in literary Hebrew lower cardinals above 'one' can occur postnominally (47), like regular adjectives and like the cardinal 'one', in regular Hebrew they are prenominal, which speaks against the adjectival analysis. In other words, cardinals higher than 'one' are distinctly more nominal than 'one', yet not fully nominal.

(47) ganavim šloša
    thief.PL   three.M
    'three thieves'
    (Shlonsky 2004, fn. 12)

Moving to simplex cardinals 'twenty' and higher (including 'hundred', 'thousand', etc.), we observe in (48) that they do not agree for gender and have no bound form.[14] They combine with a plural lexical NP, like other cardinals above 'one'. The same form of the cardinal is used in definite expressions (48b). Following the same reasoning as above, the position of the definite article (in prescriptive use) indicates the construct state.[15]

(48) a. xamišim/me'a sfarim/dlatot
      fifty/hundred.F book.M.PL/door.F.PL
      'fifty/a hundred books/doors'
   b. xamišim/me'a ha-sfarim
      fifty/hundred.F the-book.M.PL
      'the fifty/hundred books'

The gender of the syntactically simplex cardinals 'twenty' through 'ninety' is not detectable: these cardinals trigger plural agreement and do not function as multiplicands (so their gender, if any, cannot be detected from the form of the multiplier). On the other hand, the higher cardinals are specified for gender: *me'a* 'hundred' is feminine, the rest are masculine. Thus, the cardinals 'hundred' and above are relatively nominal while 'twenty' through 'ninety' exhibit neither decisively nominal nor decisively adjectival characteristics, falling somewhere along the adjective–noun continuum.

To summarize, with the exception of 'one', Hebrew cardinals are commonly divided into three clusters: those showing gender agreement ('two' through 'ten'), those having inherent gender ('hundred' and up), and the rest. The syntactic configuration in which simplex cardinals combine with lexical NPs is unambiguously that of a construct state only in definite DPs, with the lower cardinals 'three' through 'ten', which appear then in their bound form in the masculine (44c). For indefinites, there is no overt evidence of the construct state except with 'two', which occurs in its bound form then as well (44a–b). For higher cardinals, there is no bound form, and the only possible evidence of the construct state is the placement of the definite article (48b).

Table 7.1
Properties of the simplex cardinals 'one' to 'ten' in Modern Hebrew

	F, free	F, bound	F, dependent	M, free	M, bound	M, dependent
'one'	exad	—	—	axat	—	—
'two'	štayim	štey	štem	šnayim	šney	šnem
'three'	šaloš		šloš	šloša	šlošet	—
'four'	arbaʕ		—	arbaʕa	arbaʕat	—
'five'	xameš		—	xamiša	xamešet	—
'six'	šeš		—	šiša	šešet	—
'seven'	ševaʕ		švaʕ	šivʕa	šivʕat	—
'eight'	šmone		—	šmona	šmonat	—
'nine'	tešaʕ		tšaʕ	tišʕa	tišʕat	—
'ten'	eser		esre	asara	aseret	asar

Given that only lower cardinals agree for gender and occur in the bound form, Danon (2012) proposes that cardinals may combine with the lexical NP in two distinct structural configurations (cf. also the digital–nondigital distinction in Glinert 1989): the cascading structure from Ionin and Matushansky 2006 for 'two' through 'nineteen' and the specifier-head configuration for the higher cardinals. We will, however, argue (section 7.4.6) that our cascading structure is compatible with all cardinal-containing NPs in Hebrew.

7.4.3 The Dependent Form versus the Bound Form

In addition to the free and bound forms discussed in the previous section, some Hebrew cardinals ('two' and 'ten' in both genders as well as 'three', 'seven', and 'nine' in the feminine only) also occur in *dependent* forms. The full paradigm of free, bound, and dependent forms for 'one' through 'ten' is given in table 7.1. Note that 'ten' in the masculine and 'two' in both genders confirm the three-way distinction between free, bound, and dependent forms.

The feminine dependent forms occur in lower complex cardinals formed by addition ('eleven' to 'nineteen'), as in (49a), as well as before 'hundred', as in (49b). The dependent form is not triggered by definiteness, as shown in (49c). 'Three' through 'nine' in the masculine, since they lack a dependent form, use the free form in 'thirteen' through 'nineteen'; the form of 'ten' in these complex cardinals is dependent in the masculine (49d) as well as the feminine (49a).

(49) a. šloš-esre   dakot
       three.F.DPNDNT-ten.F.DPNDNT minute.F.PL
       'thirteen minutes'

b. šloš   meot
   three.F.DPNDNT hundred.F.PL
   'three hundred'

c. šaloš/*šloš   ha-dlatot
   three.F/three.F.DPNDNT the-door.F.PL
   'the three doors'

d. šloša-asar   xodašim
   three.M-ten.M.DPNDNT month.M.PL
   'thirteen months'

The only place where we can observe the dependent forms for the cardinal 'two' is inside the cardinal 'twelve' (50). The reason is that Hebrew does not allow 'two' to combine with the multiplicands 'hundred' and 'thousand', instead using the synthetic dual forms, as in (51a–b).[16] We also note that, just like *me'a* 'hundred' and *elef* 'thousand', their dual counterparts in (51a–b) do not have a separate bound form. 'Two' does combine with 'million', but in its bound rather than dependent form (51c). The same holds for 'ten' in combination with both 'thousand' and 'million' (52).

(50) a. šnem-asar   xodašim
        two.M.DPNDNT-ten.M.DPNDNT month.M.PL
        'twelve months'

     b. štem-esre   dakot
        two.F.DPNDNT-ten.F.DPNDNT minute.F.PL
        'twelve minutes'

(51) a. matayim   dakot/xodašim
        hundred.DU minute.F.PL/month.M.PL
        'two hundred minutes/months'

     b. alpayim   dakot/xodašim
        thousand.DU minute.F.PL/month.M.PL
        'two thousand minutes/months'

     c. šney/*šnem   milyon
        two.M.BOUND/M.DPNDNT million.M
        'two million'

(52) a. aseret/*asar   alafim
        ten.M.BOUND/ten.M.DPNDNT thousand.M.PL
        'ten thousand'

     b. aseret/*asar   milyon
        ten.M.BOUND/M.DPNDNT million.M
        'ten million'

To sum up, dependent forms occur only in the cardinals 'eleven' through 'nineteen' (in both conjuncts) and as the multiplier of 'hundred'.

Danon (2012) suggests that the dependent form of cardinals constitutes an argument against our approach in that it only occurs inside complex cardinals, arguing for the lexical nature of the latter. In principle, we could agree with Danon and say that the cardinals involving the dependent form, 'eleven' through 'nineteen' as well as 'three hundred' through 'nine hundred' (recall that 'two hundred' is expressed as a dual), are indeed formed in the lexicon, while all other complex cardinals are formed in the syntax. However, a lexical approach faces two types of problems. First, it requires φ-features to be present twice in one syntactic terminal, for example, on both 'three' and 'ten' in 'thirteen': compare the gender marking in (49a) to (49d). Second, for hundreds, both gender morphology on the multiplier and plural marking on the multiplicand (as in (49b)) would have to be lexical, which seems rather odd.

Importantly, the dependent form is not the only special property of the cardinals 'eleven' through 'nineteen', as discussed in the next subsection.

7.4.4 The Syntax of 'Eleven' through 'Nineteen'

As in many languages, the complex cardinals 'eleven' through 'nineteen' in Hebrew are special in that the lower cardinal precedes the higher cardinal (53a,c), while the opposite is the case for higher complex cardinals involving addition (53b,d). In both cases, those simplex cardinals that inflect for gender do so. The fact that in (53a,c) we see gender agreement on both cardinals speaks in favor of a syntactic rather than a lexical analysis. We proceed on the assumption that all Hebrew complex cardinals formed by addition are built in the syntax.

(53) a. xameš-esre   dlatot
       five.F.DPNDNT-ten.F.DPNDNT door.F.PL
       'fifteen doors'

    b. esrim   ve-xameš dlatot
       twenty and-five.F door.F.PL
       'twenty-five doors'

    c. xamiša-asar   sfarim
       five.M.DPNDNT-ten.M.DPNDNT book.M.PL
       'fifteen books'

    d. esrim   ve-xamiša sfarim
       twenty and-five.M book.M.PL
       'twenty-five books'

Does the dependent form that 'ten' takes in the cardinals 'eleven' through 'nineteen' (53a,c) have exactly the same semantics as the free form? We propose that the answer is no, that in this case a special form corresponds to a special semantics. More specifically, we propose that the dependent form of 'ten' used in 'eleven' through 'nineteen' incorporates addition as part of its semantics. The lexical entry for dependent 'ten' is given in (54a), with the corresponding extension for 'thirteen books' stated formally in (54b) and informally in (54c).

(54) a. $[\![\text{ten.DPNDNT}]\!] = \lambda g \in D_{\langle e,t\rangle} \,.\, \lambda f \in D_{\langle\langle e,t\rangle,\langle e,t\rangle\rangle} \,.\, \lambda x \in D_e \,.$
$\exists S \in D_{\langle e,t\rangle} \, [\Pi(S)(x) \wedge S = x_1 \oplus x_2 \wedge [\![\text{ten}]\!](g)(x_1) \wedge f(g)(x_2)]$

   b. $[\![\text{ten.DPNDNT-three books}]\!] = \lambda x \in D_e \,.\, \exists S \in D_{\langle e,t\rangle}$
$[\Pi(S)(x) \wedge S = x_1 \oplus x_2 \wedge [\![\text{ten}]\!](\text{book})(x_1) \wedge [\![\text{three}]\!](\text{book})(x_2)]$

   c. $[\![\text{ten.DPNDNT-three books}]\!] = x$ is a plural individual divisible into two nonintersecting parts such that one part consists of ten books and the other part consists of three books.

Note that with the special semantics for dependent 'ten' given above, there is no need to also posit a special semantics for the dependent forms of 'two' through 'nine'. As a result, the cardinals 'eleven' through 'nineteen' in Hebrew project as in (55), which is a cascading structure that we do not normally assume for addition.[17] However, if the dependent form of 'ten' is the complement of the lower cardinal, the configuration is the same for 'thirteen' as for 'three hundred'. Recall, from the previous subsection, that the dependent form of 'two' through 'nine' also occurs with 'hundred' (e.g., (49b)), which we assume has the normal cascading structure of multiplication, repeated in (56).

(55)

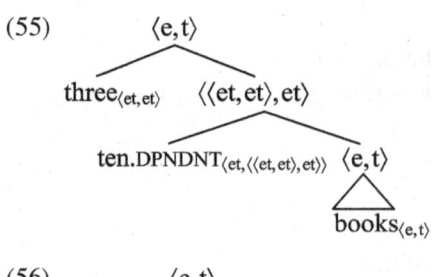

(56)

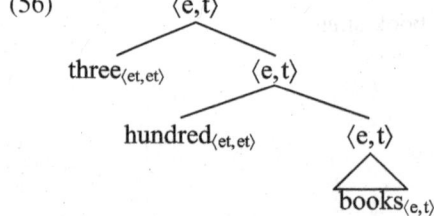

We hypothesize that the dependent forms of the cardinals below 'ten' reflect agreement with their complement: 'hundred' in (56) and dependent 'ten' in (55). The multiple agreement seen in the cardinals 'eleven' through 'nineteen' is then explained as successive agreement: in 'thirteen books', for example, first 'ten' agrees with its complement 'books', then 'three' agrees with 'ten'.

The above hypothesis requires the assumption that both 'ten' and 'hundred' differ from the highest cardinals ('thousand', 'million', etc.), which require their multiplier to be in the bound form (when one is available) rather than the dependent form (see (52)). In section 7.4.6, we will see some other differences between 'hundred' on the one hand and 'thousand', 'million', and so on on the other, which support the distinction required for the above analysis.

Thus, we see that it is quite possible to account for the distribution of the dependent form in Hebrew without postulating that complex cardinals containing this form are built in the lexicon.[18]

7.4.5 Other Complex Cardinals Involving Addition

Apart from the cardinals 'eleven' through 'nineteen', Hebrew uses the coordination strategy for addition, as illustrated above in (53b,d). When one of the conjuncts is the cardinal 'one', the lexical NP is still plural, as shown in (57). In this, Hebrew patterns with languages like English, Polish, Romanian, and Modern Greek (see chapter 5). We therefore propose the same analysis for Hebrew as for these languages: NP deletion of the lexical NP in the first conjunct, coupled with a special status for 'one' in the second conjunct.

(57) a. exad-esre   dlatot
       one.F-ten.F door.F.PL
       'eleven doors'
   b. esrim   ve-exad   dlatot
       twenty and-one.F door.F.PL
       'twenty-one doors'

Note that *exad*, which is postnominal when it is a simplex cardinal (43), is prenominal when part of a complex cardinal involving addition (57). The facts in (57) appear to argue against our NP-deletion analysis of such cardinals—which predicts *exad* to be in the same syntactic location regardless of whether it occurs on its own or as part of a complex cardinal—and in favor of treating 'twenty-one' as a unit to the exclusion of the lexical NP. However, as discussed in chapter 5, it is a general property of complex cardinals involving addition that all cardinals inside them show more noun-like behavior than when they occur on their own: for example, the cardinal *jeden* 'one' in Polish becomes indeclinable (which is more common for nouns than for adjectives), and the lower cardinals in Romanian start requiring the preposition *de*. In the case of Hebrew, *exad* 'one' inside a complex cardinal both requires a plural rather than a singular lexical NP sister and becomes prenominal, that is, less like an adjective and more like other, more nominal, cardinals. Another special property that 'one' takes on inside coordinated cardinals is that it is no longer marked for definiteness, as shown in (58).

(58) a. ha-xatula ha-exad
        the-cat.F the-one.F
        'the one female cat'

    b. esrim ve-exad   ha-xatulot
       twenty and-one.F the-cat.F.PL
       'the twenty-one female cats'

'One' is not the only Hebrew cardinal that changes its behavior when occurring inside an additive complex cardinal. 'Two' through 'ten', which have a bound form (see table 7.1) when simplex, do not appear in their bound form when inside a complex cardinal involving addition.

As discussed earlier, the simplex cardinals 'three' through 'ten' occur in their free form inside an indefinite NP (59a) but in their bound form inside a definite NP (59b); however, inside a complex cardinal involving addition, these cardinals obligatorily occur in their free form, regardless of definiteness (59c–d).

(59) a. šloša   sfarim
        three.M book.M.PL

    b. šlošet/*šloša   ha-sfarim
       three.M.BOUND/three.M the-book.M.PL
       'the three books'

    c. šloša-asar/*šlošet-aseret   (ha-)sfarim
       three.M-ten.DPNDNT/three.M.BOUND-ten.M.BOUND   the-book.M.PL
       '(the) thirteen books'

    d. esrim ve šloša/*šlošet   (ha-)sfarim
       twenty and three.M/three.M.BOUND the-book.M.PL
       '(the) twenty-three books'

'Two' is exceptional in that it occurs in the construct state in indefinite as well as definite environments (60a). However, inside an additive complex cardinal, 'two' occurs in its free form (60b).[19]

(60) a. šney/*šnayim   (ha-)sfarim
        two.M.BOUND/two.M the-book.M.PL
        '(the) two books'

b. esrim ve šnayim/*šney   sfarim
   twenty and two.M/two.M.BOUND book.M.PL
   'twenty-two books'

This is reminiscent of measure nouns. While the bound form in Hebrew is very widespread, appearing not only with nouns but also on adjectives, quantifiers, and other lexical items, measure nouns do not have it. Consider the pseudopartitive in (61a), where the container appears in the bound form indicating the complementation structure; when the regular noun is replaced by with a measure noun, such as 'liter', the free form (either singular or plural) must be used, as in (61b). We conclude that when lower simplex cardinals occur inside additive complex cardinals, they become more nominal, approximating measure nouns in their behavior.

(61) a. šloša bakbukey   yayin
        three bottle.PL.BOUND wine
        'three bottles of wine'
     b. šloša litrim/liter/*litrey   yayin
        three liter.PL/liter/liter.PL.BOUND wine
        'three liters of wine'

To sum up, we have proposed that complex cardinals involving addition in Modern Hebrew are built in two different ways: regular coordination plus NP deletion in the case of cardinals 'twenty-one' and above (with the lower simplex cardinals inside them becoming more nominal), and the uncharacteristic (for addition) cascading structure in (55) for 'eleven' through 'nineteen'.

7.4.6 Complex Cardinals Involving Multiplication

As discussed earlier, the Hebrew multiplicands *me'a* 'hundred.F', *elef* 'thousand.M', *milyon* 'million.M', and so on are specified for gender. As the examples in (62) show, the gender of the multiplier is determined by the gender of the multiplicand.

(62) a. šloš   meot   sfarim
        three.F.DPNDNT hundred.F.PL book.M.PL
        'three hundred books'
     b. šlošet   alafim   sfarim
        three.M.BOUND thousand.M.PL book.M.PL
        'three thousand books'

For those cardinals that have a distinct bound form (see table 7.1), this form obligatorily occurs before the multiplicands 'thousand', 'million', and so on (63a), as well as in definite NPs (63b).

(63) a. šlošet   alafim
        three.M.BOUND thousand.M.PL
        'three thousand'
     b. šlošet   ha-sfarim
        three.M.BOUND the-book.M.PL
        'the three books'

Danon (2012) argues that the NPs in (62) are not construct states, on the grounds that the multiplicands are not in their bound form. In particular, despite the fact that 'thousand' has a bound form in the plural, *alfey*, which, as in (64a), surfaces with a "plural of abundance" interpretation (see section 4.1.2), Danon points out that this bound form does not surface in complex cardinals, either with the bound form of the multiplier (64b) or with its free form (64c); only the combination in (62b), with a bound-form multiplier and a free-form multiplicand, is grammatical, leading Danon to argue that the entire complex cardinal is a constituent in the construct state, to the exclusion of the lexical NP—an analysis incompatible with our cascading structure.

(64) a. alfey   sfarim
        thousand.M.PL.BOUND book.M.PL
        'thousands of books'
     b. *šlošet   alfey   sfarim
        three.M.BOUND thousand.M.PL.BOUND book.M.PL
     c. *šloša alfey   sfarim
        three.M thousand.M.PL.BOUND book.M.PL

The evidence, however, is conflicting: while the plural bound form of 'thousand' is indeed unavailable in a complex cardinal, the behavior of the definite article is consistent with the entire NP ('three thousand books') being a construct state: the definite article attaches to the lexical NP, as in (65a), rather than to either of the cardinals (though see note 15). The puzzle for us is why in (65a) only the first element appears in bound form and not both, unlike in other construct states involving iteration (65b–c).

(65) a. šlošet   alafim   ha-sfarim
        three.M.BOUND thousand.M.PL the-book.M.PL
        'the three thousand books'
     b. šiša yemey   tiyul
        six.M day.M.PL.BOUND trip
        'a trip of six days'
        (Glinert 1989, 75)
     c. milxemet šešet   ha-yamim
        war.BOUND six.M.BOUND the-day.PL
        'the Six-Day War'

The question arises whether the lack of a bound form for the higher multiplicands is indicative of a noncomplementation structure (which is what Danon (2012) argues) or a morphological fact about these multiplicands. Evidence for the latter comes from the fact that measure nouns in Hebrew behave like the higher multiplicands in this respect, as discussed in the previous subsection.

We conclude that complex cardinals involving multiplication do enter the construct state, and that the lack of a bound form is a morphological quirk shared with measure nouns. Indeed, just as 'liter' can be optionally singular or plural in (61b), so *elef* 'thousand' can be either singular or plural, depending on the multiplier. While it is plural with multipliers 'three' through 'ten', as in (65a), for multipliers higher than 'ten', *elef* is used in the singular, as in (66a–b) (Glinert 1989, 83). The highest cardinals ('million', etc.) always require the singular form, even with lower cardinals (66c). The singular form is never available for *me'a* 'hundred' in the presence of a multiplier. This provides us with an additional property distinguishing 'hundred' from the highest multiplicands, and lends support to our analysis of 'hundred' as special in requiring the dependent form of the multiplier (see section 7.4.4).

(66) a. axat-asar   elef   sfarim
        one.M-ten.M.DPNDNT thousand.M book.M.PL
        'eleven thousand books'
     b. esrim-ve-šloša   elef   sfarim
        twenty-and-three.M thousand.M book.M.PL
        'twenty-three thousand books'
     c. šloša   milyon   sfarim
        three.M million.M book.M.PL
        'three million books'

In short, the surface form of *elef* changes as a function of the multiplier. While a detailed analysis of these facts would take us too far afield, treating multiplicative complex cardinals as specifiers (Danon 2012) does not bring us any closer to an explanation, as it loses us the link to measure nouns. We conclude that all complex cardinals involving multiplication in Hebrew do have the cascading structure that we propose, contra Danon.

7.4.7 Summary: Properties of Hebrew Cardinals
In this section, we have argued that cardinals in Modern Hebrew are compatible with our analysis of complex cardinals that involve multiplication, according to which they systematically require the head-complement configuration. We have argued, contra Danon (2012), that both free and bound cardinals

appear in the same syntactic configuration. Table 7.2 summarizes the properties of Hebrew cardinals we have discussed in this section.

We have proposed that Hebrew cardinals higher than 'one' are more nominal than adjectival. As table 7.2 shows, Hebrew cardinals show a mixture of nominal and adjectival properties (a mixture familiar from Russian; see chapter 6), with lower cardinals being more adjectival, given that they show (chiastic) agreement. Conversely, the highest cardinals are nominal, given their inherent gender. Having a bound form does not distinguish nominal and adjectival cardinals, just as it does not distinguish adjectives from nouns in general. Importantly, even lower cardinals are not fully adjectival: despite being heads in the construct state, they do not confer AP status on the entire cardinal-containing phrase, even though elsewhere when an adjective heads the construct state, the result is an AP (Hazout 2000; Siloni 2001; Ji-yung Kim 2002).[20] We note that this is a problem for Danon's (2012) approach as well as for ours, even though Danon assumes the construct-state analysis only where a bound form is present, while we assume it for all cardinal-containing NPs (except 'one'); either way, no cardinal-containing NPs are APs. We conclude that lower cardinals in Hebrew are not adjectives: regular adjectives in Hebrew are postnominal and show regular, not chiastic, agreement. Thus, while lower cardinals may have some properties of adjectives, we suggest that they are primarily nominal and hence able to head NP construct states. Just as we saw in Russian (chapter 6), different cardinals in a language vary in how closely they resemble prototypical nouns or adjectives. Finally, we have shown that simplex cardinals become even more nominal when occurring inside an additive complex cardinal, and we have also provided an analysis of the unusual behavior of the complex cardinals 'eleven' through 'nineteen'.

Table 7.2
Summary of the properties of Hebrew cardinals

	'one'	'two'– 'ten'	'eleven'– 'nineteen'	'twenty'– 'ninety'	'hundred'	'thousand'	'million'
Bound form of the cardinal	No	Yes	No	No	Optional	In the plural, not used	No
Lexical NP number	Agreement (SG)	Agreement (PL)					
Gender on cardinal	Chiastic agreement			None	Inherent gender		
Number marking on *elef*	—	Plural	Singular			—	—

7.5 Arabic

Cardinals in Modern Standard Arabic behave in some ways like cardinals in Modern Hebrew, discussed in the previous section; for example, they show chiastic agreement. Unlike Hebrew cardinals, however, cardinals in Standard Arabic assign case to the lexical NP, with the case varying depending on the cardinal, a situation similar to that in Russian (see chapter 6). In sections 7.5.1 and 7.5.2, we discuss the behavior of Arabic cardinals, showing that they fall on a continuum from nominal to adjectival, as in Russian. We consider the behavior of additive complex cardinals in section 7.5.3 and argue in favor of the NP-deletion analysis. In section 7.5.5, we turn our attention to Lebanese Arabic, where, as shown by Ouwayda (2014), cardinals exhibit an interesting pattern of both NP-internal and NP-external agreement.

7.5.1 Case and Number Marking with Cardinals in Standard Arabic

In Modern Standard Arabic, cardinals exhibit several distinct patterns. The cardinals 'one' and 'two' behave like adjectives: they follow the lexical NP, and they agree with it in gender and case, as shown in (67), just like adjectives (68). Masculine is the unmarked gender in Arabic, so in masculine examples like (67c) and (68b), adjectival agreement takes the form of the absence of an overt suffix corresponding to feminine -t-/-at-.[21]

(67) a. bint-un   waahid-at-un   Modern Standard Arabic
       girl-NOM one-F-NOM
       'one girl'

    b. bint-aani   iθn-at-aani
       girl-DU.NOM two-F-DU.NOM
       'two girls'

    c. walad-un waahid-∅-un
       boy-NOM one-M-NOM
       'one boy'

    d. walad-aan   iθn-∅-aani
       boy-DU.NOM two-M-DU.NOM
       'two boys'

(68) a. bint-un   jamil-at-un
       girl-NOM beautiful-F-NOM
       'a beautiful girl'

    b. rajul-un   jamil-∅-un
       man-NOM beautiful-M-NOM
       'a beautiful man'

'Three' and higher, on the other hand, precede the lexical NP and (just like lower cardinals in Hebrew: see section 7.4.2) show chiastic agreement, according to which the gender marked on the cardinal is the opposite of that of the lexical NP. This is illustrated in (69): with a masculine noun like 'man' (69a), the cardinal contains a feminine suffix, whereas with a feminine noun like 'girl' (69b), the cardinal is in the unmarked masculine form (the gender suffixes are clearly visible in Arabic, in contrast to Hebrew; cf. section 7.4.2). As shown, the lexical NP occurring with cardinals higher than 'two' is marked genitive plural. It is possible for these cardinals to occur postnominally as well, although it is a marked option. In the postnominal order, cardinals behave more like adjectives, agreeing with the NP in case rather than assigning case; however, they still exhibit chiastic agreement for number. This is shown in (69c).

(69) a. ?arbaʕ-at-u rijaal-in
         four-F-NOM man.PL-GEN
         'four men'
         (Zabbal 2005)
     b. ?arbaʕ-u banaat-in
         four-NOM girl.PL-GEN
         'four girls'
         (Zabbal 2005)
     c. banaat-un   ?arbaʕ-u
         girl.PL-NOM four-NOM
         'four girls'

Unlike (prenominal) 'three' through 'ten', which require the lexical NP to be marked genitive plural, the cardinals between 'ten' and 'hundred' require it to be marked accusative singular, as in (70a). These cardinals have only one form, and they do not show either regular or chiastic agreement. The cardinals 'hundred' and 'thousand' require the lexical NP to be marked genitive singular, as in (70b–c).

(70) a. xamsuuna rajul-an
         fifty   man-ACC
         'fifty men'
     b. miʔat-u   rajul-in
         hundred-NOM man-GEN
         'a hundred men'
     c. ?alf-u   rajul-in
         thousand-NOM man-GEN
         'a thousand men'

With complex cardinals that involve multiplication, the case and number marking on the lexical NP is determined by the immediately preceding cardinal: *rajul* is genitive singular in (71a–b), as required by 'hundred' and 'thousand', not genitive plural, as required by 'four'; compare (69a). Furthermore, 'four' shows chiastic agreement with the following cardinal, not with the lexical NP: 'hundred' is a feminine noun, so 'four' is masculine in (71a); 'thousand' is a masculine noun, so 'four' is feminine in (71b). 'Thousand' has both singular and plural forms; when following a lower cardinal, it is plural, just like any regular NP (compare the plural in (71b) to the singular in (70c)). The word for 'hundred', on the other hand, does not have a plural form, so it is singular in (71a). Finally, in 'four hundred thousand men', the case and number on 'thousand' (genitive singular), as shown in (71c), is determined by 'hundred', rather than by 'four'.

(71) a. ʔarbaʕ-u miʔat-i   rajul-in
        four-NOM hundred-GEN man-GEN
        'four hundred men'

     b. ʔarbaʕ-at-u ʔaalaaf-i   rajul-in
        four-F-NOM thousand.PL-GEN man-GEN
        'four thousand men'

     c. ʔarbaʕ-u miʔat-i   ʔalf-i   rajul-in
        four-NOM hundred-GEN thousand-GEN man-GEN
        'four hundred thousand men'

The behavior of Standard Arabic complex cardinals involving multiplication is thus consistent with our cascading structure: case and number on the lexical NP as well as on intermediate cardinals is determined by the immediately preceding cardinal.

7.5.2 Formal Properties of Individual Cardinals in Standard Arabic

The cardinals 'thousand' and 'hundred' are clearly nominal—they are inherently specified for gender, they trigger chiastic gender agreement on preceding cardinals, and they assign genitive case to the lexical NP; 'thousand', like other nominals, has both singular and plural forms, which surface according to the requirements of the preceding cardinal. Cardinals above 'ten' and below 'hundred' are less nominal: they do assign (accusative) case to the lexical NP, but they do not bear inherent gender or have a plural form. 'Three' through 'ten' exhibit a mixed nominal-adjectival pattern, reminiscent of that exhibited by the cardinals 'two' through 'four' in Russian (see chapter 6): they are adjectival in that they agree with their sister NP, but unlike adjectives they occur prenominally (as the unmarked option) and exhibit the chiastic pattern

Table 7.3
Properties of cardinals in Standard Arabic

	'one', 'two'	'three'–'ten'	'eleven'– 'ninety'	'hundred', 'thousand'
Case of lexical NP	None	GEN	ACC	GEN
Number of lexical NP	Agreement (SG for 'one', DU for 'two')	Agreement (PL)	SG	SG
Gender on cardinal	Agreement with lexical NP	Chiastic agreement (gender polarity)	None	Inherent gender
Number on cardinal	Agreement with lexical NP	None	None	None
Construct state	No	Yes	Indeterminate	Yes

of agreement. On the other hand, like nouns and unlike adjectives, they assign genitive case. Finally, 'one' and 'two' are fully adjectival and therefore do not assign case. These facts are summarized in table 7.3.[22]

For the case-assignment facts, we assume, as we did for Russian, Finnish, and Inari Sami (see chapter 3), that it is the lexical property of individual cardinals which case they assign. 'Three' through 'ten' and the higher cardinals 'hundred' and 'thousand' are sufficiently nominal to combine with their lexical NP in the construct state (Ouwayda 2014). In Standard Arabic, the construct state is manifested both on the head noun, which appears in the bound form, and on the complement, which appears in genitive case, as shown in (72a) for a cardinal-containing NP and in (72b) for a possessive construction. Following Borer (1999), we assume that the construct state, in standard Arabic and in Hebrew, involves the deletion of the feature [definite] on the head noun, plus case assignment.

(72) a. thalaath-a*t*-u   awlaad-in
        three-F.BOUND-NOM child.PL-GEN
        'three children'
    b. sayyara*t*-u   l-walad-i
        car.BOUND-NOM the-child-GEN
        'the child's car'
    (Ouwayda 2014, 21–22)

The "transdecimal" cardinals (those above 'ten' and below 'hundred'), as seen in (70a) above, do not fit into this pattern. We tentatively hypothesize that the accusative case that they assign to their lexical NPs is an impoverished version of the genitive case, much as the Russian, Ukrainian, and Belarusian paucal is an impoverished version of the genitive (see chapter 6).

Turning now to agreement, the lower cardinals 'one' through 'ten' behave like adjectives in that they agree with the lexical NP for gender, although 'three' through 'ten' are morphologically special in that they take the gender suffixes of the opposite gender. As in Hebrew, chiastic agreement is likely to be a morphological idiosyncrasy of individual lexical items, including cardinals.[23]

Turning now to number marking on the lexical NP, we propose that lower and higher cardinals bear different formal specifications for interpretable number. Lower cardinals ('three' through 'ten') are formally marked with [plural] and bear an uninterpretable feature allowing them to probe their complement for gender. The uninterpretable number feature on the lexical NP is therefore valued as a free rider. In contrast, higher cardinals are either inherently specified for gender (in the case of 'hundred' and 'thousand') or lack the gender feature altogether, and as a result they do not probe and hence do not exhibit number agreement with the lexical NP, which remains singular.

7.5.3 Complex Cardinals Involving Addition in Standard Arabic

With additive complex cardinals, we find that the form of the lexical NP is determined by the rightmost cardinal: in (73a), the lexical NP is genitive plural, as required by 'five', not genitive singular, as would be required by 'thousand'; similarly, in (73b), the lexical NP is accusative singular, as required by 'fifty'. As Zabbal (2005) points out, these data are incompatible with the right-node-raising analysis of coordination, since right-node raising is impossible when the two underlying NPs differ in number and/or case. However, we note that the data are compatible with the NP-deletion analysis that we have proposed: precisely as in Russian, in Arabic, the first instance of the NP is deleted, as schematized in (74).

(73) a. ?arbaʕ-at-u ?aalaaf-in   wa-xams-at-u   rijaal-in
         four-F-NOM thousand.PL-GEN and-five-F-NOM man.PL-GEN
         'four thousand and five men'

    b. ?arbaʕ-at-u ?aalaaf-in   wa-xamsuuna rajul-an
         four-F-NOM thousand.PL-GEN and-fifty   man-ACC
         'four thousand and fifty men'
         (Zabbal 2005)

(74) a. ?arbaʕ-at-u ?aalaaf-in   ~~rajul-in~~   wa-xams-at-u   rijaal-in
         four-F-NOM thousand.PL-GEN man-GEN and-five-F-NOM man.PL-GEN
         'four thousand and five men'

    b. ?arbaʕ-at-u ?aalaaf-in   ~~rajul-in~~   wa-xams-uuna rajul-a
         four-F-NOM thousand.PL-GEN man-GEN and-fifty   man-ACC
         'four thousand and fifty men'

Furthermore, there is evidence (Abdelaadim Bidaoui, pers. comm.) that NP deletion can apply to the NP in either conjunct, as shown in (75): the lexical NP can occur after either conjunct, and its case and number marking is determined by the cardinal that it appears with.

(75) a. ?alf-u   kitaab-in   wa-sita-t-un   ~~kutub-in~~
        thousand-NOM book-GEN and-six-F-NOM book.PL-GEN
        'one thousand and six books'

   b. ?alf-un   ~~kitaab-in~~   wa-sita-t-u   kutub-in
        thousand-NOM book-GEN and-six-F-NOM book.PL-GEN
        'one thousand and six books'

Interestingly, NP deletion in the first conjunct is impossible or at least strange when the complex cardinal ends in 'one'. Recall that 'one' in Arabic has to occur postnominally and exhibits case agreement with the lexical NP (76a) while 'thousand' occurs prenominally and requires the lexical NP to be singular (76b). Similarly, 'a thousand and one books' can have the form in either (76c) or (76d), depending on which conjunct is affected by the NP deletion (however, we do note some consultant disagreement with regard to the grammaticality of (76d)). Another way to say 'a thousand and one books' is to just repeat the lexical NP, without any cardinal 'one', as shown in (76e). Our informants report variable judgments about the case on 'book' in this example: while nominative case appears to be the prescriptively correct form, the genitive—which is rather unexpected in this context—appears to be acceptable to at least some informants.

(76) a. kitaab-un waahid-un
        book-NOM one-NOM
        'one book'

   b. ?alf-u   kitaab-in
        thousand-NOM book-GEN
        'a thousand books'

   c. ?alf-u   kitaab-in wa-~~kitaab-un~~   waahid-un
        thousand-NOM book-GEN and-book-NOM one-NOM
        'a thousand and one books'

   d. ?alf-un   ~~kitaab-in~~   wa-kitaab-un   waahid-un
        thousand-NOM book-GEN and-book-NOM one-NOM
        'a thousand and one books'

   e. ?alf-u   kitaab-in wa-kitaab-un/?in
        thousand-NOM book-GEN and-book-NOM/GEN
        'a thousand and one books'

Note that in Arabic, unlike in Hebrew (see section 7.4.5), 'one' remains postnominal inside an additive complex cardinal (76d). Thus, the NP-deletion analysis is straightforwardly applicable to Arabic, and it does not appear that 'one' undergoes any changes.

7.5.4 Interim Summary: Standard Arabic

To summarize, the behavior of simplex and complex cardinals in Standard Arabic is consistent with our proposal. While we have not discussed predicate agreement with cardinal-containing NPs in Standard Arabic, we note that it is in no way different from agreement with regular NPs: whether the verb is marked singular or plural depends on the surface order, with a postverbal subject triggering singular agreement and a preverbal subject triggering plural agreement (Aoun, Benmamoun, and Sportiche 1994).

In contrast, some dialects of Arabic, notably Lebanese and Palestinian forms of Arabic, do exhibit unique patterns of agreement with cardinal-containing NPs. We turn to this next.

7.5.5 Lebanese Arabic Agreement Facts

As discussed extensively by Ouwayda (2014), in Lebanese Arabic subject–predicate agreement and attributive-adjective agreement, both plural and singular agreement are allowed with an indefinite NP whose cardinal is higher than 'ten'. This is illustrated in (77). On the one hand, with 'three' through 'ten'—which require a plural lexical NP—plural agreement on both adjectives and verbs is obligatory (77a–c); but with cardinals higher than 'ten'—which take a singular lexical NP—singular marking becomes an option, as shown in (77d–e). When the cardinal-containing NP is definite, plural marking is again obligatory (78).

(77) a. tlat/ktiir šeby-een mnazzm-iin wešl-uu.   Lebanese Arabic
        three/many boy-PL   organized-PL arrived-PL
        'Three/many organized boys arrived.'
     b. *tlat/ktiir   šeby-een mnazzam
        three/many boy-PL   organized-∅
     c. *tlat/ktiir   šeby-een xabbar.
        three/many boy-PL   told-∅
     d. tleetiin sabi mnazzam-iin wešl-uu.
        thirty   boy organized-PL arrived-PL
        'Thirty organized boys arrived.'
     e. tleetiin sabi mnazzam   wešel.
        thirty   boy organized-∅ arrived-∅
        'Thirty organized boys arrived.'
     (Ouwayda 2014, 105–106)

(78) a.   t-tleetiin sabi l-mnazzm-iin   wešl-uu.
           the-thirty boy the-organized-PL arrived-PL
           'The thirty organized boys arrived.'
      b.   *t-tleetiin sabi l-mnazzm   wešel.
           the-thirty boy the-organized-∅ arrived-∅
           (Ouwayda 2014, 222)

Ouwayda further shows that number marking is correlated with a difference
in available interpretations. When the verb bears plural agreement, whether
with 'three' (79a), where plural marking is obligatory, or with 'thirty' (79b),
where it is optional, both collective and distributive readings are available. On
the other hand, when the verb does not bear plural marking, as in (79c), only
the distributive interpretation is available. The same effect is observed with
plural-marked and non-plural-marked adjectives, as shown in (80).

(79) a.   tlat   wleed   akal-uu ?aaleb gateau keemel.
           three child.PL ate-PL   pie   cake   whole
           'Three children ate a whole cake.'
           ←Three children each ate one whole cake (three cakes total).
           ←Three children shared one whole cake (one cake total).
      b.   tleetiin walad   akal-uu ?aaleb gateau keemel.
           thirty   child-∅ ate-PL   pie   cake   whole
           'Thirty children ate a whole cake.'
           ←Thirty children each ate one whole cake (thirty cakes total).
           ←Thirty children shared one whole cake (one cake total).
      c.   tleetiin walad   akal   ?aaleb gateau keemel.
           thirty   child-∅ ate-∅ pie   cake   whole
           'Thirty children ate a whole cake.'
           ←Thirty children each ate one whole cake (thirty cakes total).
           ↮Thirty children shared one whole cake (one cake total).
      (Ouwayda 2014, 113–114)

(80) a.   shefet   tleetiin walad   mnazzam-iin.
           saw.1SG thirty   child-∅ organized-PL
           'I saw thirty organized children.'
           ←I saw thirty children, and each was an organized person (distributive).
           ←I saw thirty children who were organized as a group (collective).
      b.   shefet   tleetiin walad   mnazzam-∅.
           saw.1SG thirty   child-∅ organized-∅
           'I saw thirty organized children.'
           ←I saw thirty children, and each was an organized person (distributive).
           ↮I saw thirty children who were organized as a group (collective).
      (Ouwayda 2014, 115–116)

Ouwayda's analysis of these facts attributes to cardinals a separate semantic type (cf. Frege 1884; Hackl 2000; Kennedy 2013, 2015; Rothstein 2013, 2016) and proposes two possible merge positions for them. In the higher position, they are inserted as specifiers of a functional head Q that introduces existential quantification over plural individuals of the relevant cardinality (there exists a plural individual of cardinality *n* such that the predicate applies to each individual part of this plural individual). In the lower position, cardinals are inserted as specifiers of the functional head #, yielding a predicate of plural individuals of the relevant cardinality. Only the latter option is compatible with a collective interpretation and the presence of the definite article.

We note that the interpretational effects observed in Lebanese Arabic are very similar to what is observed for Russian (see section 6.5). There as well, definite cardinal-containing NPs require plural marking on the predicate (cf. Pereltsvaig 2006b). Given that semantic agreement is dependent on definiteness, specificity, individuation, and so on (see chapter 4), the obligatory plural marking on definite cardinal-containing NPs, in Russian as in Lebanese Arabic, is not a surprise. Furthermore, Pesetsky (1982, 85) notes that when the predicate is singular in Russian, the cardinal-containing NP lacks a collective interpretation. This is similar to what we observed above for Lebanese Arabic.

This dichotomy is only manifested with higher cardinals, where number marking on the lexical NP fails (see section 7.5.2). With lower cardinals, where that lexical NP is specified for number, formal number agreement with adjectives and verbs is obligatory. For higher cardinals, on the other hand, where the lexical NP is not specified for number, alternative mechanisms must be employed. In other words, the presence of syntactic number features (as with lower cardinals in Lebanese Arabic, or in the presence of a DP layer) makes number agreement obligatory irrespective of the interpretation. In the absence of syntactic number (as with higher cardinals in Lebanese Arabic, in the case of indefinites), number agreement is optional and correlates with the interpretation.

While agreement failure with a cardinal-containing subject NP accounts for singular marking on the verb, it does not predict the fact that in Lebanese Arabic, pronouns bound by a cardinal-containing NP subject can be singular or plural, with the same consequences for availability of collective versus distributive interpretations. This is shown in (81).

(81) a. nejjaHet   eshriin benet baʕd-ma SellaHet   mashruuʕ-*on*
        passed.1SG twenty girl-∅ after   graded.1SG project-their
        (lli   ʕamal-uu-h lawaHd-on).
        which did-PL-it   to-self-their

'I passed twenty girls after grading their project (which they did on their own).'

←I passed twenty girls after I graded each girl's project (which she did on her own) (distributive).

←I passed twenty girls after I graded their group project (which they did on their own) (collective).

(Ouwayda 2014, 116)

b. nejjaHet   eshriin benet baʕd-ma SellaHet   mashruuʕ-*ah*
   passed.1SG twenty girl-∅ after   graded.1SG project-her

   (lli   ʕamal-uu-h lawaHd-ah).

   which did-PL-it   to-self-her

'I passed twenty girls after grading their project (which they did on their own).'

←I passed twenty girls after I graded each girl's project (which she did on her own) (distributive).

↔I passed twenty girls after I graded their group project (which they did on their own) (collective).

Following Ouwayda (2014), we propose that the number marking on the pronoun is due to agreement. As controversial as this hypothesis is for pronouns, some independent evidence for the uninterpretability of φ-features in bound pronouns comes from examples like (82), originally noted by Barbara Partee (1989) and discussed by Kratzer (1998a), Rullmann (2003), and von Stechow (2003), among others (see Kiparsky and Tonhauser 2012). On the reading in (82a), the person feature on the pronoun is not interpreted.

(82) Only I did my homework.
   a. [only I] λx . x did x's homework
   b. [only I] λx . x did my homework

An apparent argument against an analysis where number marking is linked to interpretation comes from NP-internal agreement patterns, where mixed agreement is possible, as in (83a). We concur with Ouwayda (2014) in suggesting different positions for agreeing and nonagreeing adjectives. Our analysis allows for a straightforward spellout of this intuition. While the nonagreeing adjective is contained within the lexical NP, the plural-marked agreeing adjective is merged higher than the cardinal. As a result, inside the lexical NP, the adjective shows the same number marking as the noun that it combines with, while above the cardinal, adjectives can be optionally singular or plural. Support for this analysis comes from the fact that a singular adjective must be closer to the noun than the plural one (83b), as well as from the obligatory singular marking on syncategorematic adjectives (84), which must be merged below the cardinal.

(83) a. tleetiin telmiiz   mnazzam   kesleen-iin Htajj-uu.
          thirty   student-∅ organized-∅ lazy-PL   complained-PL
          'Thirty lazy organized students complained.'
      b. *tleetiin telmiiz   mnazzam-iin kesleen Htajj-uu.
          thirty   student-∅ organized-PL lazy-∅ complained-PL
      (Ouwayda 2014, 121)

(84) a. tleetiin mhandes   madani
          thirty   engineer-∅ civil-∅
          'thirty civil engineers'
      b. *tleetiin mhandes   madaniy-iin
          thirty   engineer-∅ civil-PL
      (Ouwayda 2014, 122)

Further support for this analysis comes from the facts of Standard Arabic, where, as shown by Ouwayda and Shlonsky (2015), the adjective bears case marking and can agree in case with either the lexical noun (85a) or the cardinal (85b). This again provides evidence that an adjective can be merged either within the lexical NP or above the cardinal.

(85) a. waSala   thalaathat-u muhandisiin saabiqiin.   Standard Arabic
          arrived.3SG.M three-NOM   engineer.GEN former.GEN
          'Three former engineers arrived.'
      b. waSala   thalaathat-u muhandisiin saabiquun.
          arrived.3SG.M three-NOM   engineer.GEN former.NOM
          'Three former engineers arrived.'
      (Ouwayda and Shlonsky 2015, (19))

Consistent with this, Ouwayda and Shlonsky further show that if the adjective closest to the noun agrees with the cardinal (i.e., is marked nominative), then any adjective that is further away from the noun must also agree with the cardinal.

To conclude, the facts of Lebanese Arabic (and the corresponding facts in Standard Arabic) can be explained on our analysis of cardinals as efficiently as on any other.[24]

7.5.6 Arabic Cardinals: Summary

In this section, we have provided evidence that cardinals in Arabic, like those in other languages, are best analyzed as ɸ-defective adjectives and nouns. There are a number of properties of cardinal-containing NPs in Arabic that we have not discussed in detail, such as the construct state with cardinals, and agreement patterns with cardinal-containing NPs in Standard Arabic (see Alqarni 2015). In addition, as noted by Sadler (2010), there is some indeterminacy about the position of the cardinal and the case marking on the lexical NP in definite cardinal-containing NPs (see Ryding 2005, 335–338).

7.6 Conclusion

In this chapter, we have discussed three different languages/language families and have shown that, despite many differences in their systems of complex-cardinal formation, they are all consistent with our proposal. The head-complement agreement in Dagaare complex cardinals provides direct evidence for the cascading structure, while Welsh supports our analysis of additive complex cardinals. Welsh, Arabic, and Hebrew all provide evidence that cardinals fall on a cline from fully adjectival to fully nominal, exactly as we have discussed for cardinals in Russian and Polish (chapter 6). As discussed by Hurford (1975, 1987, 2001, 2003), lower cardinals are more likely to be adjectival while higher cardinals are more likely to be nominal. On the basis of the crosslinguistic data, we maintain the analysis, put forth in chapter 6, that cardinals do not form a special lexical class, nor do they fall into a binary nominal–adjectival divide; rather (with the exception of a few fully adjectival or fully nominal cardinals) they correspond to ϕ-defective nouns and adjectives.

8 The Modified-Cardinal Construction

In this chapter, we investigate the properties of a puzzling English construction (see Jackendoff 1977; Honda 1984; Babby 1985; Gawron 2002; Ohna 2003; Ionin and Matushansky 2004; Solt 2007; Ellsworth, Lee-Goldman, and Rhodes 2008; Maekawa 2013; Keenan 2013; Bylinina, Dotlačil, and Klockmann 2016), which we term the *modified-cardinal construction*, illustrated in (1). While this construction is most frequently found with measure phrases, as in (1a–b), it is also attested with regular nouns, as in (1c–d).[1]

(1) a. We spent a meager two hours in the alien ship.
  b. a mere one hour
    (Ohna 2003, 585)
  c. *Our Betters* ran for a healthy 112 performances.
    (Ted Morgan, *Maugham: A Biography*, 1981: New York, Simon and Schuster, 220)
  d. a career-low one touchdown in eight games
    (Maekawa 2013)

The modified-cardinal construction seems on the surface to pose a challenge to our analysis, since it looks as if the adjective is modifying the complex cardinal to the exclusion of the lexical NP: it is frequently assumed that it is the quantity (e.g., *two* in (1a)) which is meager (or amazing, or healthy, etc.). We will show in section 8.1 that this is not the case, and that both the syntax and the semantics of the modified-cardinal construction point towards the opposite conclusion, namely, that the adjective modifies the entire cardinal-containing NP. In sections 8.2 and 8.3 we will consider the two main characterizing properties of the modified-cardinal construction. The first is the obligatory appearance of the singular indefinite article *a*, even when the entire NP is plural, as shown by the ungrammaticality of (2a). The second is the correlation between the appearance of *a* and the presence of a modifier, shown by the ungrammaticality of (2b).

(2) a. *We spent *meager two hours* in the alien ship.

   b. *We spent *a two hours* in the alien ship.

In this chapter, we focus primarily on exploring the properties of the modified-cardinal construction in English, though we will also, in section 8.4, briefly consider its counterparts in Russian and French.

Our conclusion in this chapter is that the modified-cardinal construction does not present a challenge to our analysis, and that the properties of adjectival modification and indefinite-article insertion follow from independently required assumptions.

8.1 The Syntax and Semantics of the Modified-Cardinal Construction

In this section, we will show that the properties of the modified-cardinal construction are in fact not puzzling at all, but are consistent with the syntax and semantics of cardinals that we propose.

8.1.1 The Syntax of the Modified-Cardinal Construction

The first analysis of this construction in the literature, that of Jackendoff (1977), proposes that the adjective combines first with the cardinal, the adjective-cardinal combination combines with the article, and the article-adjective-cardinal combination combines with the noun. This is illustrated in (3), Jackendoff's structure for *a beautiful two weeks*.

(3)

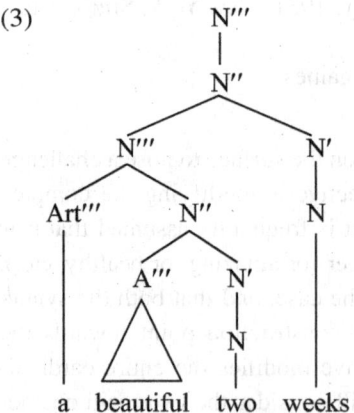

One piece of evidence against the structure in (3) comes from the semantics of examples like (4). The only analysis that attempts to build a compositional semantics of (3), that of Bylinina, Dotlačil, and Klockmann (2016), treats the cardinal as number-denoting, and as composing with the adjective prior to composing with the lexical NP. However, if the adjectives *mere* and *amazing* in (4) both modify the cardinal 'ten', this entails that ten is too few and amazingly many at the same time. Any analysis that attempts to derive the semantics of (3) has to account for examples like (4).

(4) There were a mere *ten* students and an amazing *ten* professors at the department party.

In contrast, if the cardinal combines with the lexical NP before combining the adjective, then (4) states that ten students were too few, but ten professors were amazingly many, precisely the intended reading.[2]

Furthermore, we provide independent syntactic evidence in Ionin and Matushansky 2004 that the adjective in the modified-cardinal construction combines with the entire NP: examples like those in (5). In (5a), for instance, *long* clearly does not modify just *two* but rather the entire bracketed NP.

(5) a. a long [two hours and fifteen minutes]
    b. a beautiful [two weeks and three days]
    c. an amazing [twelve performances and six hundred and two rehearsals]

This means that in *a long ten miles*, *long* combines with *ten miles* rather than with *ten*. Under our analysis of the syntax of cardinals, this is easily captured by the structure in (6) (we leave aside, for now, the location of the indefinite article). The fact that the adjective modifies the entire NP rather than just the cardinal is in fact predicted by our syntactic analysis.

(6)

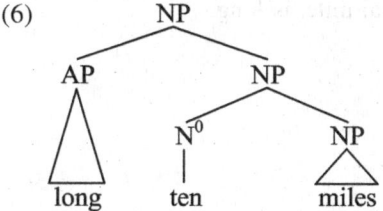

Further evidence in the same direction comes from examples like (7), about which Maekawa (2013) observes that the modifier forcing the presence of the indefinite article can be a relative clause, which does not, on the surface at least, form a constituent with the cardinal.

(7) By the end of the four days, my group and I were ready to leave, but it was [a four days [that we will all look back on with great memories]]. (Maekawa 2013)

There is furthermore independent evidence for the availability of a position for modifying adjectives above the cardinal: examples like (8), where the adjective clearly modifies not only the cardinal but the entire cardinal-containing NP. Given the existence of such definite NPs, the indefinite modified-cardinal construction is in fact expected (although, as we will see in section 8.4, in different languages the syntax of the two constructions may be not altogether the same).

(8) a. the next five years
    b. the first three presidents of the United States

Thus, our analysis solves one puzzle of the modified-cardinal construction: the appearance of an adjective in front of a cardinal. Under our analysis, this is not in fact a puzzle. Nominal cardinals head NPs, which allow adjectival modification; adjectival cardinals are also projected as part of an NP and hence are compatible with further adjectival modification. What makes some adjectives but not others compatible with this construction will be discussed in section 8.2.

8.1.2   The Semantics of the Modified-Cardinal Construction

The semantic composition of the modified-cardinal construction is also quite straightforwardly derived under our analysis. As discussed above, in *a long ten miles*, *long* combines with *ten miles*, rather than with *ten*. If cardinals are of type $\langle\langle e,t\rangle,\langle e,t\rangle\rangle$, this is easily captured, as shown by the tree in (9) and the composition in (10). (In giving *long* type $\langle e,t\rangle$, we are abstracting away from scalarity, intensionality, etc.) As shown by (10c) (and its informal variant in (10d)), this composition successfully accounts for the judgment that the entire unit of ten miles, rather than each individual mile, is long.

(9)

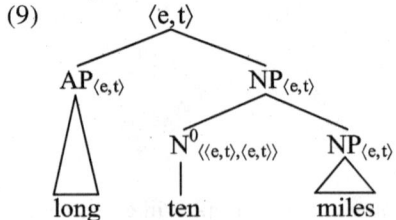

(10) a. $[\![\text{ten miles}]\!] = \lambda x \in D_e \,.\, \exists S = \Pi(x) \, [|S| = 10 \wedge \forall s \in S \, [\![\text{mile}]\!](s)]$
    b. $[\![\text{long}]\!] = \lambda x \in D_e \,.\, x$ is long
    c. $[\![\text{long ten miles}]\!] = \lambda x \in D_e \,.\, \exists S = \Pi(x) \, [|S| = 10 \wedge \forall s \in S \, [\![\text{mile}]\!](s)] \wedge x$ is long
    d. $\lambda x \in D_e \,.\, [x$ is a plural individual divisible into ten nonoverlapping individuals p_i such that their sum is x and each p_i is a mile$]$ and $[x$ is long$]$

Furthermore, the semantics of the modified-cardinal construction provide an additional piece of evidence for the view that cardinals do not have the semantic type of determiners (see chapter 2). If cardinals did have this type, namely $\langle\langle e,t\rangle,\langle\langle e,t\rangle,t\rangle\rangle$, the composition of the modified-cardinal construction would fail under any syntactic analysis. Let's take first Jackendoff's syntactic analysis, under which the cardinal is first combined with the adjective. As shown in (11), Jackendoff's approach is incompatible with giving the cardinal the semantic type $\langle\langle e,t\rangle,\langle\langle e,t\rangle,t\rangle\rangle$, since this would result in the entire NP having type t, which is not a valid type for NPs.

(11)

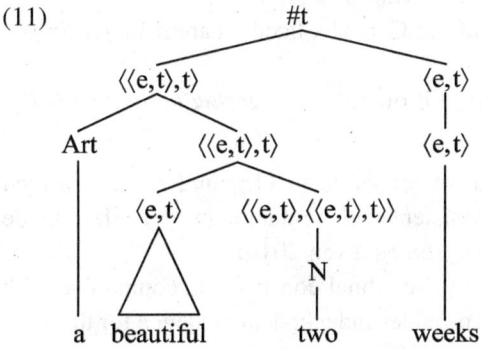

Likewise, our own syntactic analysis, on which the adjective combines with the entire NP containing the cardinal, is incompatible with the determiner type for cardinals. If cardinals have the semantic type $\langle\langle e,t\rangle,\langle\langle e,t\rangle,t\rangle\rangle$, then the NP projection containing a cardinal is a generalized quantifier:

(12) $[\![\text{ten miles}]\!] = \lambda f \in D_{\langle e,t\rangle} \, . \, \exists x \, . \, |x| = 10 \wedge [\![\text{miles}]\!](x) \wedge f(x)$

If the cardinal-noun combination has the semantic type of a generalized quantifier, $\langle\langle e,t\rangle,t\rangle$, it cannot be combined with an adjectival modifier, as shown in (13). Since *long* has the predicative type $\langle e,t\rangle$, it may not combine with something of type $\langle\langle e,t\rangle,t\rangle$ and still produce a legitimate NP.

(13)

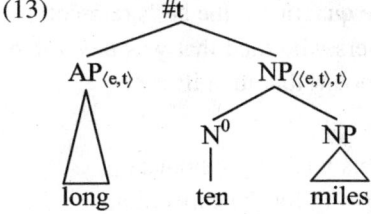

This makes an interesting prediction: NPs that have type $\langle\langle e,t\rangle,t\rangle\rangle$ should be unable to appear inside the modified-cardinal construction. This prediction receives support from NPs containing modified numerals (see chapter 9), which have been argued to be of type $\langle\langle e,t\rangle,t\rangle$ (e.g., Kamp and Reyle 1993, but see Krifka 1999 for an alternative). As shown in (14), some modified numerals cannot in fact appear inside the modified-cardinal construction.[3] At the same time, comparative numerals can appear in it, as in the examples in (15) (found on the internet by an anonymous reviewer).

(14) a.   We walked *a long ten miles.*
    b.   *We walked *a long at least/at most/exactly ten miles.*

(15) a.   The Congressional Budget Office conservatively puts those costs at
       a staggering more than half a trillion.
    b.   Expiration dates on the current Catawba licenses are *a staggering*
       more than two decades away.
    c.   Fourteen great landowners still owned *a staggering more than half*
       the entire country.

The difference between comparative numerals in (15) and other modified numerals in (14) is potentially consistent with semantic analyses that divide modified numerals into two classes (see Nouwen 2010).

    To sum up, not only is the modified-cardinal construction compatible with the semantics that we proposed, it provides independent evidence for the view that cardinals are modifiers rather than determiners.

8.2   Adjectives in the Modified-Cardinal Construction

8.2.1   Quantity versus Quality Interpretations of Adjectives: An Illusion

As noted by Honda (1984) and Solt (2007), the modified-cardinal construction can be interpreted in two ways, which Solt terms the *quality* and *quantity* interpretations, and the interpretation of the entire construction as quality or quantity is directly related to the adjective. The cases in (16) are all examples of the quality reading: here, the adjective characterizes the quality of the referent of the NP. In contrast, the cases in (17) are all examples of the quantity reading. Here, the adjective characterizes the quantity of the NP's referent: for example, in (17d), it is the number of soldiers who died that was incredible; similarly, in (17a), what is meager is the amount that the gift cost.

(16) a.   I walked a long ten miles.
    b.   ?I had to feed a very hungry five cats when I got home.
    c.   We spent a busy three weeks preparing for the expedition.
       (Solt 2007)
    d.   A lucky three students got fellowships.
       (Solt 2007)

(17) a. The gift cost a meager two dollars.
   b. A staggering one hundred chairs were piled in the corner.
   c. The play ran for an amazing three hundred performances.
   d. An incredible eight thousand soldiers died at Gettysburg.
      (Solt 2007)

As pointed out by Solt, the difference between quantity and quality modified cardinals is seen in their distinct entailment patterns, as in (18).

(18) a. I walked a long ten miles. → The ten miles that I walked were long.
   b. A lucky three students got fellowships. → The three students who got fellowships were lucky.
   c. The gift cost a meager two dollars. ↛ The two dollars the gift cost were meager.
   d. A staggering one hundred chairs were piled in the corner. ↛ The one hundred chairs that were piled in the corner were staggering.

Some adjectives, notably *amazing*, allow both readings: for example (19a) can mean that the ten performances were of amazing quality (the quality reading), or that running for as many as ten performances is an amazing feat (the quantity reading). Only on the quality reading does (19a) entail (19b).

(19) a. The play ran for an amazing ten performances.
   b. The ten performances for which the play ran were amazing.

As the example in (20) shows, *good*, like *amazing*, is compatible with both the quantity reading (20a) and the quality reading (20b), with only the latter entailing that the three days were good. However, given the distinct meanings of *good* in (20a–b), it is not clear that we are dealing with the same lexical item.

(20) a. A: How long did it take to drive to Canada?
      B: It was a good three days.
   b. A: What kind of weather did you have on your vacation?
      B: We had a good three days.
      (Honda 1984)

Another difference Solt notes between the two construction types is that quality modified cardinals, but not quantity ones, require the lexical NP to denote a single unit (the "grouping effect" of Bylinina, Dotlačil, and Klockmann 2016). In (21b), unlike its variant in (21a), the five songs have to be sequential. In contrast, in the quantity modified-cardinal construction in (21c), the twelve hours need not be sequential.

(21) a. He played five boring songs, but in between he played one really
         good one.
   b. ??He played a boring five songs, but in the middle he played a
         really good one.
   c.   I spent an incredible twelve hours grading the final exam.

When the lexical NP denotes animate beings, such as people, the quality modified-cardinal construction requires them to be interpreted as a group, as in (22a); if the group interpretation is not possible, the modified-cardinal construction is odd, as shown in (22b). Again, there is no such constraint on the quantity modified-cardinal construction, as seen in (22c–d).

(22) a. We served dinner to a hungry twenty hikers.
   b. ??At various points during the day, a hungry twenty hikers showed
         up wanting to be fed.
   c.   We served dinner to a staggering one hundred and ninety-two
         hikers.
   d.   At various points during the day, a staggering one hundred and
         ninety-two hikers showed up wanting to be fed.

Solt proposes four different structures for the modified-cardinal constructions, depending on whether the meaning is that of quality or quantity and on whether the NP denotes a measure phrase or not. Bylinina, Dotlačil, and Klockmann (2016) propose only two structures, one for quantity and one for quality; in both, the modifier forms a constituent with the cardinal to the exclusion of the lexical NP. We believe that a proliferation of structures is not in fact necessary and that the difference between the quality and quantity constructions can ultimately be traced back to the semantics of the adjectives.

Evidence for a nonsyntactic explanation comes from the fact that adjectives like *amazing* and *staggering* can have the quantity interpretation even when modifying an NP that does not contain cardinals. While the quality interpretation is more common, NPs that denote quantities also allow the quantity reading with such adjectives. This is shown by examples with measure nouns, such as in (23) and (24), as well as with group nouns, as in (25).

(23) a. Over the summer the three groups lost a *staggering ton* in weight
        between them.
   b. Yet I truly feel that, as cliche-ish as it may sound, what they
        contributed to this benchmark epic which ran for *an amazing dozen*
        original years and is …
   c. Bertha the elephant has shed *an incredible TON* of flab.
   (Found via Google)

(24) a whole bunch of bananas
   (Honda 1984)

(25) a. In the past five years Ultimate Fighting Championship has gained
      enormous popularity and sales. Its audience is now *a staggering
      group* of all ages.
   b. A *staggering crowd* of more than four thousand strong braved the
      rain to watch the Holiday Train band in Hamilton last night.
   c. The Bible prophesies that *a staggering army* of two hundred million
      demons will be released from the "bottomless pit" in the last days.
   d. The Cobras started as *a meager group* of five.
   e. What has started as *a meager army* of twelve thousand men (three
      battalions) ...
   f. Porn for Mobile Devices Achieves *an Amazing Crowd*
   g. What *an AMAZING crowd*—thirteen thousand six hundred and
      thirty-six fans cheering our ladies on, including an incredible two
      thousand five hundred and six UA students.
   h. Yesterday it was reported that Cooper witnessed *an incredible
      crowd* of Mubarak supporters surge across a no-man's land.
   (Found via Google)

In the case of group nouns, the quality and quantity readings cannot be
attributed to different structures without postulating some invisible functional heads, for which there is no independent motivation. This means that
the emergence of the quantity reading with cardinals should be attributed
to semantics rather than to syntax. Evaluative adjectives ultimately evaluate properties of the individual denoted by the NP. The individuals in the
extension of a group noun like *crowd* have many properties: a crowd is big,
it is energetic, it is loud, and so on. As a result, *amazing* can evaluate any of
these properties: the crowd might be amazing because of its size, because of
its energy, or because of the noise level that it produces. Conversely, *meager*
can only evaluate the size of the crowd, which makes examples like (25e)
unambiguous, unlike (25g), for example. The choice of the noun also matters:
measure nouns, as in (23), strongly prefer the quantity interpretation, probably because their denotata do not have any properties besides their size (or
volume/length/etc.).

If we do not postulate structural ambiguity for group and measure nouns
modified by adjectives, there is no reason to postulate such ambiguity for
the modified-cardinal construction. Indeed, cardinal-containing NPs like *two
hundred hikers* can be treated along similar lines. Two hundred hikers can
be amazing either because the individual hikers bear some amazing property

(e.g., they have a great deal of endurance), or else because of their number (two hundred hikers is a lot to have around). Conversely, a measure phrase such as *twenty meters*, just like the measure noun *ton*, preferentially gives rise to the quantity interpretation. Adjectives like *mere* or *meager* force the quantity reading for cardinal-containing NPs, as in (26), just as they do for group nouns.

(26) a. a mere forty-six years
     b. a miserable seventy-five pounds
     c. a bare six inches
     (Honda 1984)

Ultimately, we argue that the distinction between quality and quantity interpretations of the modified-cardinal construction is not structural: *amazing* modifies some property of the individual, and quantity happens to be one such property. Note that other evaluative adjectives, like *staggering* and *meager*, can only evaluate quantity: *a staggering army* is staggering because of its size, and the same is true about *a staggering twenty hikers*. There is no "quality" reading in this case. Whether a given adjective has only a "quality" reading (e.g., *hungry*), only a "quantity" reading (e.g., *staggering*), or both (e.g., *amazing*) is determined by the semantics of the adjective. Separate structures for the quality and quantity readings are redundant.

8.2.2 Modification of Groups versus Individuals

We turn next to Solt's (2007) observation that the quality reading, but not the quantity reading, requires a sequential or group interpretation (see (21)–(22)). In line with our intuition that the source of the two readings is the adjective rather than the structure, we propose to rely on the fact that properties can characterize atoms, groups, or plural individuals. As illustrated in (27), there is a three-way distinction between (i) quantity adjectives such as *staggering*, which cannot combine with sets of atomic individuals; (ii) quality adjectives such as *hungry*, which can apply to atomic individuals and to groups; and (iii) stubbornly distributive adjectives (Schwarzschild 2011) such as *tall*, which only apply to atomic individuals.

(27) a. a staggering crowd/*person/two hundred people
     b. a hungry crowd/person/two hundred people
     c. a tall person/*crowd/*two hundred people

We propose that the difference between (27a) and (27b) is that quantity adjectives like *staggering* must apply to plural individuals, while quality adjectives like *hungry* apply to atoms or groups. Group nouns such as *crowd* denote sets of groups (viewed here as a type of atom; see Landman 2000), but can also be interpreted as sets of plural individuals (via Landman's↓ operator). This makes them compatible with both *hungry* and *staggering*. On the other hand, *two hundred people* denotes a plural individual and hence is compatible with quantity adjectives with no further stipulations. It can also be converted to a group reading via Landman's ↑ operator, which allows it to appear with adjectives like *hungry*. The grouping effect seen in (21)–(22) arises because conversion to a group is only possible for those sets of plural individuals that are contiguous in space and/or time (cf. Bylinina, Dotlačil, and Klockmann 2016). We note here that bare plurals are incompatible with adjectives like *staggering*; if bare plurals are kind-denoting (Carlson 1977a; Chierchia 1998; etc.), this incompatibility follows.

Stubbornly distributive adjectives, however, cannot normally combine with either groups or plural individuals, as illustrated in (28). Such adjectives denote properties such as size, color, and shape, which are properties of atomic individuals.

(28) a. ??I met a tall five people the other day.
   b. *She bought a blue six pencils.
   c. ??I bought a rectangular five tables.
   d. ??We fed a big five hikers the other day.

To sum up, what we are proposing is the following. A cardinal-containing NP normally denotes a set of plural individuals. When such an NP is combined with a quantity adjective like *staggering*, no complications arise. But when it is combined with a quality adjective like *hungry* or *tall*, the question arises whether groups are in the extension of the denotation of the adjective. If they are, as with *hungry*, the plural individual is converted into a group, and the grouping effect noted by Solt (2007) arises; if not, as with *tall*, the outcome is ungrammatical. In other words, the quantity versus quality readings are traceable to the semantics of the adjective and are all captured via the structure in (9).

8.3 Article Insertion in the Modified-Cardinal Construction

8.3.1 Articles with Cardinal-Containing NPs

A major puzzle concerning the modified-cardinal construction is article insertion in the presence of modification (*a long ten miles*). The appearance of an article per se is not a puzzle at all: like other nouns, cardinals co-occur with determiners. In fact, cardinals can co-occur with a variety of determiners (cf. Zamparelli 2004):

(29) a. the twenty people
     b. every seven days
     c. these five books

However, it is also a fact that (except for semilexical cardinals like *hundred*, discussed below) unmodified cardinals do not occur with the indefinite article (**a twenty books*). As discussed in chapter 2, on our proposal, the existential force of a cardinal-containing indefinite NP results from existential closure or from a choice-function variable; the head introducing it is null with most cardinals but surfaces as *a* with semilexical cardinals as well as in the modified-cardinal construction. In other words, *a* is inserted for purely syntactic reasons, it does not introduce a semantic operation. We also note that there are other areas of language in which absence versus presence of an article appears to have no effect on semantics. For example, in some dialects of Italian, female proper names can appear with or without a definite article: *Maria* or *la Maria* (cf. Longobardi 1994, 1999, et seq.; Matushansky 2008c). Assigning different semantics to proper names depending on whether they require a phonologically overt article or not seems rather unwelcome. We see no reason why the overt/covert character of the indefinite article shouldn't also be treated as a purely syntactic phenomenon.

As discussed in chapter 2, the indefinite article appears obligatorily when the leftmost cardinal is one of the so-called semilexical cardinals (*hundred, dozen*, etc.), exemplified in (30), as well as in the modified-cardinal construction, as in (31a). Nothing comparable happens with English bare plurals, as (31b) shows (but see Bennis, Corver, and Den Dikken 1998 for indefinite-article insertion in Dutch plurals).

(30) a. *(a) hundred/thousand/million/dozen books   Semilexical cardinals
     b. (*a) twenty/thirty/five/twelve/one thousand
        books

(31) a. a stunning one thousand/twenty-five books   Modified cardinals
     b. (*a) (stunning) books   Bare plurals

We note furthermore that whereas a semilexical cardinal can optionally appear with either *a* or *one* (32a), it can no longer appear with *a* inside the modified-cardinal construction (32b), inside a complex cardinal (32c), or after determiners in general (32d) (see also Ellsworth, Lee-Goldman, and Rhodes 2008). This provides evidence that *a* does not form a constituent with the modified cardinal and is inserted above the NP level, possibly in D^0. Further evidence for this is that the modified-cardinal construction can take other determiners besides *a*, as shown in (33).

(32) a. a/one hundred books
    b. a stunning one/*a hundred books
    c. one thousand one/*a hundred books
    d. the one/*a hundred books that I read

(33) a. the/that long five miles that I walked
    b. my lucky two friends who won the fellowship
    c. those busy nine days of snow we had
      (Maekawa 2013)

To simplify matters, we assume that all determiners, including *a*, are inserted in the D position. We have already argued in chapter 2 that cardinal-containing NPs like *three books* combine with a choice-function operator in D. Its realization as an overt indefinite article or as a null element depends on the nature of the cardinal, with only "semilexical" higher cardinals like *hundred, thousand*, and so on requiring an overt D.

A potential problem for the view that *a* is inserted in D comes from the correlation between the presence of *a* with a coordinated cardinal and the presence of a semilexical cardinal inside it, as shown in (34).

(34) a. a hundred and two books
    b. a thousand and five hundred thirty-five books
    c. *a twenty-two books
    d. *a thirty-five books

The appearance of *a* here is determined by the leftmost simplex cardinal: as we saw in (30), *a* appears with *hundred, thousand*, and so on and not with *twenty, thirty*, and so on, and (34) shows the same pattern. Now, if *a* is merged at a level where the cardinal-containing NPs are no longer interpreted as predicates, then it must be merged above the coordinated NP *hundred and two books* in (34a) (see chapter 5 for more discussion of coordination). This means that a particular property of one of the conjuncts (the requirement that *a* be inserted) affects the entire coordinated NP. This is expected on the analysis of case assignment in Inari Sami and Romanian complex cardinals that we argued for in chapter 5. Since, as we proposed, the lower cardinal inherits the properties of the higher cardinal, the entire additive complex cardinal behaves like the higher cardinal. In English, this is manifested in *a* insertion only when the left conjunct is a cardinal that requires *a* insertion, such as *hundred* or *thousand*. As a side note, we point out that the facts in (34) provide another argument against a theory that treats complex cardinals as morphological compounds. Such a theory would have to explain why, despite the fact that English compounds have their heads on the right, it is properties of the leftmost lexical item inside the compound that determine whether article insertion takes place. We now turn to *a* insertion in the presence of adjectival modification.

8.3.2   The Effect of Modification on Article Insertion

One possible account for the absence of the indefinite article with lower cardinals and its reemergence with modification can be drawn from Matushansky 2006. Matushansky proposes the operation of *m-merger*, which rearranges the syntactic tree by collapsing two heads into one when they are found in a local configuration with no intervening elements, as illustrated in (35).

(35)

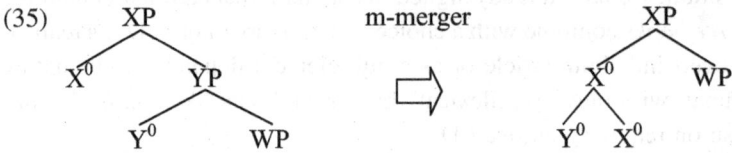

If D^0 and N^0 are not structurally separated by a modifier, m-merger can occur; like other morphosyntactic operations, it can be conditioned by individual lexical items (in this case, applying to *twenty* but not *hundred*). Further support for this approach comes from examples like (7), repeated here, where a relative clause optionally triggers the indefinite-article insertion.

(7)   By the end of the four days, my group and I were ready to leave, but it was [a four days [that we will all look back on with great memories]]. (Maekawa 2013)

What is puzzling, however, is the fact that this indefinite-article insertion is generally *not* triggered by postmodification by relative clauses or PPs, as shown in (36). Even when a relative clause clearly attaches above the NP level, as shown by the availability of cumulative readings in (37), indefinite-article insertion is still not triggered (37b). We leave this as an open question.

(36)   a.   We spent (*a) two hours *that were amazing* in the alien ship.
       b.   I walked (*a) five miles, *which were long*.
       c.   She forgot to return (*a) three books *from the library*.

(37)   a.   The two hunters *who shot three pheasants* won a prize.
       b.   (*A) two hunters *who shot three pheasants* won a prize.

We also note that there is one context where *a* surfaces with a cardinal-containing NP despite the fact that the modifier does not on the surface intervene between the article and the cardinal: with degree fronting, as in (38b) (see Matushansky 2002).

(38) a. More arduous a road I have never walked.

   b. More arduous a ten miles I have never walked.

Finally, we note that article insertion in the presence of adjectival modification is part of a larger phenomenon of article insertion in the presence of modification. This phenomenon is also seen in such places as direct quotations, as the examples in (39) show.

(39) a. He responded with *(an) absent, "What do you want, Grew?"

   b. He responded with (??a), "What do you want, Grew?"

We now propose that the insertion of the indefinite article both with coordinated cardinals (34a–b) and in the modified-cardinal construction reflects the more nominal nature of the complex NP resulting from coordination and modification, respectively. The same intuition accounts for the use of the indefinite article with direct quotations, as in (39). Following Hurford (1975), we hypothesize that higher cardinals are more "nominal" than lower cardinals (see also chapter 6 on Slavic languages). For English, the hypothesis that *hundred*, *thousand*, *million*, and so on are more nominal is circular: the only diagnostic that we have for this property in the English case is the insertion of the indefinite article with unmodified-cardinal-containing NPs. From the syntactic point of view, being more nominal should correspond to a syntactic feature, which can be lexically specified or potentially assigned by the syntactic environment. We propose therefore that NPs headed by less nominal elements can become more nominal because of the features introduced by their modifiers. Adjectives have to agree (covertly in English) in some ϕ-features that nouns bear; in the case that the head of the NP does not have a feature that the adjective is specified for, agreement failure ensues, and the adjective acquires the default value of the feature (cf. Preminger 2011). As a result, the entire NP containing the adjective then becomes specified for the default value of the feature. As an example, let's suppose that the relevant feature in English is [αNumber], and that cardinals are not specified for this feature with the exception of *hundred*, *thousand*, *million*, and so on. Suppose that *a* insertion is sensitive to the presence of the Number feature. For *hundred*, *thousand*, and so on, *a* insertion therefore takes place. For all other cardinals, *a* insertion occurs only in the presence of modification. Assuming that adjectives in English covertly agree for number, by the above reasoning, a modified-cardinal-containing NP bears the default (singular) value of the [αNumber] feature, and this results in *a* insertion.[4]

This analysis explains why a singular article is used with a plural cardinal-containing NP; it also formalizes the intuition that *a* insertion correlates with the cardinal being more "nominal": more "nominal" means, in the case of English, bearing a Number feature.

8.4 Modified Cardinals in Other Languages

While in English the modified-cardinal construction is highly productive, this is not the case in other languages, including Russian and French.

In Russian, there is a small class of adjectives, termed *prequantifiers* by Babby (1985, 1987; see also Crockett 1976; Corbett 1979), that can appear in front of a cardinal-containing NP, as shown in (40). Unlike demonstratives, possessives, quantifiers, ordinals, sequence adjectives, and specificity-inducing adjectives, which agree in case with the entire NP (41), prequantifiers obligatorily appear in genitive case with cardinals higher than 'one'. (A prior stage of the Russian language allowed nominative case on the prequantifier, but that is no longer the case in modern Russian.) All prequantifiers give rise to quantity readings only, which clearly makes them a special case of the modified-cardinal construction.

(40) a. On s'el   celyx/dobryx/*porazitel'nyx   Russian
        he eat.M.PAST whole/kind/amazing.PL.GEN
        dvadcat'   kotlet.
        twenty.ACC/NOM patty.PL.GEN
        'He ate a good/a whole twenty meat patties.'
    b. On s'el   kakix-to/žalkix/*ničtožnyx
        he eat.M.PAST some/pitiable/miserable.PL.GEN
        tri   kotlety.
        three.ACC/NOM patty.PAUC
        'He ate a mere/a meager three meat patties.'
    c. ne-polnyx/bityx   tri   časá
        NEG-full.PL.GEN/beaten.PL.GEN three.ACC/NOM hour.PAUC
        'an incomplete/a damned three hours'

(41) èti/každye/pervye/predyduščie/kakie-to/opredëlennye
     this/every/first/previous/some/specific.PL.NOM
     dvadcat'   kotlet
     twenty.ACC/NOM patty.PL.GEN
     'these/every/the first/the previous/some/the specific twenty meat patties'

Babby (1987) proposes that prequantifiers form a constituent with a cardinal to the exclusion of the lexical NP. This is not compatible with our analysis of complex cardinals. Two questions arise: how do prequantifiers compose with cardinal-containing NPs, and why are they in the genitive case? With regard to the second question, we refer the reader to Matushansky and Ruys 2015a. With regard to the first, the Russian construction in (40) is much less productive than the modified-cardinal construction in English and receives only the quantity reading. We propose that this is because in Russian, the modified-cardinal construction is only possible when the cardinal-containing NP functions as a measure phrase. There are two pieces of evidence for this analysis. First, only a very small list of adjectives can function as premodifiers, and when combined with a measure noun in the singular, they receive the same interpretation as in the modified-cardinal construction (42). Second, the indefinite adjective *kakoj-to* 'some' does not require a quantity reading (43b–d) except with cardinal-containing NPs (43a) and measure phrases (42b).

(42) a. Ona provela   tam celyj   den'.
        she spend.F.PAST there whole.NOM day.NOM
        'She spent a whole day there.'

    b. Za kakuju-to nedelju   v gorod prišlo   nastojaščee leto.
       for some.ACC week.ACC in city   come.N.PAST real   summer
       'In a mere week, real summer had come to the city.'

    c. My proždali   v priemnoj u doktora   bityj   čas.
       we wait.PL.PAST in reception at doctor.GEN beaten.GEN hour.GEN
       'We waited in the doctor's waiting room for a whole damned hour.'

    d. Ona vylila   na nego dobryj   litr   duxov.
       she pour.F.PAST on him kind.NOM liter.NOM perfume.PL.GEN
       'She poured a good liter of perfume on him.'

(43) a. On s'el   kakix-to   tri   kotlety.
        he eat.M.PAST some.PL.GEN three.ACC/NOM patty.PAUC
        'He ate a mere three patties.'

    b. On s'el   kakie-to   tri   kotlety.
       he eat.M.PAST some.PL.NOM three.ACC/NOM patty.PAUC
       'He ate an unspecified three meat patties.'

    c. On s'el   kakie-to   kotlety.
       he eat.M.PAST some.PL.NOM patty.PL.NOM
       'He ate some patties.'

    d. On poel   kakix-to   kotlet.
       he eat.M.PAST some.PL.GEN patty.PL.GEN
       'He ate part of some patties.'

We propose that the Russian facts are directly tied to the special syntactic status of measure-denoting cardinal-containing NPs in Russian, as discussed in Matushansky and Ruys 2015a,b (see section 6.5). Measure-denoting cardinal-containing NPs have two distinct syntactic properties. First, they fail to trigger agreement on the verb, as discussed in section 6.5 and illustrated again in (44).

(44) a. Prošlo/*prošli pjat' let.
        went.N.SG/PL five years
        'Five years passed.'

   b. Na ètot pirog nužno/*nužny   tri   stakana muki.
        on this pie   necessary.N.SG/PL three glasses flour.GEN
        'This pie requires three cups of flour.'

Second, measure-denoting cardinal-containing NPs are obligatorily inanimate. Russian has grammaticalized the animacy distinction, which is reflected in accusative-case realization for masculine nouns ending in a consonant (a.k.a. the second declension) and for all plurals. Lacking a dedicated accusative form, these nouns appear with surface genitive case if they are animate and with surface nominative case if they are inanimate. The animacy distinction can be detected in cardinal-containing NPs if they are headed by 'two', 'three', or 'four', as illustrated in (45a). In contexts where cardinal-containing NPs clearly denote measures, exemplified in (45b) by a differential, they uniformly behave as inanimates even when the lexical NP is headed by an animate noun.

(45) a. uvidet' tri   čexla   / trëx   čelovek
        see.INF three.ACC/NOM cover.PAUC three.ACC/GEN person.PL.ACC/GEN
        'to see three covers/people'

   b. Apel'siny končilis'   za četyre   čeloveka do   menja.
        orange.PL finish.PL.PAST for four.ACC/NOM person.PAUC before me
        'Oranges ran out four people before my turn.'
        (Mel'čuk 1980)

Following Matushansky and Ruys (2015a,b), we predict that Russian cardinal-containing NPs occurring with prequantifiers should pattern with measure-denoting NPs on both counts: they should obligatorily appear with default (neuter-singular) agreement, and they should exhibit the inanimate pattern of accusative-case syncretism. However, a search of the Russian National Corpus for *celyx* 'whole.GEN.PL' followed by a form of 'two'/'three'/'four' followed by an animate noun proved inconclusive on this point. The search turned up extremely few relevant examples; among those few examples, we found both patterns of accusative-case syncretism, as shown in (46).

(46) a. Da bud'   ona xot' čertova poddannaja i   imej
yes be.IMPER she even devil's.F subject.F   and have.IMPER
celyx   *tri*   *materi,*   ona budet moeju ...
whole.PL.GEN three.ACC/NOM mother.PAUC she be.FUT my.F.INS
to est' ona budet kamer-jungferoj   moim   plemjannicam,
that is she be.FUT lady-in-waiting.INS my.PL.DAT nephew.PL.DAT
slyšal?
hear.M.PAST
'Even if she were the devil's own subject, and even if she had *a
whole three mothers*, she will be mine ... that is, she will be the
chambermaid to my nephews, you hear?'
(Nikolaj Gejnce, *Knjaz' Tavridy*, Moscow: Terra, 1994; first
published 1898)

b. Putin imeet   v Bannom   pereulke *celyx*   *dvux*
Putin have.3SG.PRES in of-baths.LOC lane.LOC whole.PL.GEN two.GEN
predstavitelej.
representative.PL.GEN
'Putin has in the Lane of the Baths *a whole two representatives*.'
(Mihail Rostovskij, "Rossija-rodina klonov," *Moskovskij
Komsomolec*, January 14, 2003)

The reason for so few cases of animate lexical NPs inside cardinal-containing NPs with *celyx* is that the vast majority of lexical NPs in such cases are headed by measure nouns such as 'year', 'minute', and so on. This provides further confirmation of the intuition that prequantifiers yield (or are combined with) measure NPs.

In order to investigate agreement with prequantifiers, we searched the Russian National Corpus for the sequence singular verb + *celyx* + numeral and the sequence plural verb + *celyx* + numeral. Unfiltered results yielded 445 and 144 documents, respectively, with the two agreement patterns, suggesting that while default singular marking (47a) is more common, plural marking on the verb (47b) is also possible. The availability of both agreement patterns was previously noted by Mel'čuk (1985). We hypothesize that plural agreement arises when the NP is specific in some relevant sense; we leave this issue open for further research.

(47) a. V karmane *ležalo*   celyx   sorok   šest'
in pocket.LOC lie.N.PAST whole.PL.GEN forty.NOM six.NOM
polnovesnyx   sovetskix   rublej.
sterling.GEN.PL Soviet.GEN.PL ruble.GEN.PL
'In the pocket there lay a whole forty-six sterling Soviet rubles.'
(Tat'jana Solomatina, *Bol'šaja sobaka, ili "Èklektičnaja živopisnaja vavilonskaja povest' o zarytom,"* Moscow: Eksmo, 2009)

   b. V mašine *sidjat*   celyx   dva   djad'ki,   a   ne
in car.LOC sit.3PL.PRES whole.PL.GEN two.NOM uncle.PAUC and not
odin   novyj   naš   avtoslesar'   Andrej.
one.NOM new.NOM our.NOM auto-mechanic.NOM Andrew
'In the car there sit a whole two guys, and not our one car mechanic Andrew.'
(Tat'jana Solomatina, *Akusher-HA! Bajki*, Moscow: Eksmo, 2010)

Given the above facts, we hypothesize that the Russian modified-cardinal construction is only possible when the cardinal-containing NP is interpreted as a measure. Given the existence of syntactic correlates of this interpretation (agreement and animacy) as well as the genitive-case marking on the adjective, we need to assume that the measure interpretation of Russian cardinal-containing NPs is itself syntactically encoded rather than achieved as a result of a meaning shift.

A similar effect can be observed in French, where the adjective *bon* 'good' receives a quantity interpretation when it precedes a cardinal or a measure noun (48b). The same is true for *good* and *healthy* in English (48a).

(48) a. a good liter of wine/a good twenty people
   b. Certaines personnes peuvent   être   contagieuses   French
certain.PL person.PL can.3PL.PRES be.INF contagious.F.PL
un bon   deux semaines après leur rétablissement.
a   good two week.PL   after their recovery
'Certain persons can be contagious a good two weeks after their recovery.'

Importantly, the modified-cardinal constructions in Russian and French are not structurally different from the one in English. The reason for the more restricted distribution in Russian and French is the requirement that the lexical NP be measure-denoting.

8.5 Conclusion

In this chapter, we have shown that the modified-cardinal construction provides further evidence in favor of the syntax and semantics that we propose for complex cardinals. After demonstrating that neither the adjective nor the indefinite article forms a constituent with the cardinal to the exclusion of the lexical NP, we discussed the sources of the quantity interpretation of the adjective as well as of indefinite-article insertion. We have also shown that the modified-cardinal construction is not unique to English but exists in other languages as well. While in Russian it requires a measure interpretation, this is not the case for English, where the construction is fully productive.

9 The Syntax of Modified Numerals

In this chapter, we consider NPs that contain so-called *modified numerals*: comparative numerals (as in *more/fewer than two books*), superlative numerals (as in *at least/at most two books*), and prepositional numerals (as in *about ten books*, *between two and four books*). As we do not treat the combination of comparative/superlative/preposition and numeral as a unit to the exclusion of the lexical NP, we use the above terms as mere shorthand for "constructions containing a comparative/superlative/prepositional element followed by a cardinal numeral and a lexical NP."

As mentioned in chapter 1, there is a rich body of literature on modified numerals in both theoretical linguistics and psycholinguistics, going back at least to Krifka 1999. This literature includes such topics as the scopal interactions of modified numerals and their interactions with modal operators (e.g., Heim 2000; Hackl 2000; Büring 2008); collective versus distributive interpretations of them (e.g., Buccola and Spector 2016); and the ignorance implicatures and other inferences generated by them (e.g., Geurts and Nouwen 2007; Geurts 2010; Cummins and Katsos 2010; Geurts et al. 2010; Nouwen 2010, 2015; Schwarz, Buccola, and Hamilton 2012; Coppock and Brochhagen 2013; Westera and Brasoveanu 2014; Alexandropoulou 2015; Kennedy 2015; Marty, Chemla, and Spector 2015; Alexandropoulou et al. 2016; Schwarz 2016a,b). It is beyond the scope of the present work to consider the consequences of our proposal for all of these linguistic phenomena: doing so would probably result in an additional monograph.

Here, we focus on the internal constituency of modified numerals, since this is the issue that relates most directly to our proposal. As our focus is on the structure of modified numerals, we do not have anything to say concerning the entailments and/or implicatures generated by them. In section 9.1, we will discuss the syntax of modified numerals, and show that they provide much independent crosslinguistic evidence in favor of our analysis of cardinals. We show that while some modified numerals, including superlative numerals, appear in an adjunction structure, comparative and prepositional numerals require complementation. In sections 9.2 and 9.3 we will discuss the structure of comparative and prepositional numerals, respectively, in more depth.

9.1   Internal Structure of NPs Containing Modified Numerals

In Generalized-Quantifier Theory (Barwise and Cooper 1981; Keenan and Stavi 1986), modified numerals are analyzed as complex determiners, with the bracketing in (1).

(1) *Complex-determiner analysis*
    a. [[more than ten] books]   Comparative numeral
    b. [[at least ten] books]   Superlative numeral
    c. [[about ten] books]   Prepositional numeral

However, the semantics that we have proposed for cardinals requires them to have the bracketing structure in (2): the cardinal must combine with the lexical NP before combining with the modifier. The bracketing in (2) has been argued for on independent grounds by Geurts and Nouwen (2007) and Arregi (2010) (cf. Kadmon 1992).

(2) *Modified-NP analysis*
    a. [more than [$_{NP}$ ten books]]   Comparative numeral
    b. [at least [$_{NP}$ ten books]]   Superlative numeral
    c. [about [$_{NP}$ ten books]]   Prepositional numeral

In this section, we provide syntactic evidence in favor of the bracketing in (2) and against the bracketing in (1): this evidence comes from word-order facts in Hebrew and in Basque (based on Arregi 2010) and from case-assignment facts in Russian.

9.1.1   Word Order with Modified Numerals
Arregi (2010) provides syntactic evidence in favor of the modified-NP bracketing for comparative numerals in (2a) over the complex-determiner bracketing in (1a), drawn from facts of word order in Hebrew and Basque. In Hebrew, the numeral *exad* 'one' follows the lexical NP (3a), while other numerals precede it (3b) (see also section 7.4). This paradigm is preserved in comparative numerals, as shown in (3c–d).

(3) a. Dani kana   sefer exad.   Hebrew
        Dani buy.PAST book one
        'Dani bought one book.'

   b. Dani kana   shney sfarim.
      Dani buy.PAST two   book.PL
      'Dani bought two books.'
   c. Dani kana   yoter mi-sefer   exad.
      Dani buy.PAST more from-book one
      'Dani bought more than one book.'
   d. Dani kana   yoter mi-šney   sfarim.
      Dani buy.PAST more from-two book.PL
      'Dani bought more than two books.'
  (Arregi 2010, (13)–(14))

Arregi argues that the paradigm in (3) provides two separate arguments in favor of the bracketing in (2a) over the one in (1a). First, the fact that word order in comparative numerals parallels that in unmodified numerals shows that the syntactic relationship between the numeral and the lexical NP (which in our analysis is a head-complement relationship) is preserved in comparative numerals. Second, the bracketing in (1a) would have to account for why the complex determiner 'more than one' is discontinuous in (3c). The same problems arise for the complex-determiner bracketing of superlative numerals in (1b), as shown in (4).

(4) a. Dani kana   le-faxot sefer exad.
      Dani buy.PAST to-less book one
      'Dani bought at least one book.'
   b. Dani kana   le-faxot šloša sfarim.
      Dani buy.PAST to-less three book.PL
      'Dani bought at least three books.'
  (Nora Boneh, pers. comm.)

Basque further supports the argument made on the basis of Hebrew. First, since Basque is a head-final language, comparatives have the order shown in (5a). With unmodified numerals, *bat* 'one' follows the lexical NP (5b), while other numerals precede it (5c).

(5) a. Jonek   Patxik   baino liburu gehiago irakurri   Basque
      John.ERG Patxi.ERG than book more.ABS read
      d-u-∅.
      3.ABS-have-3SG.ERG
      'John has read more books than Patxi.'
   b. Liburu bat   erosi d-u-t.
      book one.ABS bought 3.ABS-have-1SG.ERG
      'I have bought one book.'

c. Hiru liburu   erosi   d-u-t.
three book.ABS bought 3.ABS-have-1SG.ERG
'I have bought three books.'
(Arregi 2010, (15)–(16))

Turning to comparative numerals, we find that the word order is exactly the
same as with unmodified numerals: *bat* follows the lexical NP (6a), while other
numerals precede it (6b). Once again, the bracketing in (2a) can capture the
facts, whereas the complex-determiner bracketing in (1a) cannot.

(6) a. Liburu bat   baino gehiago erosi d-u-t.
       book   one.ABS than more.ABS buy 3.ABS-have-1SG.ERG
       'I have bought more than one book.'
    b. Hiru liburu   baino gehiago erosi d-u-t.
       three book.ABS than   more.ABS buy 3.ABS-have-1SG.ERG
       'I have bought more than three books.'
    (Arregi 2010, (17))

Similar facts hold for prepositional numerals, as shown in (7a–c): the preposition *inguru* 'around' follows the cardinal-containing NP and does not form a
unit with the cardinal. In the case of *gutxienez* 'at least' (7d–e), the word order
is flexible: a focus particle, *gutxienez* can appear in different parts of the sentence, with (7d) representing the more neutral word order. Once again, *gutxienez* does not form a unit with the cardinal.

(7) a. Kilo   bat azukre   inguru erosi d-u-t.
       kilogram one sugar.ABS around buy 3.ABS-have-1SG.ERG
       'I have bought around/about one kilogram of sugar.'
    b. Azukre kilo   bat   inguru erosi d-u-t.
       sugar kilogram one.ABS around buy 3.ABS-have-1SG.ERG
       'I have bought around/about one kilogram of sugar.'
    c. Hiru kilo   azukre   inguru erosi d-u-t.
       three kilogram sugar.ABS around buy 3.ABS-have-1SG.ERG
       'I have bought around/about three kilograms of sugar.'
    d. Hiru liburu   erosi d-it-u-t   gutxienez.
       three books.ABS buy 3.ABS-PL-have-1SG.ERG at-least
       'I have bought at least three books.'
    e. Hiru liburu   gutxienez erosi d-it-u-t.
       three books.ABS at-least   buy 3.ABS-PL-have-1SG.ERG
       'I have bought at least three books.'
    (Itxaso Rodríguez, pers. comm.)

To sum up, the Hebrew and Basque word-order facts provide compelling evidence that a cardinal does not form a unit with the modifier or preposition to the exclusion of the lexical NP (for more evidence, see Arregi 2010). Further evidence for the modified-NP analysis (the bracketing in (2)) comes from case-assignment facts in Russian. We turn to these next.

9.1.2 Case Assignment in Russian

In this section we will use evidence from the behavior of various modified numerals in Russian to argue against the complex-determiner analysis in (1) and for the modified-NP analysis in (2).

Our argument will proceed as follows: we will first demonstrate that various Russian "modifiers" that can appear in cardinal-containing NPs can function as adjuncts (as is the case with 'exactly' or 'at least') or can take their sister as a complement (which is what happens with 'more than' or with 'about'). Crucially, the syntactic status of such a "modifier" is reflected in the case marking on the lexical NP, which is impossible to account for with the bracketing in (1) but turns out to be expected with the structure in (2).

As shown in (8) (see also chapter 6), in direct cases in Russian, the case marking on the lexical NP inside a cardinal-containing NP depends both on the cardinal and on the case assigned to the entire NP. The verb *kupit'* 'to buy' assigns accusative case, as shown by the example without a cardinal in (8a). If its object is instead a cardinal-containing NP, as in (8b), the accusative case shows up on the cardinal and the lexical NP is marked with the case assigned by the cardinal (in this example it is paucal case, the case assigned by the cardinals 'two' through 'four'). Finally, if the cardinal-containing NP is assigned an oblique case, such as genitive, that oblique case must be assigned to both the numeral and the lexical NP, as in (8c).

(8) a. Maša kupila šar.   Russian
       Mary bought balloon.ACC/NOM
       'Mary bought a balloon.'
    b. Maša kupila dva   šará.
       Masha bought two.ACC balloon.PAUC
       'Mary bought two balloons.'
    c. Maša ne kupila i   dvux   šarov.
       Masha NEG bought EMPH two.GEN balloon.PL.GEN
       'Mary didn't even buy two balloons.'

In the context of a phrasal comparative (for the difference between phrasal and clausal comparatives in Russian, see 9.2.1), the cardinal-containing NP behaves precisely as in (8c): the cardinal no longer bears accusative case but instead is obligatorily genitive, as shown in (9a). The same holds if the comparative contains an NP headed by a measure noun (9b) or a regular lexical NP (9c). On the account of Pancheva (2006, 2010), the genitive case in these environments is assigned not by 'more' itself but by a null preposition corresponding to English *than* (this preposition is overt in other Slavic languages, e.g., Polish and Czech).

(9) a. Maša kupila   bol'še dvux   Comparative + numeral
       Mary buy.PAST.F more two.GEN
       šarov.
       balloon.PL.GEN
       'Mary bought more than two balloons.'
    b. Maša kupila   bol'še funta   Comparative + measure noun
       Mary buy.PAST.F more pound.GEN
       saxara.
       sugar.GEN
       'Mary bought more than a pound of sugar.'
    c. Ljudjam   nužno   bol'še xleba.   Comparative + lexical noun
       people.DAT necessary more bread.GEN
       'People need more bread.'

Such spatial prepositions as *okolo* 'near/around/about', *ot* 'from', and so on, which appear in modified numerals with completely predictable meanings, behave in the same fashion: the genitive case that they assign to lexical NPs in spatial PPs (10a) appears on both the cardinal and the lexical NP in a cardinal-containing NP (10b).

(10) a. Ja živu okolo školy.   P + lexical NP
        I live near school.GEN
        'I live near the/a school.'
     b. Maša kupila   okolo dvux   jaščikov   P + numeral
        Mary buy.PAST.F about two.GEN box.PL.GEN
        knig.
        book.PL.GEN
        'Mary bought about two boxes of books.'

These facts are expected if the comparative 'more than' and the preposition 'about' take the cardinal-containing NP as a complement (the modified-NP analysis in (2)) and assign case to it (see also Arregi 2010). However, under the assumption that the comparative and the preposition combine with the cardinal to the exclusion of the lexical NP (the complex-determiner analysis in (1)), the question arises how the lexical NP receives oblique-case marking. Given that the comparative-cardinal combination or preposition-cardinal combination cannot be a syntactic head, under the complex-determiner analysis in (1) it must be merged as a specifier, and genitive-case assignment then has to be attributed to an independent factor (e.g., to a null Num^0 head agreeing with the specifier into which the modified cardinal is merged). Crucially, therefore, a modified-cardinal specifier under this approach must correlate with genitive case. We will now show that this conclusion leads to incorrect predictions for adverbial "modifiers" in cardinal-containing NPs.

Indeed, as is easy to see, both the superlative adverbial *po krajnej mere* 'at least' in (11) and the approximative adverbial *primerno* 'approximately' in (12) are transparent to case assignment, irrespective of whether they appear in a cardinal-containing NP, a measure phrase, or a lexical NP. In environments with direct-case assignment, the direct case is exhibited by the cardinal and the lexical NP is assigned case by the cardinal, regardless of the presence or absence of the superlative modifier, as shown in (11a). In oblique environments, the entire NP is in the oblique case, as in (11b), again regardless of whether the superlative modifier is present or absent. The same facts hold for regular nouns (11c–d). All of this behavior is replicated by the approximative modifier, as shown in (12).

(11) a. Maša kupila (po krajnej mere) dva shará.
   Mary buy.PAST.F along extreme measure two.ACC balloon.PAUC
   'Mary bought (at least) two balloons.'

   b. Maša ulybnulas' (po krajnej mere) dvum mal'čikam.
   Mary smile.PAST.F along extreme measure two.DAT boy.PL.DAT
   'Mary smiled at (at least) two boys.'

   c. Maša kupila (po krajnej mere) kilogram saxara.
   Mary buy.PAST.F along extreme measure kilogram.ACC sugar.GEN
   'Mary bought (at least) a kilogram of sugar.'

   d. Maša ulybnulas' (po krajnej mere) Pete.
   Maru smile.PAST.F along extreme measure Peter.DAT.
   'Mary smiled at (at least) Peter.'

(12) a. Ètot saxar vesit (primerno) dva kilogramma.
   this sugar weighs approximately two.ACC kilogram.PAUC
   'This sugar weighs (approximately) two kilograms.'

   b. Slon naelsja (primerno) desjatju tonnami sena.
   elephant eat.till.full.PAST approximately ten.INS ton.PL.INS hay.GEN
   'The elephant became full by eating (approximately) ten tons of hay.'

   c. Ètot saxar vesit (primerno) kilogram.
   this sugar.NOM weighs approximately kilogram.ACC
   'This sugar weighs (approximately) a kilogram.'

    d. Slon   naelsja   (primerno)   tonnoj sena.
       elephant eat.till.full.PAST approximately ton.INS hay.GEN
       'The elephant became full by eating (approximately) a ton of hay.'

Once again, if we follow the modified-NP analysis in (2) and assume that the superlative adverbial 'at least' and the approximative adverbial 'approximately' are adjuncts to the entire NP, the pattern in (11) and (12) falls out: adjuncts don't interfere with case assignment. Conversely, if the two adverbials had been merged with the cardinal in Spec,NumP to the exclusion of the lexical NP, we would have expected the same genitive case on the cardinal-containing NP that we have seen with modified numerals containing the comparative and the preposition.

9.1.3 Numeral Modifiers as Adjuncts

The above facts argue in favor of treating the superlative modifier in Russian as an adjunct, as sketched in (13), rather than as a head.

(13) [$_{DP}$ [$_{AdvP}$ at least/at most/approximately] [$_{NP}$ ten books]]

The behavior of superlative and approximative modifiers in Russian cannot be captured on the bracketing in (1), which treats the modifier and the numeral as a unit, to the exclusion of the lexical NP. If 'at least two' or 'approximately two' were a complex determiner, we would expect the lexical NP to be assigned case by the complex determiner. Once again, we have no independent evidence of what case assignment by a complex determiner would be like. Furthermore, there is no reason to expect the complex determiner to assign different cases in (11a) and (11b), nor is it at all clear why the cardinal inside the complex determiner would receive different cases in those examples. Any analysis that tries to treat *po krajnej mere dva* 'at least two' as a complex determiner would need to make a number of stipulations in order to account for these case-assignment facts. Further evidence against viewing superlative modifiers as heads is their internal composition: *po krajnej mere* is literally 'at extreme.DAT measure.DAT', with the preposition *po* 'along' assigning dative case to *mera* 'measure' exactly as in other environments (e.g., *po stolu*, lit. 'along table.DAT').

We also saw that the behavior of superlative modifiers in Russian parallels the behavior of unambiguously adverbial expressions like *primerno* ('approximately'), as shown in (12) above. Further evidence that the superlative modifier is an adjunct is that it can be displaced (14a), again like an adverbial (14b), and unlike the comparative modifier (14c). The same facts are found in English, as shown in (15), which suggests that the adjunct analysis of the superlative modifier holds for English as well.

(14) a. Maša kupila   pjat'   knig,   po   Superlative numeral
         Mary buy.PAST.F five.ACC book.PL.GEN along
         krajnej mere.
         extreme measure
         'Mary bought five books, at least.'

     b. Maša kupila   pjat'   knig,   Adverbial numeral
         Mary buy.PAST.F five.ACC book.PL.GEN
         primerno.
         approximately
         'Mary bought five books, approximately.'

     c. *Maša kupila   pjati   knig,   Comparative numeral
         Mary buy.PAST.F five.GEN book.PL.GEN
         bol'she.
         more

(15) a. Mary bought five books, at least.   Superlative numeral
     b. Mary bought five books, approximately.   Adverbial numeral
     c. *Mary bought five books, more than.   Comparative numeral

In other words, case marking on the lexical NP in modified numerals is fully determined by the properties of the "modifier," which is difficult to capture if this "modifier" is contained in a specifier in the extended projection of the lexical NP. Conversely, if the "modifier" combines with the entire cardinal-containing NP, it will assign case to the entire NP if it is a head and will not do so if it is an adjunct. We conclude that the modified-NP analysis in (2) is more compatible with case assignment in Russian modified cardinals than the complex-determiner analysis in (1).[1] Now that we have sketched an analysis of numeral modifiers that are adjuncts, we consider in more depth the syntax of modified numerals where the alleged modifier is actually the head of the construction. The next two sections address, respectively, the syntax of comparative and prepositional numerals.

9.2   The Syntax of Comparative Numerals

In this section, we consider the syntax of comparative numerals, focusing on two types of comparative numerals in Russian. The Hebrew and Basque facts and Russian case-assignment facts discussed above provide evidence that comparative numerals have the bracketing in (2a), in which the cardinal forms a unit with the lexical NP, and not in (1a), where the cardinal forms a unit with the comparative expression. We now address in more depth the exact syntactic structure of comparative numerals.

Comparative constructions are descriptively divided into two types, depending on the constituent that functions as the standard of comparison: a noun phrase in phrasal comparatives and a (potentially elided) clause in clausal comparatives. In Russian (as in many other Slavic languages, as well as in French and German), the two comparatives are expressed by different means: with a genitive NP, as in (16a), and with a *wh*-word followed by a nominative NP, as in (16b) (Heim 1985; Lechner 1998; Pancheva 2006).

(16) a. Maša vyše Ivana.   Russian phrasal comparative
        Mary taller Ivan.GEN
        'Mary is taller than Ivan.'
     b. Maša vyše, čem Ivan.   Russian clausal comparative
        Mary taller than Ivan.NOM
        'Mary is taller than Ivan is.'

Both phrasal and clausal comparatives can be used with cardinal-containing NPs, as (17) shows.

(17) a. bol'še pjati   knig   Russian phrasal comparative
        more   five.GEN book.PL.GEN
        'more than five books'
     b. bol'še, čem pjat'   knig   Russian clausal comparative
        more   than five.NOM book.PL.GEN
        'more than five books'

We take the English *more* and the Russian *bol'še* 'more' as the synthetic comparative form of *many/much* and *mnogo* 'many/much', respectively (Matushansky and Ionin 2014; see also Bresnan 1973).

9.2.1 Differences between Phrasal and Clausal Comparative Numerals

We observe two differences between phrasal and clausal comparative numerals. The first is the availability of 'many' versus 'much' readings. The two types of readings are both available for English comparatives, as shown in (18).

(18) more than five sandwiches
     a. 'Many' reading ≈ 'six or more sandwiches'
     b. 'Much' reading ≈ 'five sandwiches plus something else'

These readings can be brought out by the continuations in (19a–b), respectively. The same effect is observed with measure nouns, as shown in (20).[2]

(19) I ate more than five sandwiches …
     a. 'Many' reading: … I ate six!
     b. 'Much' reading: … I ate five sandwiches plus a bowl of soup!

(20) I bought more than a pound of apples …
    a. 'Many' reading: … I bought a pound and a half.
    b. 'Much' reading: … I also bought some bananas.

Russian phrasal comparative numerals have only the 'many' reading available to them, while clausal comparative numerals are compatible with both 'many' and 'much' readings, as shown in (21); the same facts hold for French (22).

(21) a. bol'še pjati   knig   Russian phrasal comparative
      more five.GEN book.PL.GEN
      ≈ 'at least six books'
    b. bol'še, čem pjat'   knig   Russian clausal comparative
      more   than five.NOM book.PL.GEN
      ≈ 'at least six books' *or*
      ≈ 'five books and something else'

(22) a. plus de dix livres   French phrasal comparative
      more of ten book.PL
      ≈ 'at least eleven books'
    b. plus que dix livres   French clausal comparative
      more that ten book.PL
      ≈ 'at least eleven books' *or*
      ≈ 'ten books and something else'

The second difference between phrasal and clausal comparatives has to do with the ability of the comparative to combine with a referential (type e) expression. As shown in (23), the English comparative is compatible with referential expressions. In Russian, only the clausal comparative, not the phrasal, is compatible with referential expressions, as shown in (24).

(23) a. I invited more than (just) Peter and Mary.   English
      (I also invited Tom.)
    b. I read more than these five books.
      (I read six other books also.)

(24) a. Ja priglasila   bol'še, čem Petju   i   Mašu.   Russian
      I invite.PAST.F more   than Peter.ACC and Mary.ACC
      'I invited more than Peter and Mary.'
    b. Ja pročitala   bol'še, čem èti   pjat'   knig.
      I read.PAST.F more   than this.PL five.ACC book.PL.GEN
      'I read more than these five books.'
    c. *Ja priglasila   bol'še Peti   i   Maši.
      I invite.PAST.F more Peter.GEN and Mary.GEN
    d. *Ja pročitala   bol'še ètix pjati   knig.
      I read.PAST.F more these five.GEN book.PL.GEN

Thus, an account of the semantics of phrasal and clausal comparatives has to explain why, in Russian, phrasal comparatives lack 'much' readings and cannot combine with referential expressions, while clausal comparatives are able to do both.

9.2.2 Against Different Bracketing for Phrasal and Clausal Comparatives

At first blush, the above differences between phrasal and clausal comparatives might suggest that phrasal comparatives have the bracketing in (1a)—on which the comparative element combines directly with the cardinal—while clausal comparatives have the bracketing in (2a). This would account for the distribution of 'many' and 'much' readings, as sketched in (25), and it would also explain why phrasal comparatives cannot combine with referential expressions (since the comparative would be combining directly with the cardinal, not with an entire NP or DP). On this analysis, the reason that (25b) would allow for both 'many' and 'much' readings is that the former is subsumed in the latter: 'five books plus something else' could in fact be six books.

(25) a. [[more than five] books] ≈ six or more books
     b. [more than [five books]] ≈ five books plus something else

However, there are a number of problems with this explanation. First, as discussed in section 9.1.2, the pattern of case assignment with phrasal comparatives provides strong evidence in favor of the bracketing in (2a) for phrasal comparatives. Putting the semantic difference between phrasal and clausal comparatives into the bracketing, as in (25), would leave these case-assignment facts unexplained.

Second, we observe that the use of a comparative with a regular NP produces the same difference between phrasal and clausal comparatives as its use with a numeral, as shown in (26) and (27) for Russian and French, respectively (compare (21) and (22)). A comparative with a degree clause (introduced by the complementizer *que/čem* 'that') leads to the interpretation where 'more' applies to the denotation of its sister—'X and something else'—whereas a comparative with a degree phrase (introduced by *de* or marked by genitive case) leads to the interpretation where 'more' applies to the quantity denoted by its sister.

(26) a. Ljudjam   nuzhno   bol'še xleba.   Russian phrasal
        people.DAT necessary more bread.GEN   comparative
        'People need more bread.'

   b. Ljudjam   nuzhno   bol'še, čem xleb.   Russian clausal
      people.DAT necessary more   than bread.ACC   comparative
      'People need something besides bread.'

(27) a. Il faut   plus de Baudelaire dans ce livre.   French phrasal
        it necessitates more of Baudelaire in   this book   comparative
        'We should have more Baudelaire in this book.'
     b. Il faut   plus que Baudelaire dans ce livre.   French clausal
        it necessitates more that Baudelaire in   this book   comparative
        'We should have something besides Baudelaire in this book.'

These data show that the pattern of interpretation of phrasal comparatives is not specific to NPs containing cardinals. Furthermore, there is evidence that both phrasal and clausal comparatives combine with an entire NP, as shown in (28).

(28) a. bol'še ukazannogo   količestva   knig   Russian phrasal
        more   indicated.GEN number.GEN book.PL.GEN   comparative
        'more than the indicated number of books'
        ≈ 'a greater number of books than was indicated'
     b. bol'še, čem ukazannoe   količestvo   Russian clausal
        more   than indicated.NOM number.NOM   comparative
        knig
        book.PL.GEN
        'more than the indicated number of books'
        ≈ 'a greater number of books than was indicated', *or*
        ≈ 'the indicated number of books plus something else'

The data in (27) and (28) suggest that in French and in Russian, both types of comparatives combine with a constituent containing the NP rather than with just the cardinal, as predicted on the bracketing in (2a). The difference between 'many' and 'much' readings cannot therefore stem from differential bracketing, as sketched in (25). Instead, we analyze both phrasal and clausal comparatives as having the bracketing in (2a). Following Pancheva (2006, 2010; cf. Arregi 2010), we assign to phrasal and clausal comparatives the structures in (29a–b), respectively.

(29) a. Phrasal comparatives: $[_{DegP}$ more/bol'še $[_{PP}$ than/$\varnothing_{GEN}$ $[_{NP}$ five books]]]
     b. Clausal comparatives: $[_{DegP}$ more/bol'še $[_{CP}$ than/čem $[_{NP}$ five books]]]

A compositional analysis of the structures in (29) is provided in Matushansky and Ionin 2014, where the analysis is framed in terms of degrees, following Pancheva. We believe that somewhat simpler derivations can be proposed for comparative numerals if an entity-based analysis of measure nouns (Matushansky and Zwarts 2017) is adopted, but we leave the details for future research.

9.3 The Syntax of Prepositional Numerals

Corver and Zwarts (2006) discuss four potential structures for prepositional numerals, as spelled out in (30). The two options in (30a–b) are compatible with the bracketing in (2c) that we have argued for, on which the cardinal forms a unit with the lexical NP. In fact, we have proposed a structure akin to (30a) for comparative numerals (see (29)) and a structure akin to (30b) for superlative numerals (see (13)). In contrast, the options in (30c–d) are two variants of the structure in (1c), on which the modifier/preposition and the cardinal form a unit to the exclusion of the lexical NP. We have argued against the complex-D variant in (30c) on the basis of word-order facts in Hebrew and Basque; those word-order facts also provide an argument against the PP-specifier option in (30d).

(30) a. $[_{PP}$ around/above $[_{NP}$ ten books]]   Prepositional head
      b. $[_{DP} [_{AdvP}$ around/above] $[_{NP}$ ten books]]   Adverbial specifier
      c. $[_{DP} [_{D}$ around/above ten] $[_{NP}$ books]]   Complex D
      d. $[_{DP} [_{PP}$ around/above ten] $[_{NP}$ books]]   PP specifier

We now examine case assignment with Russian prepositional numerals, applying the same tests as we applied to comparative and superlative numerals; we show that these facts provide evidence for the structure in (30a) and against the ones in (30b–d). We then discuss the behavior of Dutch prepositional cardinals, which is what led Corver and Zwarts (2006) to argue for the PP-specifier structure in (30d). We provide an alternative analysis of the Dutch facts, in favor of the structure in (30a).

We set aside here the distinction between specifiers and adjuncts, concentrating mostly on the constituency.

9.3.1 Case Assignment with Russian Prepositional Numerals

The Russian preposition *okolo* 'near/around/about' assigns genitive case to its lexical-NP sister, as shown in (31): even though the verb *vesit'* 'to weigh' assigns accusative case (31a–b), the lexical NP that follows *okolo* is obligatorily genitive (31c–d). As noted earlier, the preposition *okolo* assigns genitive also when it is used as a spatial preposition (31e).

(31) a. Etot saxar   vesit   funt.   Russian
    this sugar.NOM weighs pound.ACC/NOM
    'This sugar weighs a pound.'
  b. Etot saxar   vesit   pjat'   funtov.
    this sugar.NOM weighs five.ACC pound.PL.GEN
    'This sugar weighs five pounds.'
  c. Etot saxar   vesit   okolo funta.
    this sugar.NOM weighs near   pound.GEN
    'This sugar weighs about a pound.'
  d. Etot saxar   vesit   okolo dvux   funtov.
    this sugar.NOM weighs near   two.GEN pound.PL.GEN
    'This sugar weighs about two pounds.'
  e. Ja živu okolo školy.
    I   live near   school.GEN
    'I live near the/a school.'

The case-assigning properties of *okolo* are naturally captured on the prepositional-head structure in (30a); *okolo* forms a minimal pair with the adverbial *primerno* ('approximately/about'), which has a very similar meaning, but, as discussed in section 9.1.2, behaves like an adjunct (30b) rather than a head. This is consistent with the fact that *okolo* is a preposition in other contexts, while *primerno* is an adverb. For example, *primerno* can modify a PP, whereas *okolo* cannot (32a): instead, *okolo* must head the PP and assign genitive case to its sister NP, as in (32b).

(32) a. Ja pridu   primerno/*okolo v vosem'   časov.
    I   come.1SG.PRFV approximately/near at eight.ACC hour.PL.GEN
    'I will come approximately at eight o'clock.'
  b. Ja pridu   okolo vos'mi   časov.
    I   come.1SG.PRFV near   eight.GEN hour.PL.GEN
    'I will come at about eight o'clock.'

The behavior of prepositional numerals with *okolo* cannot be captured on the complex-D structure in (30c), for the same reasons discussed with respect to comparative numerals: the complex-D account would have to stipulate that the complex determiner *okolo dvux* in (31d) assigns genitive case, and it would have to explain how the cardinal inside the complex determiner gets its case assignment.

Consider next the PP-specifier structure in (30d): on this structure, *okolo dvux* is in the specifier of the DP, and should have no effect on case assignment to the lexical NP. We would then expect the lexical NP to be assigned accusative case by the verb: since the verb *vesit'* assigns accusative case, as shown in (31a–b), this would result in the ungrammatical sentence in (33).

(33) *Etot saxar vesit   [okolo dvux]   funty.
      this sugar weighs near two.GEN pound.ACC.PL

9.3.2   Internal Structure of Prepositional Numerals

The facts reported in the previous section provide evidence in favor of (30a) as the structure for prepositional numerals. We note further that on this analysis, the preposition in *around twenty books* or *between ten and twenty books* combines with an NP (as schematized in (30a)), not with a DP. There are two pieces of evidence in favor of this analysis. First, as shown by Matushansky and Zwarts (2017), the entire prepositional numeral can be embedded inside a DP. The fact that the determiner in (34) (examples taken from the web) appears above the preposition indicates that the entire PP is merged below the DP level.

(34) a. the [*over 9 million liters* of water and 50,000 filters distributed by FEMA]
     b. for the duration of those *up to ten minutes*
     (Matushansky and Zwarts 2017)

The second piece of evidence comes from the fact that prepositional numerals cannot contain true DPs, as shown in (35). (Thanks to an anonymous reviewer for the example in (35).) In this they behave like phrasal comparatives and unlike clausal comparatives (see section 9.2).

(35) a.   We invited around (*the) twenty guests.
     b. *John invited between [Ann and Bill] and [Ann, Bill, Carol, and Dan].

We conclude that in prepositional numerals, the preposition combines with an NP, which consists of the cardinal plus the lexical NP.

A potential piece of evidence against this proposal (pointed out by an anonymous reviewer) comes from the behavior of approximative *about* PPs. In an approximative PP, approximation must affect the quantity (36a), not the property denoted by the lexical NP (as in (36b), where caramels might be viewed as "approximately" chocolates). This might seem to speak in favor of the bracketing [about ten], rather than, as we propose, [about [ten chocolates]]. However, we already saw in the case of comparatives (see section 9.2) that the existence of 'many' versus 'much' readings does not necessitate different bracketing. We note that *about* PPs behave just like phrasal comparatives, in that they both disallow 'much' readings (36b) and cannot combine with a referential expression (36c).

(36) They ate about ten chocolates.
    a. 'Many' reading ≈ nine or eleven chocolates
    b. 'Much' reading ≈ ten caramels
    c. *They ate about the ten chocolates.

For an account of why prepositional cardinals do not allow the type of reading in (36b), please see Matushansky and Zwarts 2017.

9.3.3 Conjoined Prepositional Numerals: Evidence from English and Russian

We next consider conjoined prepositional numerals such as *between two and three feet* or *between ten and twenty books*. Corver and Zwarts (2006), who argue for the PP-specifier structure in (30d) for prepositional numerals (see the next section), analyze conjoined prepositional numerals as having the structure in (37a). They argue that conjunctions like *ten and twenty* are licensed by the presence of *between*: it is unnatural to talk about *ten and twenty books*. Corver and Zwarts argue that this is expected on the PP-specifier structure in (37a), where *between ten and twenty* is a PP (i.e., the conjunction *ten and twenty* cannot occur in the specifier of the DP unless it first combines with *between*). In contrast, if *ten and twenty books* forms a unit, it is predicted to occur regardless of the presence of *between*.

(37) [$_{DP}$ [$_{PP}$ between ten and twenty] [$_{NP}$ books]]   PP specifier

We suggest that the reason *ten and twenty books* does not normally occur without *between* is pragmatics: it is strange to talk about *ten and twenty books* when one could instead say *thirty books* (see also chapter 5 for more discussion of the effects of pragmatics on conjunction).

The analysis in (37) runs into difficulties with conjoined prepositional numerals that have the lexical NP spelled out in both conjuncts, such as *between ten books and twenty books* or *between two ounces and three pounds*. These require the fairly bizarre and otherwise unmotivated structures in (38), with *between ten books and twenty* occurring in the specifier of *books* and *between two ounces and three* occurring in the specifier of *pounds*, which is clearly not the right analysis.

(38) a. [$_{DP}$ [$_{PP}$ between ten books and twenty] [$_{NP}$ books]]   PP specifier
    b. [$_{DP}$ [$_{PP}$ between two ounces and three] [$_{NP}$ pounds]]

In contrast, the analysis of prepositional numerals on which the cardinal forms a unit with the lexical NP, as in (30a), has no problem deriving all types of conjoined prepositional numerals. We analyze prepositional numerals like *between ten and twenty books* as involving NP deletion of the lexical NP in the first conjunct (39a), exactly as in the case of complex cardinals formed by addition (see chapter 5). Support for the structure in (39a) comes from the fact that it is possible to spell out the lexical NP in both conjuncts (39b). Furthermore, the lexical NPs in the two conjuncts do not even need to be identical (39c) or to contain numerals (39d).

(39) a. [$_{PP}$ between [$_{NP}$ [$_{NP}$ ten ~~books~~] and [$_{NP}$ twenty   Prepositional head
          books]]]
      b. [$_{PP}$ between [$_{NP}$ [$_{NP}$ ten books] and [$_{NP}$ twenty books]]]
      c. [$_{PP}$ between [$_{NP}$ [$_{NP}$ two ounces] and [$_{NP}$ three pounds]]]
      d. [$_{PP}$ between [$_{NP}$ [$_{NP}$ a foot] and [$_{NP}$ a yard]]]

Further evidence for our proposal comes from case assignment with conjoined prepositional numerals in Russian, like *ot dvux do trex futov*, literally 'from two to three feet', as in (40a). An analysis of the entire conjoined numeral as a complex determiner, as in (30c), is untenable, since it cannot account for the case-assignment facts: *ot* and *do* behave exactly like *okolo* (31), in that each preposition requires the element following it (be it a numeral, as in (40a), or a lexical NP, as in (40b)) to bear genitive case. This is unexpected if the sequence *ot dvux do trex* is a complex determiner. Furthermore, neither the complex-D structure (30c) nor the PP-specifier structure (30d) can account for the facts in (40b–c). As shown by (40b), *ot ... do* can occur with two lexical NPs: the PP-specifier structure in (30d) would have to say that the entire sequence *ot futa do yarda* occurs in the specifier of a null lexical NP, which is not an independently motivated analysis. Another problem for the PP-specifier analysis is that *ot ... do* can have the meaning 'between', as in (40a), or 'interval from ... to ...', as in (40c). The PP-specifier analysis can capture the former meaning but not the latter.

(40) a. Stebel' etogo   rastenija byvaet dlinoj   ot   dvux   Russian
          stalk   this.GEN plant.GEN is.HAB length.INS from two.GEN
          do trex   futov.
          to three.GEN foot.PL.GEN
          'The stalk of this plant can be between two and three feet in length.'
      b. Stebel' etogo   rastenija byvaet dlinoj   ot   futa
          stalk   this.GEN plant.GEN is.HAB length.INS from foot.GEN
          do yarda.
          to yard.GEN
          'The stalk of this plant can be between a foot and a yard in length.'

   c. V vozraste ot   dvux   do trex   let,   rebenok
      in age   from two.GEN to three.GEN years.GEN child.NOM
      poznaet mir.
      knows world
      'In the interval from two to three years of age, the child gains
      knowledge of the world.'

The facts in (40) are, once again, fully captured on the prepositional-head structure in (30a), with NP deletion of the lexical NP that follows the first numeral. The examples in (40a–c) would then have the underlying structures in (41a–c), respectively. These structures receive further support from the fact that it is possible to pronounce the deleted NP, as in *ot dvux futov do trex futov* (see (39) for the same facts in English).

(41) a. [PP [PP ot   dvux   ~~futov~~] [PP   do trex   Prepositional head
         from two.GEN foot.PL.GEN until three.GEN
      futov]]
      foot.PL.GEN
      'from two to three feet'
   b. [PP [PP ot   futa] [PP do   yarda]]
         from foot.GEN until yard.GEN
      'from a foot to a yard'
   c. [PP [PP ot   dvux   ~~let~~] [PP   do   trex   let]]
         from two.GEN year.PL.GEN until three.GEN years.GEN
      'from two to three years'
   d. [PP [PP ot   goda] [PP do trex   ~~let~~]]]]
         from year.GEN until three.GEN years.GEN
      'from one year to three'

Finally, as (41d) shows, it is not necessarily the linearly first lexical NP that is deleted, arguing further for the analysis and constituency that we assume.

9.3.4 Prepositional Numerals in Dutch

Potential counterevidence for our analysis of prepositional numerals comes from the work of Corver and Zwarts (2006), who argue for the PP-specifier structure in (30d) for prepositional numerals. Their argument is based primarily on data from Dutch.

The examples in (42) through (46) below (Corver and Zwarts's (29) through (33)) show that expressions containing prepositional numerals behave like DPs rather than PPs on five different tests. First, PPs can be displaced to the right of the verb, but DPs and expressions containing prepositional numerals cannot, as shown in (42). Second, topicalization is possible out of a DP or an expression containing a prepositional numeral, but not out of a regular PP, as shown in (43). Third, extraction of the pronoun *er* is possible out of a DP or an expression containing a prepositional numeral, but is not so good out of a PP, as shown in (44). Finally, expressions containing prepositional numerals occur in two positions, (45) and (46), that subcategorize for DPs but not for PPs.

(42) a. *Ik heb __ uitgenodigd [twintig kinderen].   Dutch, DP
        I   have   invited   twenty children
        'I invited twenty children.'

     b. Ik heb __ gedanst [rond   de twintig tafels].   PP
        I   have   danced   around the twenty tables
        'I danced around the twenty tables.'

     c. *Ik heb __ uitgenodigd [rond   de twintig   Prepositional
        I   have   invited   around the twenty   numeral
        kinderen].
        children
        'I invited approximately twenty children.'

     d. *Ik heb __ verstuurd [tegen   de twintig kaarten].   Prepositional
        I   have   sent   against the twenty postcards   numeral
        'I have sent close to twenty postcards.'

(43) a. Volgelingen heeft Jomanda [tweeduizend __].   DP
        followers   has   Jomanda two.thousand
        'Jomanda has two thousand followers.'

     b. *Volgelingen heeft Jomanda [met [tweeduizend __]]gepraat.   PP
        followers   has   Jomanda with two.thousand   talked
        'Jomanda talked with two thousand followers.'

     c. Volgelingen heeft Jomanda [ruim over de   Prepositional
        followers   has   Jomanda wide over the   numerals
        tweeduizend __].
        two.thousand
        'Jomanda has well over two thousand followers.'

     d. Kaarten   heb ik [tegen   de tweeduizend __]   Prepositional
        postcards have I   against the two.thousand   numeral
        'I sent close to two thousand postcards.'

(44) a. Jomanda heeft er   [tweeduizend __].   DP
        Jomanda has   there two.thousand
        'Jomanda has two thousand of them.'

     b. ?Ik heb er   met [tweeduizend __] gepraat.   PP
        I have there with two.thousand   talked
        'I have talked with two thousand of them.'

    c.   Jomanda heeft er   [ruim over de tweeduizend __]. Prepositional
        Jomanda has   there far   over the two.thousand   numeral
        'Jomanda has far over two thousand of them.'

    d.   Ik heb er   [tegen   de tweeduizend __]   Prepositional
        I have there against the two.thousand   numeral
        verstuurd.
        sent
        'I have close to two thousand of them sent.'

(45) a.   Jan ontmoette [de kinderen].   DP
        Jan met   the children
        'Jan met the children.'

    b.   *Jan ontmoette [rond   de kinderen].   PP
        Jan met   around the children

    c.   Jan ontmoette [rond   de twintig kinderen].   Prepositional
        Jan met   around the twenty children   numeral
        'Jan met approximately twenty children.'

    d.   Jan ontmoette [tegen   de twintig kinderen].   Prepositional
        Jan met   against the twenty children   numeral
        'Jan met close to twenty children.'

(46) a.   Ik reken op [de kinderen].   DP
        I count on the children
        'I count on the children.'

    b.   *Ik reken op [rond   de kinderen].   PP
        I count on around the children

    c.   Ik reken op [rond   de twintig kinderen].   Prepositional
        I count on around the twenty children   numeral
        'I count on approximately twenty children.'

    d.   Ik reken op [tegen   de twintig kinderen].   Prepositional
        I count on against the twenty children   numeral
        'I count on close to twenty children.'

On the basis of these tests, Corver and Zwarts argue that expressions containing prepositional numerals are DPs rather than PPs, which rules out the prepositional-head structure in (30a).

At the same time, Corver and Zwarts argue that the preposition-numeral combination has PP properties, which cannot be captured on an analysis that treats the preposition as adverbial, as in the adverbial-specifier structure in (30b), or as a complex D (30c). The prepositional numeral can be preceded by expressions like *somewhere*, which also co-occur with regular PPs. This is illustrated in (47).

(47) a. Er   hebben zich [ergens   in de twintig deelnemers]
        there have   REFL   somewhere in the twenty participant.PL
        aangemeld voor het spelletje.
        registered for   the game
        'Somewhere around twenty participants registered for the game.'
     b. Er   hebben zich [iets boven de twintig mensen] aangemeld voor
        there have   REFL   a.bit above the twenty people   registered for
        het spel.
        the game
        'A little over twenty people registered for the game.'
     (Corver and Zwarts 2006, (34))

Furthermore, the occurrence of the definite article *de* between the preposition
and the numeral is obligatory (48b) or at least optional (48a), depending on
the preposition; in contrast, expressions that are unambiguously adverbial
cannot be followed by *de* (48c–d), indicating that the preposition that occurs
with prepositional numerals is not an adverb.

(48) a. Er   waren [rond   (de) twintig deelnemers].
        there were   around   the twenty participant.PL
        'There were around twenty participants.'
     b. Er   waren [tegen   *(de) twintig deelnemers].
        there were   against   the twenty participant.PL
        'There were close to twenty participants.'
     c. Er   waren [ongeveer   (*de) twintig deelnemers].
        there were   approximately   the twenty participant.PL
        'There were approximately twenty participants.'
     d. Er   waren [hoogstens (*de) twintig deelnemers].
        there were   at.most   the twenty participant.PL
        'There were at most twenty participants.'
     (Corver and Zwarts 2006, (36)–(37))

Finally, the fact that complex prepositions like *in de buurt van* 'in the neigh-
borhood of' can also occur with numerals speaks against their status as adverbial or as complex determiners. All of these considerations lead Corver and
Zwarts to treat the preposition in prepositional numerals as truly prepositional,
not adverbial, and opt for the structure in (30d).

   While we agree that the prepositions occurring with prepositional numerals
are indeed prepositional, we argue that the prepositional-head structure in
(30a), rather than the PP-specifier structure in (30d), is the right analysis of
prepositional-numeral constructions. Evidence for this comes from the behavior of measure nouns used with prepositional constructions. Measure nouns

like *kilo* can be used with prepositions, as shown in (49a). Such "prepositional measure nouns" behave just like prepositional numerals on Corver and Zwarts's tests: like DPs, and unlike true PPs, they cannot be displaced to the right of the verb (compare (49b) to (42)); they allow topicalization of the lexical NP (compare (49c) to (43)); they allow extraction of the pronoun *er* (compare (49d) to (44)); and they occur with verbs that subcategorize for DPs (compare (49e) to (46)).

(49) a. Ik heb [rond   de kilo suiker] gegeten.
        I have around the kilo sugar   eaten
        'I have eaten about a kilogram of sugar.'
    b. * Ik heb __ gegeten rond   de kilo suiker.
        I have   eaten   around the kilo sugar
    c. %Hooi heeft de olifant   rond   de ton gegeten.
        hay has   the elephant around the ton eaten
        'The elephant has eaten around a ton of hay.'
    d. Jomanda heeft er   [ruim over de ton __].
        Jomanda has   them far over the ton
        'Jomanda has more than a ton of them.'
    e. %Ik reken op rond   de kilo suiker.
        I count on around the kilo sugar
        'I count on around a kilogram of sugar.'

Given these parallels, Corver and Zwarts should analyze prepositional measure nouns as having the same structure as prepositional numerals: in both cases, the preposition–numeral/measure noun sequence should occur in the specifier of the DP, as shown in (50a–b). Corver and Zwarts furthermore assume that when the numeral occurs without a preposition, it still occurs in the specifier of the lexical NP, as shown in (50c); extending this analysis to measure nouns, we get (50d).

(50) a. [$_{DP}$ [$_{PP}$ rond de twintig] kinderen]
    b. [$_{DP}$ [$_{PP}$ rond de kilo] suiker]
    c. [$_{DP}$ [$_{NP}$ twintig] kinderen]
    d. [$_{DP}$ [$_{NP}$ kilo] suiker]

However, the above analysis of measure nouns runs into problems with gender assignment. Dutch has two definite determiners, *de*, which is used with singular common-gender nouns and with all plural nouns, and *het*, which is used with singular neuter-gender nouns. Since *kilo* is a common-gender noun, it is preceded by *de* (51a). If the diminutive form of *kilo* is used, as in (51b), the definite determiner *het* is used, since all diminutives in Dutch are neuter.

(51) a. rond   de   kilo suiker
         around the.c kilo sugar
         'around a kilo of sugar'
     b. rond   het/*de kilootje suiker
         around the.N/C kilo.DIM sugar
         'around a kilo of sugar'

In order to account for (51) on their PP-specifier analysis, Corver and Zwarts would need to say that the definite determiner is agreeing with the measure noun, which is inside the PP in the specifier of the lexical NP, as in (50b). Consider next what happens in the absence of *rond*: as shown in (52), the demonstrative has to agree with the measure noun (*deze* is common, *dit* is neuter), rather than with the lexical noun *suiker* (which is common, and should therefore take *deze* regardless of the form of the measure noun).

(52) a. Ik heb deze/*dit kilo suiker thuis gebracht.
         I   have this.C/N kilo sugar home brought
         'I have brought this kilogram of sugar home.'
     b. Ik heb dit/*deze kilootje suiker thuis gebracht.
         I   have this.N/C kilo.DIM sugar home brought
         'I have brought this kilogram of sugar home.'

In order to preserve the analysis on which the measure noun is in the specifier of the lexical NP (50d), and to account for the gender agreement between the demonstrative and the measure noun in (52), Corver and Zwarts would be forced to adopt the structure in (53a); however, this structure does not yield the right semantics, since it turns the substance NP into a nonrestrictive modifier. In contrast, our structure in (53b) has no problem capturing the agreement between the demonstrative and the head of the NP.

(53) a. $[_{DP} [_{DP}$ deze kilo$]$ suiker$]$ Measure noun inside PP specifier
     b. $[_{DP}$ deze $[_{NP}$ kilo $[_{NP}$ suiker$]]]$ Measure noun as head

We have argued that prepositional numerals have the structure in (30a), where the cardinal and the lexical NP form a unit to the exclusion of the preposition. As schematized in (30a), the preposition combines with an NP rather than with a DP; in section 9.3.2, we presented several pieces of evidence in favor of this analysis. However, in Dutch, as shown by examples (42) through (49) above, the definite article *de* appears between the preposition and the lexical NP, suggesting the presence of a DP level. For our purposes, both the NP and DP analyses work equally well. Yet there is reason to believe that *de* in these constructions is not a normal definite article; as indicated by the translations, *de* inside Dutch prepositional numerals does not signal uniqueness or identifiability or familiarity, as definite articles normally do. Furthermore, Plank (2004), analyzing a similar use of a definite article in German, observes that it does not behave like a normal definite article from the morphological standpoint: it does not reflect the gender or the case of the NP. It is reasonable to suppose, then, that *de* in the above examples does not signal the presence of a DP.

To sum up, the analysis on which the cardinal combines with the lexical NP prior to combining with a preposition is better able to account for crosslinguistic data than an analysis on which the preposition and the cardinal form a unit to the exclusion of the lexical NP.

9.3.5   Prepositional Numerals: PPs or NPs?

We are left with a conundrum: there is much evidence both from Dutch (gender-agreement facts) and from Russian (case-assignment facts) that prepositional numerals have the prepositional-head structure in (30a), that is, that they are PPs. At the same time, the Dutch facts in (42) through (46) indicate that structures containing prepositional numerals behave like noun phrases rather than PPs. Nor are these facts specific to Dutch: while many of the Dutch syntactic tests (topicalization, rightward movement) are not applicable to English, the subcategorization facts of English are like those in Dutch. Verbs that subcategorize DPs rather than PPs are nevertheless compatible with prepositional numerals, as shown in (54a–b); the same holds for prepositional measure nouns (54c).

(54) a. John met *twenty children/around twenty children/*around the children*.

   b. I count on *twenty children/around twenty children/*around the children*.

   c. John bought *a pound of apples/around a pound of apples/*around the apples*.

One explanation could be that the PP structure that we argue for in (30a) is further embedded inside a noun phrase, as schematized in (55a) for prepositional numerals and in (55b) for prepositional measure nouns. The constituency facts (gender, case assignment) are captured, because the cardinal forms a unit with the lexical NP rather than with the preposition. Yet the entire construction behaves like a DP rather than a PP, hence the subcategorization facts.

(55) a. $[_{DP} [_{NP} e [_{PP} \text{around} [_{NP} \text{twenty children}]]]]$

   b. $[_{DP} [_{NP} e [_{PP} \text{around} [_{NP} \text{a pound of apples}]]]]$

However, Matushansky and Zwarts (2017) present evidence against the analysis in (55). In particular, they show that in both Dutch and Russian, the number/gender specification of a prepositional numeral is not fixed, contrary to what we would expect under the structure in (55): since gender and number are core properties of any noun, the postulated null noun in (55) should have a certain constant number/gender specification. Consider Matushansky and Zwarts's examples (56)–(57). In (56), the measure noun and the substance noun are both singular and neuter, and each on its own would take the neuter-singular definite article *het*. However, the entire prepositional measure does not denote a singular entity, making it incompatible with a singular determiner, hence the ungrammaticality of (56a). Example (56b) is also ungrammatical because neither the measure noun nor the substance noun is compatible with the definite article *de* (which can appear with plurals as well as with common-gender nouns). In contrast, (57) is grammatical: the definite article *de* is compatible with the substance noun, which is plural in (57a) and common in (57b), as well as with the nonsingularity of the entire prepositional measure.

(56) a. *het   rond   een pond   meel   dat   ik gekocht heb   Dutch
         DEF.N.SG around a   pound.N flour.N that.N I bought have
     b. *de   rond   een pond   meel   die   ik gekocht heb
         DEF.PL/C.SG around a   pound.N flour.N that.PL/C.SG I bought have

(57) a. de   rond   een pond   aardappeltjes die   ik gekocht heb
        def.PL around a   pound.N potatoes.PL   that.PL I bought have
        'the around a pound of potatoes that I bought'
     b. de   rond   een kilo   cocaïne   die   ik gekocht heb
        def.PL around a   kilogram.C.SG cocaine.C.SG that.PL I bought have
        'the around a kilogram of cocaine that I bought'

The Russian facts in (58) also illustrate the lack of a null noun in prepositional measures: the only determiner that is marginally compatible with the prepositional measure in (58a) is the plural one; its acceptability is much improved when the measure phrase is plural, as in (58b). Once again, the plurality of the determiner must be licensed by something inside the prepositional measure.

(58) a. ??èti/*ètot/*èta   okolo   litra   vody   Russian
        this.PL/M.SG/F.SG around liter.M.GEN water.F.GEN
        '??these over a liter of water'
     b. èti   okolo   pjatidesjati litrov   vody
        this.PL around fifty.GEN   liter.M.PL.GEN water.F.GEN
        'these over fifty liters of water'

The above facts are not expected on the null-noun analysis in (55): if the null noun were plural, it should license *de* in Dutch and the plural determiner in Russian regardless of the internal composition of the prepositional measure; and if the null noun were singular, gender agreement on the determiner would be expected. Matushansky and Zwarts conclude, therefore, that the prepositional measure is not embedded inside an NP.

Assuming that approximation in prepositional measures and prepositional numerals works the same, it would be odd to expect a null noun in the latter but not in the former. We conclude that the structure in (55) is not correct, and that prepositional numerals like *around twenty children* are PPs, not NPs. We propose that their NP-like behavior (e.g., the ability to appear as an argument of a transitive verb) is explained by their denotation (they denote entities rather than locations) and not by their lexical category. While a semantic analysis of prepositional numerals is beyond the scope of this work, proposals for a compositionally correct interpretation include Nouwen 2008, 2010; Schwarz, Buccola, and Hamilton 2012; Blok 2013, 2016a,b, for the semantics of directional numeral modifiers; and Matushansky and Zwarts 2017 for prepositional measure phrases.

9.4 Summary

In this chapter, we saw that there is much independent evidence for our analysis that a cardinal combines directly with the lexical NP prior to combining with a preposition, comparative operator, or modifier. We have proposed that so-called numeral modifiers actually fall into two broad categories: those that are modifiers, or adjuncts, such as 'at least' and 'exactly', and those, such as 'more than' and prepositions, that take the cardinal-containing NP as their complement. We showed that the behavior of all types of modified numerals supports our analysis that the cardinal does not form a unit with the so-called modifier to the exclusion of the lexical NP.

10 Partitives and Fractions

A discussion of cardinals, and complex cardinals in particular, would be incomplete without an analysis of fractions. On the one hand, fractions appear to be very similar to cardinals in that they denote quantities, and on the other hand, they contain cardinals as part of their internal structure. In this chapter, we build the syntax and semantics of fractions (section 10.2) on the basis of the syntax and semantics of partitives (section 10.1), while also accounting for the differences between the two. We will argue here for an analysis of English fractions as a type of relational noun whose lexical semantics incorporates the part-whole relation, rather than treating them as cardinal-containing partitives (contra Ionin, Matushansky, and Ruys 2006), thereby predicting certain syntactic distinctions between partitives and fractions.

10.1 Partitives and Pseudopartitives

There is syntactic similarity between fractions and partitives, especially if the former are viewed as a subtype of cardinals (1).

(1) a. all/some/*three* of my friends
    b. *half*/a quarter/two-thirds of my friends

Apart from the fact that both (1a) and (1b) involve the preposition *of* and the *part-of* relation between the denotation of the entire DP (*X of my friends*) and that of the embedded DP (*my friends*), it seems unquestionable that fractions and partitives form an integral part of a spectrum of related constructions, given in (2).[1]

(2) a. two of these three students   Regular count partitives
    b. a liter of water/five feet of snow   Measure partitives
    c. three-quarters of the cake/the beans/my   Fractions
       friends
    d. a number/lot/bunch of cats/my friends   Vague measure partitives

Given this fact, it would be certainly desirable to provide an analysis that would account for all of the constructions in (2). The first prerequisite for such an analysis would be a unified semantics for the preposition *of*. Whereas for measure partitives and vague measure partitives it has been suggested (De Hoop 1998; Vos 1999; Schwarzschild 2002a; Chierchia 2010; Martí Girbau 2010; and Rothstein 2011b, among others) that the preposition *of* is semantically vacuous and present for case reasons only, in regular partitives it is generally taken to encode the *part-of* relation (Ladusaw 1982; Hoeksema 1984; Barker 1998; Ionin, Matushansky, and Ruys 2006; etc.). We will follow the latter strategy, arguing that in all partitive constructions the *part-of* relation is contributed by the preposition. Our evidence for such an approach comes from the so-called *Partitive Constraint* and from fractions with a value greater than one.

10.1.1   The Partitive Constraint

As noted by Jackendoff (1977), with refinements by Barwise and Cooper (1981), Ladusaw (1982), Hoeksema (1984), and others (see De Hoop 1997, 1998 for an overview), the overt lexical NP in regular partitives is heavily constrained (3). The empirical generalization describing which NPs are allowed as complements of the *of* PP has been dubbed the Partitive Constraint.

(3) a. *two of every student
    b. *five of many/most linguists
    c. *four of students
    d. *six of a student

While earlier analyses (Jackendoff 1977; Selkirk 1977) treated the Partitive Constraint as a kind of definiteness effect, later work (Reed 1991; Abbott 1996; Hoeksema 1996a,b) converged on the hypothesis that the constraint is due to specificity (Reed 1991): indefinite NPs are also, as in (4), allowed in partitives on the condition that they are specific/referential in the sense of Fodor and Sag (1982), denoting the unique entity x such that the speaker intends to refer to x (cf. Heim 1991).[2] The Partitive Constraint is therefore restated requiring the lexical NP inside the *of* PP to be of type e (i.e., definite or specific).

(4) a. One of *three things* will happen: the aliens will be friendly, the aliens will be hostile, or the aliens won't care.
    b. Two of *some students that I know* are having a party.

A further refinement of Reed's (1991) clarification came from Abbott (1996), who demonstrated that existentially quantified NPs are in fact allowed in partitives (5) and argued for a general pragmatic principle, one that prohibits mentioning entities unless there is some reason for it, as the source for the Partitive Constraint.

(5) a. Every year only *one of many applicants* is admitted to this program.
   (Abbott 1996)
   b. John was *one of several students* who arrived late.
   (Ladusaw 1982)

A purely pragmatic approach cannot, however, explain the grammaticality contrast between the existentially quantified NPs in (4) and (5) and the universally quantified or bare NPs in (3). Conversely, as shown by Ionin, Matushansky, and Ruys (2006), the assumption that the partitive *of* is a regular preposition denoting the *part-of* relation (6) immediately and correctly predicts that its complement has the semantic type e. Importantly, we follow Barker (1998) in assuming that *of* generalizes over material *part-of* and individual *part-of* (cf. Link 1983) relations: if *of* combines with a singular NP, mass or count, it yields the set of material parts; if it combines with a plural NP, *of* yields the set of individual parts.

(6) $[\![of]\!] = \lambda x \in D_e . \lambda y \in D_e . y \leq x$
   (Ionin, Matushansky, and Ruys 2006)

Given that cardinals and most quantifiers can only combine with atomic sets (see chapter 2 for discussion), we correctly predict that cardinals and quantifiers cannot appear in partitives whose lexical NP is singular and thus denotes an entity or a kind (7a). However, we also incorrectly predict that cardinals and most quantifiers cannot appear in partitives whose lexical NP is a plural (7b): the set of individual parts of a plurality that the *of* PP denotes is not atomic, since it may contain plural individuals.

(7) a. *three/each of gold/this gold/child
   b. three/each of my friends/some students that I know

To deal with this issue, in Ionin, Matushansky, and Ruys 2006 we propose (see also Jackendoff 1977; Milner 1978; Abney 1987; Cardinaletti and Giusti 1992, 2005; Barker 1998; Zamparelli 1998) that in regular partitives the cardinal or the quantifier does not combine directly with an *of* PP but rather with an elided singular NP, to which the *of* PP is adjoined. In (8) we schematize this.

(8)

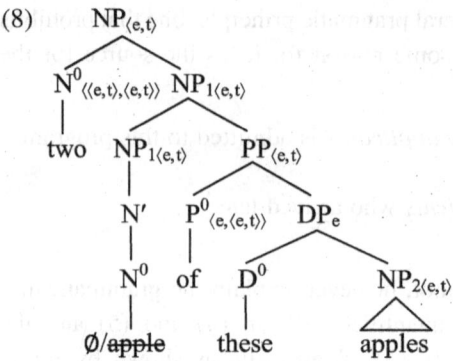

(Ionin, Matushansky, and Ruys 2006)

As is easy to see, the Partitive Constraint can now be derived. First of all, the structure in (8) and the meaning of *of* in (6) derive the meaning of (9a) and the impossibility of (9b) and (9c).

(9) a.   three of my friends
    b.   *three of this apple
    c   *three of this water

In (9a) the *of* PP returns the set of (plural) individuals that are individual parts of *my friends*. Combined with a null singular count noun (*friend*) it yields a set of atomic friends (10a). In (9b), the *of* PP returns the set of material parts of *this apple*. When combined with a null singular count noun (*apple*), which denotes in the count domain, it yields the entire apple (10b)—it is impossible to construct a plural individual consisting of that single apple. The same is true for (9c) because the sole atomic part of *water* is its totality (see Chierchia (2004)), and counting fails for the same reason as in (9b), as shown in (10c).

(10) a.   [[three ~~friend~~ of my friends]] $\approx \lambda x \in D_e$. x is a plural individual
            divisible into three nonintersecting nonempty individuals p_i such
            that their union is x and each p_i is an atomic friend and is *an*
            individual part of my friends
     b.   *[[three ~~apple~~ of this apple]] $\approx \lambda x \in D_e$. x is a plural individual
            divisible into three nonintersecting nonempty individuals p_i such
            that their union is x and each p_i is an atomic apple and is *a*
            material part of this apple
     c.   *[[three ~~water~~ of this water]] $\approx \lambda x \in D_e$. x is a plural individual
            divisible into three nonintersecting nonempty individuals p_i such
            that their union is x and each p_i is the totality of water and is *a*
            material part of water

The same reasoning accounts for the fact that bare plurals are impossible in count partitives (11): assuming that bare plurals denote kinds (see Carlson 1977a,b; Chierchia 1998; etc.), they are atomic and therefore incompatible with count partitives.[3]

(11) a. *one of children
b. *several/all of books

Support for this approach comes from the fact that kind-denoting singular definites are also incompatible with count partitives, under any interpretation, as (12) shows.

(12) a. *Ten of whales/the whale are extinct.
b. *Ten of whales/the whale are swimming in the ocean.

The ungrammaticality of partitives built on a singular universal, as in (3a), is due to the fact that it undergoes Quantifier Raising, leaving behind an e-type trace, that is, an atom. The situation is analogous for the quantifiers *all*, *each*, and *most*, which quantify over atomic parts (C. Roberts 1987). Conversely, if a partitive contains a universal quantification over pluralities (13), it is correctly predicted to be grammatical.

(13) ?She interviewed two of every seven people that walked down the street.

Given that universal quantification over cardinal-containing NPs requires the formation of a plural individual, when such a universal undergoes Quantifier Raising, the trace it leaves behind also denotes a plural individual, which does have atomic parts, as shown in (14).

(14) $\forall x$ [x is seven people that walked down the street $\rightarrow$ there exists a plural individual y divisible into two nonintersecting nonempty individuals p_i such that their union is y and each p_i is an atomic person and is *part of x* and she interviewed y]

The same reasoning derives the behavior of existentials, irrespective of whether existential force is achieved by existential closure over a choice function (Reinhart 1997; Winter 1997), or by Quantifier Raising of the plural existential: in both cases the complement of the preposition *of* denotes a plural individual, yielding the LFs in (15) for (5).

(15) a. every year $\exists x$ [x is many applicants $\wedge$ only one *of x* is admitted]
b. $\exists f$ [John was one *of f (several students)*]

Plural entities can have atomic individual parts, so an NP containing a vague cardinal (*few*, *several*, *many*, etc.) is possible inside a partitive even though it is quantificational. This correctly predicts that, as illustrated in (16), a nonspecific pure existential NP is possible as long as quantification is over plural individuals.

(16) a. Not being *one of six children*, I have nothing to say on the matter.
   b. What about the interaction between *any two of three intelligent species*?
   c. I would hate it for my boyfriend and me to be *two of seventeen housemates*—we would never be able to kiss in private.

To summarize, the fact that cardinals must combine with an atomic set makes count partitives incompatible with a singular NP_2, be it count or mass.

10.1.2 One or Two Nouns?

As discussed above, the approach to cardinals advocated here requires them to combine with atomic sets. The hypothesis that the preposition *of* returns a set of parts of its complement is therefore insufficient: to make count partitives work, we need to further restrict that set to a set of atomic parts. Following Ionin, Matushansky, and Ruys (2006) we suggested that count partitives involve the adjunction of an *of* PP to a null singular noun. An alternative solution would be to assume an optional insertion of the atomizing function At, which returns a subset of those members of a set that have no proper parts (17).

(17) $[\![At]\!] = \lambda f \in D_{\langle e,t \rangle} . \lambda y . f(y) \wedge \forall z [[f(z) \wedge z \leq y] \rightarrow z = y]$

The introduction of At makes it possible for an *of* PP to function as a complement of a cardinal, as schematized in (18).

(18)

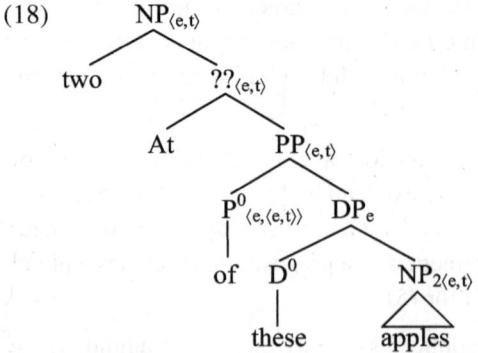

Given that the structure in (18) has the distribution of an NP, and given that cardinals and determiners don't normally combine with PPs, as illustrated in (19a), At should be viewed as a nominal element. However, the introduction of At does not explain why, as (19b) shows, partitives are possible only with those determiners that allow NP ellipsis. Assuming that At is actually the copy of the nominal head (*apples* in (18)) accounts for both of these properties.

(19) a. I congratulated each of {my colleagues/*in my department}.
   b. When faced with several problems, deal with each/*every (of them) in turn.

Further evidence for the presence of a null noun in count partitives comes from superlatives and ordinals. The fact that both are morphologically adjectival leads us to expect that they should be unable to combine with PPs (20), which in turn suggests that the grammaticality of superlative and ordinal partitives is due to an NP before the preposition *of*.[4] Independent crosslinguistic evidence for the presence of a null nominal head in superlatives is provided by Matushansky (2008a).

(20) a. Geraldine chose the most expensive of {the dresses/*in the shop}.
   b. *X-Ray* was the last of {today's movies/*in Jordan's schedule}.

The presence of gender marking on determiners and adjectives inside partitives in languages that have gender agreement (21) also supports the presence of a null noun, since PPs are generally not assumed to bear ϕ-features. It likewise suggests that the null noun in partitives is an elided copy of the overt noun rather than some dummy element—the latter hypothesis might lead us to expect a constant gender marking in partitives.

(21) a. une/*un de mes filles   French
      one.F/M of my daughters
      'one of my daughters'
   b. un/*une de mes neveux
      one.M/F of my nephews
      'one of my nephews'

There are, however, some problems for the hypothesis that partitives contain an elided noun. First, this hypothesis cannot explain why an overt noun in the same position is impossible, as (22) shows.[5]

(22) a. *two linguists of my friends
   b. *every woman of my relatives

Second, when NP$_2$ is a pronoun, as in (23) and (24), there is no overt NP that can serve as antecedent to the elided noun. However, some approaches to pronouns (Postal 1969; Panagiotidis 2002; Elbourne 2002) assume a richer syntactic structure to them, including a null NP, which can function as such an antecedent.

(23) a. *One thousand two hundred and eight of you* pledged to support The
        LINGUIST List.
     b. *The two of us* spent a romantic evening together.

(24) *Uno di noi* pensa que ...   Italian
     one of us thinks that
     (Cardinaletti and Giusti 2005, (115b))

To summarize, there exists some syntactic evidence for an elided noun in
partitives, which can be semantically singular, yielding a structure that denotes
an atomic set, necessary for composition with cardinals and singular quantifiers. However, this proposal is not without its problems.

10.2 Fractions

Among the issues to be discussed in this section are the obligatory use of the
preposition *of* in English fractions (25a), the alternation between the indefinite
article (25a) and the cardinal *one* (25b), fractions that amount to greater than
one unit (25c), and combinations of a cardinal and a fraction, with the lexical
NP following the cardinal (25d) or the fraction (25e).

(25) a. a third *(of) the wedding cake/the marbles/the wine
     b. a seventy-fifth/one-seventy-fifth of the population
     c. seven-fifths of the GNP
     d. two cakes and three-quarters
     e. two and three-quarters of a cake/*the cake

We begin by providing a compositional semantics for fractions and then
move on to a discussion of the syntax and semantics of complex fractions (e.g.,
two-fifths), combined fractions (e.g., *two and three-fifths*), and fractions greater
than one (e.g., *five-fourths*).

10.2.1 Compositional Semantics of Fractions

The semantics of fractions in (26) proposed by Ionin, Matushansky, and Ruys
(2006) correctly derives the truth conditions for fraction partitives, but is otherwise far from straightforward. The hypothesis that fractions, like cardinals,
are modifiers (semantic type $\langle\langle e,t\rangle,\langle e,t\rangle\rangle$) requires the condition (i) in (26),
whose purpose is to undo the work that the preposition *of* has done: (i) takes
a set of parts of a particular object (i.e., the denotation of NP_2) and returns the
largest such part, that is, the object itself.

(26) $[\![\text{third}]\!] = \lambda P \in D_{\langle e,t \rangle} . \lambda x \in D_e$.

(i) $\exists y \in D_e [y \in P \land \forall z [z \in P \to z \le y] \land$

(ii) $\exists S \in D_{\langle e,t \rangle} [\Pi(S)(y) \land |S| = 3 \land x \in S \land$

(iii) $\exists \mu [\mu \in M \land \forall s_1, s_2 [(s_1 \in S \land s_2 \in S) \to \mu(s_1) = \mu(s_2)]]]]$,

where M is a contextually determined set of measure functions.

On the one hand, the redundancy in the semantic combination of the preposition *of* and a fraction allows us to achieve uniformity of meaning for the preposition *of* across count and fraction partitives. On the other hand, it can also be reasonably argued that this unification is unnecessary. Given that the preposition *of* (and its equivalents in other languages) also functions as a case marker, in which case it is semantically vacuous, the analysis of fractions can be simplified as follows.[6]

(27) $[\![\text{third}]\!] = \lambda y \in D_e . \lambda x \in D_e$.

(ii) $\exists S \in D_{\langle e,t \rangle} [\Pi(S)(y) \land |S| = 3 \land x \in S \land$

(iii) $\exists \mu [\mu \in M \land \forall s_1, s_2 [(s_1 \in S \land s_2 \in S) \to \mu(s_1) = \mu(s_2)]]]$,

where M is a contextually determined set of measure functions.

Since (27) applies to an entity rather than a set, it naturally predicts that, as shown in (28), fractions (unlike cardinals) cannot combine with bare NPs, with or without the preposition *of* to assign case.

(28) a. *one-third (of) apples
    b. *two-thirds (of) cake

The next adjustment that we undertake here deals with the fact that the conditions (ii) and (iii) in (27) are (naturally) extremely similar to the meaning of the cardinal *three*, repeated for convenience in (29); however, the way they are stated does not permit us to derive the meaning of a fraction from the meaning of the corresponding cardinal.

(29) $[\![\text{three}]\!] = \lambda P \in D_{\langle e,t \rangle} . \lambda x \in D_e . \exists S \in D_{\langle e,t \rangle} [\Pi(S)(x) \land |S| = 3 \land \forall s \in S \ P(s)]$

The issue is easy to resolve: while the definition in (27) says that the whole is divisible into three parts that are identical with respect to some measure function (27iii) and one of which is the external argument of *third* (27ii), saying that all of these three parts are identical to the external argument of *third* with respect to that measure function (30) allows a simple reformulation of the meaning of a fraction in the terms of the corresponding cardinal (31).

(30) $[\![\text{third}]\!] = \lambda y \in D_e \cdot \lambda x \in D_e \cdot$

     (ii) $\exists S \in D_{\langle e,t \rangle} \left[\Pi(S)(y) \wedge |S| = 3 \wedge x \in S \wedge \right.$

     (iii) $\exists \mu \left[\mu \in M \wedge \forall s \in S \left[\mu(s) = \mu(x) \right] \right] \right],$

where M is a contextually determined set of measure functions.

(31) $[\![\text{third}]\!] = \lambda y \in D_e \cdot \lambda x \in D_e \cdot \exists \mu \in M \left[x \leq y \wedge [\![\text{three}]\!](\mu(x))(y) \right],$

where M is a contextually determined set of measure functions.

As a result of these amendments the lexical meaning for the derivational suffix -*th* becomes completely straightforward:

(32) $[\![\text{-th}]\!] = \lambda f \in D_{\langle\langle e,t \rangle,\langle e,t \rangle\rangle} \cdot \lambda y \in D_e \cdot \lambda x \in D_e \cdot \exists \mu \in M \left[x \leq y \wedge f(\mu(x))(y) \right],$

where M is a contextually determined set of measure functions.

As shown in the definition in (32), we assume that the morphological similarity between fractions and ordinals, exemplified in (33), is coincidental and most probably due to crosscultural influence. Support for this view comes from the fact that in some languages fractions and ordinals are morphologically distinct: for instance, while lower Spanish ordinals and fractions (from two to ten) have the same morphological realization (33c), for higher numerals they are clearly distinct (34).[7]

(33) a. un   troisième   French
        a/one third.M
        'one-third'

    b. dve   šestyx   Russian
        two.F sixth.F.PL
        'two-sixths'

    c. tres séptimos   Spanish
        three seven.ORD/FRACT.M.PL
        'three-sevenths'

(34) a. veintiún   trentavos   Spanish
        twenty.one.M thirtieth.M.PL
        'twenty-one-thirtieths'

    b. trigésimo/*treintavo   lugar
        thirty.ORD/thirty.FRACT place
        'thirtieth place'

To summarize, we depart from the analysis proposed by Ionin, Matushansky, and Ruys (2006) in assuming that fractions have the standard type of relational nouns ($\langle e, \langle e, t \rangle \rangle$) and that the role of the preposition *of* is that of a case marker. As a result, we get rid of the redundancy implicit in the definition of a fraction, thereby facilitating straightforward morphological derivation of a fraction from the corresponding cardinal.

10.2.2 Complex Fractions

Following Ionin, Matushansky, and Ruys (2006), we assume that complex fractions involving multiplication, as in (35), should be analyzed just like complex cardinals involving multiplication: they do not contain a constituent consisting of the multiplier (a cardinal) and the fraction to the exclusion of the lexical NP. Rather, as suggested by the semantics of *third* in (31), we hypothesize that complex fractions involving multiplication project a cascading structure, as in (36).

(35) a. three-sevenths of your income
     b. one-fifteenth of the GNP

(36)

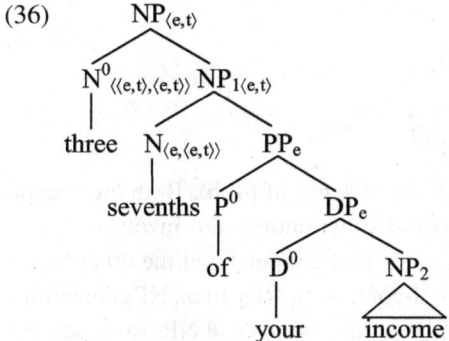

The semantic composition being totally straightforward, we now turn to cardinal-fraction combinations involving addition.

10.2.3 Combined Fractions

NPs containing simultaneously cardinals and fractions, as in (37a), provide additional support for our treatment of complex cardinals involving addition, as in (37b). Indeed, as discussed in chapter 5, the derivation of the correct truth conditions for (37b) under our assumptions about the semantics of cardinals (cf. (29)) requires the presence of two copies of the lexical NP, one in each conjunct of the cardinal. Exactly the same assumptions have to be made for an NP containing simultaneously a cardinal and a fraction, as illustrated in (38).

(37) a. two and three-fifths of *a cake*
     b. two hundred and three *cakes*

(38)

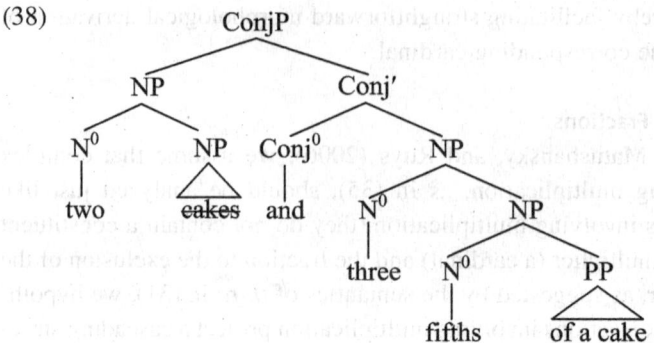

Evidence for the presence of the two copies of the lexical NP comes from the fact that either of them can remain overt, as in (39), although not both of them at once (40a).

(39) a. two *cakes* and three-fifths
     b. two and three-fifths of *a cake*

(40) a. #two *cakes* and three-fifths of *a cake*
     b. #twenty *people* and two *people*

The oddness of (40a) is on a par with the oddness of (40b). Both are acceptable under the interpretation where two distinct entities are involved, whose cardinality is specified: one denoted by the first conjunct and the other by the second conjunct. Unlike regular cardinal-containing NPs, then, NPs containing both a cardinal and a fraction allow either of the two lexical NPs to be deleted. We hypothesize that this difference is due to the fact that the lexical NP has a different semantic type in the two conjuncts of a complex fraction: it is a predicate when it combines with a cardinal and an entity when it combines with a fraction, and as a result, the choice is not without effect on the information structure and therefore is not grammaticalized.

The assumption that the fraction noun combines with a constituent of semantic type e, might lead us to expect in examples like (41) either that a particular cake is asserted to exist that everyone (counterintuitively) got two-thirds of (the existential taking wide scope over the universal, or a specific interpretation of the indefinite) or that for each individual a cake is asserted to exist that they received two-thirds of (the existential taking scope below the universal, yet again incorrectly predicting as many separate cakes in existence as there are individuals receiving fractions of them).

(41) Everyone got two-thirds of a cake.

To handle this issue, as well as a similar one arising with fractions greater than one, it becomes necessary to assume yet another scope position for the existential, which, as we will now show, provides further support for our cascading analysis of complex cardinals in general and complex fractions in particular.

10.2.4 Fractions Greater Than One

The scope of the existential quantifier in examples like (42a), as well as the interpretation of the complex fraction in (42b), further supports our cascading structure for fractions. To see this we need to determine the status of Anti-Uniqueness (Jackendoff 1968; Barker 1998) in complex fractions.

(42) a. If you have seven cakes and six people, each person gets
    seven-sixths of a cake.
  b. We cut each apple into three pieces, and in the end we had
    six-thirds of an apple.

It is easy to see why fractions with a final value greater than one should be incompatible with a definite or universally quantified NP_2, as in (43a): no object can have more than two halves, three thirds, four quarters, etc. What is unexpected is the contrast between (43a) and (43b): (43a) is perfectly grammatical, even though it is nonsensical (as indicated by #, a marker of infelicity), while (43b) is clearly ungrammatical. This is the case even though the denotation of the two NPs it the same: why is the partitive so much worse than the corresponding fraction?[8]

(43) a. #sixteen-eighths of the blueberry pie/every donation
    b. *two of the blueberry pie/every donation

Given that the complex fraction *sixteen-eighths* doesn't form a constituent to the exclusion of the lexical NP, we can derive the contrast by positing that the infelicity of (43a) is due to pragmatics (it requires nonlinguistic, arithmetical computation to find out that 16/8 gives you more than one) while the ungrammaticality of (43b) results from semantics: any attempt to count entities belonging to a singleton set is doomed to failure. The latter constraint also explains the ungrammaticality of (44a), but not the grammaticality of (44b)—to explain this we need to consider the scope of the existential quantifier in both configurations.

(44) a. *If you have six cakes and three people, each person gets *two of a cake.*
    b. If you have six cakes and three people, each person gets *six-thirds of a cake.*

Besides the infelicitous interpretation assimilating (44b) to (43b), (44b) can also entail that each person gets six objects, each of which is a third of a (different) cake—in other words, the indefinite NP can take scope below *six*. The cascading structure in (36) makes it possible for the existential quantifier to adjoin to the NP *thirds of a cake* on the assumption (see Heim and Kratzer 1998) that the subject position of this NP can be saturated by a null operator PRO (45).

(45)

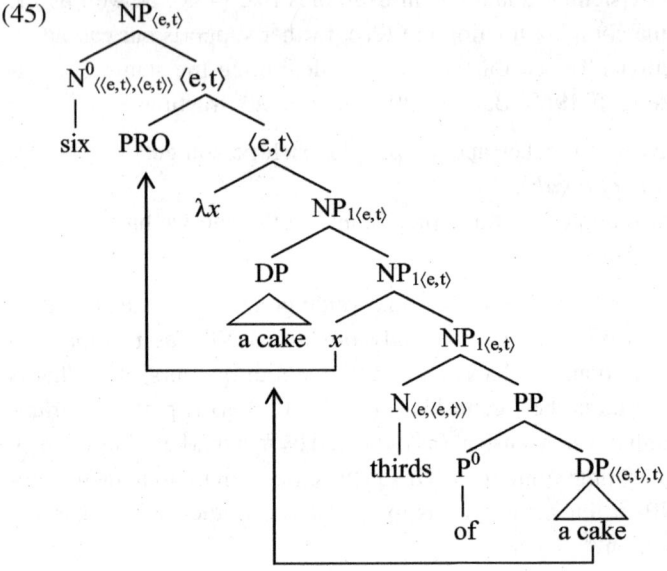

As is easy to ascertain, even if the same analysis were applied to a count partitive, it would still not yield the same effect; nor would it do so if the complex fraction were a constituent to the exclusion of the lexical NP.

10.3 Summary

We have argued for distinct syntactic structures for count partitives and fractions. Following Ionin, Matushansky, and Ruys (2006), we hypothesize that the complement of a cardinal in a count partitive is an NP with a null head, and the preposition *of* introduces the *part-of* relation. Conversely, we have demonstrated that the syntax and semantics of fractions necessitate a different structure, where the partitive PP is the complement of the relational fraction noun and the preposition *of* is semantically vacuous and appears for case reasons only, contra Ionin, Matushansky, and Ruys (2006).

We have simplified the semantics of fractions proposed by Ionin, Matushansky, and Ruys (2006) and proposed a lexical entry for the fraction-deriving suffix *-th*, clearly distinguishing it from the (frequently homophonous) ordinal suffix (on the semantics of ordinals, see, e.g., Bylinina et al. 2015). We have also shown that the resulting semantic composition correctly accounts for the interpretation of cardinal-fraction combinations involving addition.

The contrasting felicity of count partitives and complex fractions in cases where the resulting NP violates Anti-Uniqueness provides further evidence for our treatment of fractions, as does the possibility for the existential quantifier to scope between the multiplier and the fraction.

11 Conclusion

In this work, we have proposed an analysis of the syntax and semantics of simplex cardinals, and of the internal composition of complex cardinals, and we have also considered the implications of this analysis for constructions containing cardinals.

An important generalization that arises from our work is that the construction of complex cardinals in language uses entirely linguistic means. Building on our prior work in Ionin and Matushansky 2006, we have demonstrated that multiplication in cardinal-containing NPs is achieved as a combination of the semantics of simplex cardinals that we propose and the complementation structure. Turning to complex cardinals involving addition, we have updated our earlier proposal in several ways. First, we have discussed the adpositional strategy of complex-cardinal formation. Second, we have provided evidence that all additive complex cardinals involve NP deletion. And third, we have reexamined our proposal regarding the semantics of additive cardinals and have put forth a proposal for a novel interpretation of *and*.

In this work, we have expanded our coverage of complex cardinal systems crosslinguistically, which has enabled us to consider how language-specific characteristics (such as the construct state in Semitic languages, case assignment in Slavic languages, and the existence of two different numeral systems in Modern Welsh) play out in the construction of complex cardinals. Moving beyond the syntax and semantics of complex cardinals themselves, we have also examined the consequences of our proposal for related phenomena, including modified numerals and fractions.

A number of topics falling beyond the immediate scope of this work remain open for further investigation. These include a range of morphological, syntactic, and semantic issues. The similarities detected between the syntax and semantics of counting and of measuring (see also Rothstein 2013, 2016, 2017) indicate a need for a more unified analysis bringing together cardinals, vague cardinals, and measure nouns. Also relevant for further investigation is the formation and interpretation of ordinals (Bylinina et al. 2015) and other numeral derivatives. A topic left out of our current work is the syntax and semantics of *zero* (Bylinina and Nouwen 2017), which, though a cardinal, does not participate in complex-cardinal formation. As mentioned in chapter 9, there is a large body of literature on the semantics of modified numerals: while we have argued that within modified numerals, cardinals combine with the lexical NP before combining with the modifier, we have not examined the consequences of this proposal for all the different theories of modified-numeral semantics and pragmatics. And finally, there is always room for more cross-linguistic coverage of cardinal systems. In particular, interesting cardinal systems briefly touched upon in this work but not examined in their entirety include those of Scottish Gaelic (where number marking on the lexical NP is conditioned by both the cardinal and the head noun; see Acquaviva 2006), different varieties of Arabic, and Slavic languages other than Russian and Polish.

While no research is ever complete, this work in its present state makes a twofold contribution. On the one hand, it moves complex cardinals from a peripheral topic to the core of generative linguistics by demonstrating their relevance to linguistic theory. On the other hand, we provide both extensive empirical coverage and detailed syntactic analyses of a number of cardinal systems crosslinguistically. Thus, this work is intended as a resource for both generative linguists and typologists.

Notes

Chapter 2

1. Another way of formulating that $\Pi(S)(x)$ is true is $x \in \Pi(S)$. We maintain the formulation in (4), since it is the one used in our original proposal. The conventional way of indicating that a formula is true involves "= 1," which leads to confusion given that the only existing way of indicating that the cardinality of a particular set is three is by "= 3." This is why we indicate that the predicate Π applies to a partition and an individual without recourse to the "= 1" strategy.

2. Defined in terms of sets (cf. Gillon 1984; Schwarzschild 1994), a family of sets C is a cover of the set X iff

i. C is a set of subsets of X
ii. Every member of X belongs to some set in C
iii. $\varnothing$ is not in C

The first two conditions amount to claiming that X is the union of all members of C ($\cup C = X$). The last condition is superfluous for the definition in terms of (plural) individuals, since we do not assume empty individuals in the domain.

3. An anonymous reviewer objects that the behavior of "semilexical" cardinals like *hundred* should not be taken to be indicative of the behavior of all cardinals, given that *hundred* has a number of special properties (it must obligatorily combine with another cardinal or a determiner—**(a/one) hundred*—can appear in the plural, etc.). We come back to this issue in section 2.5, where we discuss the possibility that so-called semilexical cardinals have a different semantic type than other cardinals, and argue against it and in favor of a uniform semantics for all cardinals.

4. Once the notion of partition is introduced, the question arises whether it could be built into the predicative analysis of cardinals. As shown in section 2.1.1, the standard predicative analysis of cardinals (1) cannot derive the meaning of complex cardinals. What if instead simplex cardinals have the semantics in (i), modified from (4)?

(i) a. $[\![\text{three}]\!] = \lambda x \in D_e \,.\, \exists S \,[\Pi(S)(x) \wedge |S| = 3]$
  b. $[\![\text{hundred}]\!] = \lambda x \in D_e \,.\, \exists S \,[\Pi(S)(x) \wedge |S| = 100]$
  c. $[\![\text{three hundred books}]\!] = \lambda x \in D_e \,.\, \exists S \,[\Pi(S)(x) \wedge |S| = 3]$
$$\wedge \; \exists S' \,[\Pi(S')(x) \wedge |S'| = 100 \wedge [\![\text{books}]\!](x)]$$

The result is still problematic: in order to simultaneously be divisible into one hundred nonoverlapping individuals and three nonoverlapping individuals, it is sufficient for a plural individual denoted by *three hundred books* to consist of just one hundred books. This is clearly unsatisfactory. The issue would not be solved by having *hundred* combine with *three* before it combines with *books*: since all three elements are predicates, the order in which they combine is irrelevant. Thus, the use of partitions cannot salvage the predicative analysis of cardinals.

5. Postulating some kind of a [±multiplicand] feature runs into a problem for such complex cardinals as the French *quatre-vingt* 'eighty', where *vingt* 'twenty' can be a multiple with *quatre* 'four' only. However, this issue arises irrespectively of which linguistic analysis is adopted for complex cardinals.

6. This said, in some languages being a multiplicand does appear to correlate with certain such properties, but not in any way that promises a straightforward solution, as will be discussed in section 2.5.2 with regard to the proposal of Rothstein (2013). For example, in Russian, both *sto* 'hundred' and *tysjača* 'thousand' serve as multiplicands, yet the latter behaves much more like a noun with respect to case assignment and agreement (see chapter 6).

7. It is completely irrelevant for our argument which syntactic levels are projected above the NP in these expressions. The choice-function operator could just as easily be located in another functional head; what matters for us is the semantic type of the indefinite.

8. This analysis derives the collective interpretation of the subject; the distributive interpretation is derived via the DIST operator, as on any standard analysis (see section 2.6). The pragmatic nature of the 'at least' reading is convincingly argued for by Koenig (1991) and Buccola and Spector (2016), among others.

9. Arithmetical uses of cardinals should be distinguished from the use of cardinals in counting (which may involve morphologically distinct forms, especially for lower cardinals; see Hurford 2003 for the details). We will not discuss counting uses here.

10. While we maintain that NP-internal uses of cardinals require the semantics of modifiers, in some languages (such as Seri: see chapter 3), cardinals denote predicates in the lexicon (with the NP-internal semantic modifier use being derived), which is why we provide the derivation in (34a) as well as in (34b).

11. As with NP-internal cardinals, we abstract away here from intensionality, while noting the fact that names of numbers are surely rigid designators (Kripke 1980).

12. Rothstein (2013) notes that *ten* is an exception to this generalization (although *tens of thousands of* is a lot more frequent than simply *tens of*; for discussion, see "Why do we not say 'tens of'?" *English Language and Usage Stack Exchange*, http://english.stackexchange.com/questions/44975/why-do-we-not-say-tens-of, accessed February 16, 2017); yet she implies that the existence of the bound morpheme *-ty* can somehow explain this.

13. Even in English, the multiplier can be dropped in measure phrases combining with attributive adjectives: *a nearly (one)-hundred-foot-long jellyfish* and *the (one)-hundred-foot-long journey* are perfectly fine.

14. Sjef Barbiers (pers. comm.) suggests that the obligatory article in the Dutch approximative use is due to the presence of the null equivalent of the lexical suffix *-tal* 'number'.

15. It must be noted, however, that just like lexical powers, measure nouns in Hebrew do not exhibit construct-state morphology:

(i) a. *šloša [alfey   sfarim]
       three thousand.PL$_{\text{BOUND}}$ book.PL
    b. šlošet [alafim   sfarim]
       three$_{\text{BOUND}}$ thousand.PL book.PL
       'three thousand books'
    (Danon 2012)

(ii) šloša litrim/*litrey   vodka
     three liter.PL/liter.PL$_{\text{BOUND}}$ vodka
     'three liters of vodka'
     (Danon 2012)

We propose that this is a morphological phenomenon related to the special status of measure nouns with respect to the definiteness feature (see section 7.4). In Biblical Hebrew both lexical powers and measure nouns behave as expected in this respect:

(iii) 'ăśeret ṣimdêy   kerem
      ten$_{\text{BOUND}}$ acre.PL$_{\text{BOUND}}$ vineyard
      'ten acres of vineyard'
      (Isaiah 5:10)

16. We are primarily concerned here with relational NPs denoting quasi-equivalence relations (Barker 1999). However, adjectives like *mutual*, *equal*, and *reciprocal* and adverbs derived from them need not be contained in a plural NP (cf. *mutual distrust*, *reciprocal obligation*), though a plural licensor, explicit or implied, is nonetheless necessary. We believe that the analysis presented below can account for such uses as well.

17. There exist alternative proposals (for instance, Le Bruyn, De Swart, and Zwarts 2013) suggesting that relational nouns may not always have an internal argument. The treatment of a noun like *sister* as a simple sortal, however, does not explain its reciprocal interpretation in *two sisters*, which is what concerns us here.

18. There is some variability of judgments on this point: for some speakers, examples like (66) are felicitous even if there are some group members who are not each other's good friends (provided they are good friends of other members of the group). Such weakly reciprocal readings are particularly likely to arise if the group under discussion is very large: for example, with four people, (66) seems to require that they are all each other's very good friends, whereas with fifty people, some exceptions are allowed. This may be purely pragmatic (with a group of very large size, it is hard to imagine everyone being good friends of everyone else), or there may be genuine speaker variability in whether relational plurals are interpreted as strongly versus weakly reciprocal. If indeed relational plurals are *not* strongly distributive, then we need to use covers (Gillon 1984, 1987; Schwarzschild 1994, 1996) instead of the distributive operators in (67) in our derivations. Ultimately, this does not affect our proposal.

19. For ease of exposition we represent the $dist_0$ operator as a syntactic node; the syntactic treatment of λ-abstraction is as in Heim and Kratzer 1998. Nothing crucial depends on these assumptions.

20. On the crosslinguistic morphosyntactic collapse of reflexive, reciprocal, and anticausative verbs see, among others, Geniušienė 1987; Haspelmath 1987, 1990, 1993; and Embick 1997a,b.

21. Needless to say, this presupposition only concerns atomic arguments. It might seem at first sight that some relational nouns permit their two arguments to be identical (*I'm my own boss*, *He is his own worst enemy*, etc.). However, the fact that the presence of the emphatic adjective *own* is obligatory suggests that these apparent exceptions involve proxy interpretations (see Safir 2004; Reuland 2005; Reuland and Winter 2009) and therefore are not reflexive. An alternative hypothesis (Hackl 2002) is to appeal to individual concepts/guises.

Some evidence in favor of postulating irreflexivity as an inherent property of all nouns comes from the interpretation of reflexive nouns and adjectives formed with the prefix *self-/auto-*. Setting aside deverbal nouns and adjectives, such as *self-loader* and *self-destructive*, all the rest are anticausative rather than reflexive (e.g., *self-adhesive* paper does not adhere to itself, but rather *by* itself, without an external agent, and a *self-portrait* is not a portrait depicting the portrait itself). Interestingly, comparatives are also inherently irreflexive: for example, *Sabine is smarter than anyone* does not mean that Sabine is smarter than herself. Though the standard treatment of superlatives (see, e.g., Heim 1995) has irreflexivity as part of the truth conditions of the superlative morpheme, it should probably be taken as providing additional evidence for the hypothesis that superlatives are morphologically derived from comparatives (Stateva 2002; Bobaljik 2012).

22. It is tempting to hypothesize that verbs like *embrace* or *kiss* are inherently irreflexive, which is both eminently compatible with their semantics and provides a straightforward explanation for why the presence of a reflexive clitic on them gives rise to a reciprocal rather than reflexive interpretation in the plural (cf. Hackl 2002, 177). Since the issue is very far removed from the topic of this section, we set this hypothesis aside here.

23. The configuration under discussion does not fall under the purview of the *i*-within-*i* filter for two reasons: the QRing reflexive operator cannot be taken as the subject of the NP, and the subject of the NP does not bind its internal argument. As noted by Hackl (2002), his analysis should be excluded by the *i*-within-*i* filter.

24. Cappelletti, Butterworth, and Kopelman (2001), Cappelletti et al. (2005), and Domahs et al. (2005), among others, claim to provide evidence for a double dissociation between numerals and other words in certain aphasics. However, these studies appear to have concentrated on the number words rather than on the syntax or compositional semantics of cardinals, and they therefore can be argued to point merely to a loss of lexical knowledge of a particular lexical-semantic class.

Chapter 3

1. These proposals are quite different from one another, and make different predictions. For instance, Giusti (1991, 1997) argues that (in languages like Italian) vague cardinals

like *many* as well as cardinal numerals can function as ether Q^0 (when occurring with a bare noun) or an adjectival specifier (when occurring with D). On the other hand, Zamparelli (1995, 2002a) provides arguments for treating cardinal numerals as heads but vague cardinals, as well as modified cardinals (*exactly three*), as maximal projections. These differences are not directly relevant for our discussion (for our treatment of modified cardinal numerals, see chapter 9; for our view on the relationship between cardinals and determiners, see chapter 2). We concur with the proposals of Giusti, Zamparelli, and Ritter that cardinals are heads, so we can accommodate some of their predictions. Our concern, however, is with the internal composition of complex cardinals, something none of these proposals address.

2. For now, we focus on the case-assignment properties of cardinals in order to argue in favor of the complementation structure in (3). In section 3.4, we will provide evidence that many (though by no means all) cardinals in familiar languages behave morphologically like nouns, which makes them compatible with the structure in (3). While in (3), both cardinals are labeled as nouns, as we will show in section 3.4, nonbase cardinals can be adjectival rather than nominal, depending on the language. Base cardinals such as *hundred*, however, uniformly behave as nouns.

3. The fact that *liter* 'liter' appears in the singular is due to the fact that Dutch measure nouns generally do not inflect for plural in numeral NPs. We will return to this issue in chapter 4.

4. A number of languages including Albanian (Giusti and Turano 2004) and Estonian (Norris 2015) have been claimed to allow case agreement in pseudopartitives. However, as it turns out, it is observed only in some case positions. In Estonian case agreement is only observed in oblique cases, whereas in direct-case positions the substance NP is marked partitive. Conversely, in Albanian case agreement is only observed in direct cases, while in oblique-case positions a case different from that assigned to the whole pseudopartitive appears, showing that the substance NP is in fact assigned some case. Case agreement in pseudopartitives is also observed in Modern Greek (Stavrou 2003), where it alternates with genitive marking, and in Armenian (Koptjevskaja-Tamm 2001).

5. This choice seems at first glance not to be an innocent one, since the hypothesis that case assignment reflects the presence of an underlying preposition can also be made for cardinal-containing NPs. Given that in a number of languages cardinals combine with a plural NP in the genitive or partitive case, the question arises whether they should be treated as pseudopartitives, that is, as taking a PP complement headed by a null preposition assigning the proper case. We propose a pseudopartitive structure for Welsh cardinal-containing NPs in chapter 7 and show that it requires no change in the semantics of cardinals or in the constituency proposed. Therefore, in the absence of any evidence to the contrary, we will continue to assume that the lack of an overt preposition involves case assignment by the noun itself.

6. Note that the lexical NP in Inari Sami appears in the singular; we will return to this issue in section 3.5. It should be noted that the partitive case in Inari Sami is deficient in exactly the same way Russian paucal case is: it has only singular forms and is highly unproductive syntactically in that it is only used after higher cardinals and (for some speakers) in comparatives (Nelson and Toivonen 2000). However, unlike the Russian paucal case, which could potentially be treated as a number rather than a case (see

chapter 6), the Inari Sami partitive case is not amenable to such an analysis, due to the existence in the language of a real dual, on the one hand, and the use of partitive in nonnumerical contexts, on the other.

7. Attempts to reduce all agreement to Spec-head or c-command configurations have been repeatedly made, but require so many additional assumptions that we have disregarded them here, as their end result would have amounted, for our purposes, to the same simple fact: morphological agreement can take place whatever the underlying configuration. See Chung 2013 and Matushansky 2013 for some discussion.

8. Objections to Zweig's analysis are twofold: on the one hand, it is crosslinguistically the case that ordinals are transparently derived from cardinals, which is not captured on his approach; and on the other hand, different cardinals show different morphosyntactic behavior, as we discuss throughout this book, and therefore multiple varieties of Card would be required. With no independent justification, there is no reason to prefer this proposal over ours. In addition, this proposal faces all of the same problems as other proposals treating cardinals as specifiers, as will be discussed below.

9. A search of the Corpus of Contemporary American English (Davies 2008–) yielded not a single example of the (23a) type of construction. Occasional examples of the types in (23b–c) can be found on Google, but they are extremely rare.

10. Under most views of syntactic case, nouns cannot assign case and an additional head (a null equivalent of the English preposition *of*) is thus thought to be required. While this hypothesis can also be made compatible with our analysis, below we will provide evidence that both cardinals and nouns assign case themselves, and that the cases assigned by different cardinals may be different.

11. The plural of *unus* is regular. The cardinal *duo* sometimes has a genitive *duum*. The cardinals *duo* and *tres* do not inflect when used in complex cardinals formed by addition (see chapter 6 for similar facts in Polish and their treatment).

12. Allen et al. (1903, 60) provide some examples where it behaves as a noun even in the singular: "The singular *mille* is sometimes found as a noun in the nominative and accusative: as, *mille hominum misit*, he sent a thousand (of) men; in the other cases rarely, except in connection with the same case of *milia*: as, *cum octo milibus peditum, mille equitum*, with eight thousand foot and a thousand horse."

13. The question arises whether paucal is a case or a number. We will argue for the former option in chapter 6.

14. Franks (1994) proposes that cardinals assign *structural* genitive (or paucal) case, whereas other nouns assign *inherent* genitive. Inherent case is stronger than structural case, so when an inherent oblique case is assigned (by the verb or preposition), it wipes out structural case but not another inherent case. We will return to this in chapter 6.

15. The Seri glosses given are based on the Spanish approximations in Moser and Marlett 1994 and may not be as precise as one would wish them to be.

16. On the surface, however, a cardinal might have a morphologically plural complement, as in *two books*. We return to this issue in chapter 4.

17. It could be argued that a third structure is available, where the classifier is a head, as in (47a), but the cardinal is a specifier, as in (47b). In our semantic approach, such a structure can be excluded on the grounds of semantic redundancy: the cardinal in a

specifier position requires a complement denoting an atomic set, that is, another classifier.

18. The traditional approach to plurality associates plural morphology with the marked meaning of plurality. The alternative, proposed by Sauerland (2004), suggests that plural is the semantic default while singular is actually the marked choice, indicating the presence of a presupposition of atomicity:

(i) $[\![\text{SG}]\!](x)$ is only defined if x is an atom.

. $[\![\text{SG}]\!](x) = x$, where defined.

In Sauerland's view, the presence of singular marking in languages like Finnish, Welsh, Turkish, Hungarian, and Inari Sami is a straightforward morphological reflection of the CL head in (47a). The structure in (47b) does not seem to be compatible with these languages at all.

19. Note that the verb is singular in (54) and (55) and the determiner is plural in (53). For the effects of cardinals on number marking inside and outside the NP, see chapter 4.

Chapter 4

1. The behavior of suppletive plurals in Russian seems at first glance to suggest that the problem is purely morphological, as suppletive plurals are possible in the kind of adjectives exemplified in (4)–(5):

(i) a. mnogo-ljud-n-yj   (cf. *čelovek* 'person' vs. *ljudi* 'people')   Russian
        many-people-ADJ-AGR
        'having many people'
   b. mnogo-det-n-ij   (cf. *rebënok* 'child' vs. *deti* 'children')
        many-children-ADJ-AGR
        'having many children'

A careful investigation shows, however, that the situation is more complex. In particular, the stem that appears in adjectives derived from cardinal-containing NPs is not the plural stem, but rather the adnumerative stem, as shown by the following two examples:

(ii) a. trëx-let-n-ij   (cf. *god* 'year' vs. *gody* 'years'
        three-year.ADN-ADJ-AGR   vs. *tri goda* 'three years')
        'three-year'
   b. pjati-let-n-ij   (cf. *god* 'year' vs. *gody* 'years'
        five-year.ADN-ADJ-AGR   vs. *pjat' let* 'five years')
        'five-year'

For the measure noun 'year' the plural form is distinct from the form used with cardinals (see chapter 6): the former is regular, while the latter is not. The result can still be argued to be inconclusive, since for the noun *celovek* 'person' the pattern is reversed: the true plural (*ljudi* 'people') is suppletive and the adnumerative form is regular. However, (ia) turns out to be misleading: it is impossible to form adjectives on the basis of this noun and any cardinal; only the vague numeral in (ia) can do so.

2. A potential counterexample is provided by the irregular plural *pence*, since both *a three-pence coin* and *a three-penny coin* are allowed. Strikingly, *penny* is the only English measure noun whose plural clearly distinguishes the measure interpretation (*pence*) and the individuated interpretation (*pennies*).

3. It has been suggested (see Sauerland 2003 and Sauerland, Anderssen, and Yatsushiro 2005, among others) that the plural can (or even must) be underspecified for number (i.e., that general number can be realized as plural). The question therefore arises whether the plural marking in cardinal-containing NPs in languages such as English is an instance of general number or (as we propose) a result of agreement. The former hypothesis, however, cannot readily deal with the singular number with cardinals higher than 'one' in Hungarian, Welsh, Finnish, and so on.

4. To be more precise, while the lower cardinals prefer or require the direct singular variant, the higher cardinals only allow the plural one, with the cutoff point generally assumed to be *deg* 'ten' (King 2003, 111). We return to this matter in chapter 7.

5. As noted by Dal Pozzo (2007), the two kinds of vague numerals do not have the same syntax: the case-assigning vague numerals cannot be preceded by a demonstrative, while the case-agreeing ones can, strongly suggesting that only the latter function as modifiers rather than quantifiers. Note, however, that this difference in semantics does not correlate with number marking (unlike with *every* vs. *all*), showing that the two must be kept separate.

6. Schuh (1998, 233) also provides an example where a bare (singular) direct-object NP introduces a new discourse referent (also singular), which is picked up by a personal pronoun in a subsequent clause, thus showing that bare NPs need not be incorporated.

7. An alternative gloss for this example, 'meat of bush' (Schuh 1998, 211), suggests that some apparent plurals rendered as singulars could be due to the translator's error.

8. The same pattern also obtains NP-externally. While Schuh treats person-number-gender preverbal markers as "personal pronouns" (i.e., as subject clitics), their analysis as subject agreement is not excluded. In either case, Miya obeys Corbett's Agreement Hierarchy.

9. Schuh (1998, 259) glosses the example in (33) as 'his twenty chests of gowns', but the possessor must in fact form a constituent with 'gowns', as is clear from the discussion and examples on p. 250, where the possessive pronouns are clearly postnominal, and the alternative translation of the same example in context on p. 168.

10. Much of the material in this section was presented as Matushansky and Ruys 2014.

11. The same pattern is reported for Armenian (Donabédian 1993, 185–187): while plural marking in Armenian is conditioned by specificity, measure nouns are obligatorily singular even in definite NPs. We will return to this matter in section 4.3.3.

12. While some nouns in the list below (e.g., *vrouw* 'woman') are not typically classified as measure nouns, Klooster (1972) nevertheless treats them as such.

13. In modern Dutch, however, *bit* and other information-theory-derived measure units (*kilobyte*, *megabyte*, etc.) obligatorily appear in the singular in cardinal-containing NPs.

14. If a bare plural is used instead of a mass noun in (41a), plural agreement becomes possible in some idiolects. Singular agreement in (41b) is likewise possible in some idiolects.

15. A more serious problem for Bale, Gagnon, and Khanjian's analysis is posed by languages with no evidence for general number, such as Finnish, as discussed in section 4.2.2.

16. In all the Western Armenian examples below, we follow Sigler (1996) in using the # sign in front of a plural morpheme to indicate that use of this morpheme would not make the sentence ungrammatical but rather would force it to have a different interpretation, not the one indicated by the translation.

17. According to Ó Maolalaigh (2013), there are in fact forty distinct nouns that occur in the singular with the cardinals 'three' to 'ten' in the corpus that he examines, and for some of them, the plural form with these cardinals is also attested.

18. As noted by Greene (1992, 532), in Lewis Gaelic a few nouns also distinguish between the special adnumerative form, used only after these adjectival cardinals, and the plural form.

19. Interestingly, a search of the British National Corpus turns up only one example of the string *there were a group of* compared to fifteen examples of *there was a group of*. Given that the numbers are rather low (sixteen instances total, compared to ninety-one instances of the same alternation in COCA), it would be premature to draw strong conclusions about dialectal differences.

20. We note that instances of plural determiners with group nouns do exist, as in the following examples:

(i) a. We deserve these bunch of promising liars because we vote them into office.
     (http://home.earthlink.net/~netquest/zap.html)
   b. These team of specialists have hit the ground running to assist the industry with immediate needs.
     (http://www.agrapoint.ca/secondary/news_release_may72001.shtml)
   c. These group of patients are frequently younger and do not have atherosclerotic disease.
     (http://imaging.birjournals.org/cgi/content/full/15/1/45)

However, such examples are extremely infrequent: a search for the sequence *these group of* on COCA yields only two hits, in contrast to *this group of*, which yields 870 hits. Thus, it is highly likely that examples like those in (i) constitute performance errors.

One might object that a search of a corpus of American English is not informative, given that the phenomenon of plural agreement with group nouns is supposed to be a feature of British English. However, as discussed above, COCA searches show no shortage of plural predicate agreement with group nouns in American English. Furthermore, a search of the British National Corpus yields results that are no different from those of COCA: the search for *these group of* yielded zero hits, compared to 170 hits for *this group of*.

21. It is generally assumed that singular marking on the predicate reflects agreement with the syntactic head (the group noun), while plural marking reflects semantic

agreement with the subject. However, there is evidence that singular marking can also correspond to agreement failure, which is correlated with the interpretation of the subject as a measure. Addressing this issue here would take us too far afield. For more information on how it works in the case of Russian, we refer the reader to Matushansky and Ruys 2015a,b.

Chapter 5

1. An anonymous reviewer points out that deriving *twenty-five books* from *twenty books and five books* incorrectly predicts the availability of (ia), given the availability of (ib).

(i) a. *John and Mary read twenty-five books, respectively.
   b. John and Mary read twenty books and five books, respectively.

However, while (ia), an instance of asyndetic coordination, is indeed ungrammatical, the corresponding cases with overt conjunction are perfectly fine, as shown in (ii).

(ii) a. John and Mary read one hundred and five books, respectively.
    b. John and Mary read one hundred books and five books, respectively.

We propose to link this contrast to the interpretation of *respective(ly)*, which, according to Gawron and Kehler (2002), depends on pragmatic sequencing. In the absence of an overt conjunction, no reasonable sequencing can be established for cardinal-containing NPs, given that the order of cardinals is always the same. The introduction of an overt *and* allows for sequencing into two parts.

2. The third possibility is that a (null) classifier is projected as a sister of both cardinals in conjunction, making it possible for the overt lexical NP to be adjoined to the entire coordination phrase, as in *[[two hundred Cl and twenty Cl] books]*. We will not examine this possibility here, but see chapter 3 for arguments against positing a null classifier as sister to a cardinal in nonclassifier languages.

3. The cardinal 'one' in Polish behaves differently, as will be discussed in section 5.3.2. For a more detailed exposition of the syntax of Polish numeral NPs, see chapter 6 and Matushansky and Ionin, to appear.

4. NP deletion specifically in cardinal-containing NPs is also attested outside of complex cardinals involving addition, but only under highly restricted conditions. As discussed by Kayne (2003), in certain age and time-of-day specifications in English and some other languages, the lexical NP can be omitted, as illustrated in Kayne's examples below.

(i) a. At the age of seven (?years), John ...
   b. At the age of three *(months), their newborn daughter already weighed twelve pounds.
   c. Five *(years) later ...

(ii) Sono le sei.   Italian
     are   the six
     'It's six o'clock.'

(iii) bez   pjati   (minut)   sem' (časov)   Russian
without five.GEN minute.PL.GEN seven hour.PL.GEN
'five (minutes) to seven (o'clock)'

This effect is restricted to a very small set of measure nouns that varies from language to language: for example, in English, omission is possible (indeed, preferred) for years of age, but not for months of age, or years in general, as shown in (i). Note that in Welsh, as discussed in chapter 7, a special allomorph is used to distinguish years of age from years in general. In light of this, it is not clear whether there is any relationship between the phenomenon observed by Kayne and NP deletion in complex cardinals, which is not restricted in these ways.

5. We are grateful to Ida Toivonen and to Elsi Kaiser for information about Inari Sami and Finnish, respectively.

6. Where not indicated otherwise, the Romanian examples are due to Roxana Girju (pers. comm.).

7. The same effect is attested in Italian, where *uno* 'one', which agrees for gender as a simplex cardinal, is invariant inside an additive complex cardinal. This is illustrated in (i).

(i) vent-uno/*vent-una ragazze   Italian
twenty-one.M/F   girl.F.PL
'twenty-one girls'

8. The contrast between the adpositional strategy and the coordination strategy does not do justice to the full range of cognitive concepts used to express addition. Greenberg (1978) contrasts the comitative strategy (which for him comprises both comitative prepositions and coordination) with the superessive strategy (the preposition 'on') and notes the existence of three rarer ways of encoding addition: possession, excess, and remainder. Hanke (2010) extends the range of available rare options, adding other spatial relations, and observes that motion verbs can also be used to encode these. We will not address these patterns here.

9. One might expect a partitive reading for (68a) (as in *one boy out of twenty*). However, not only is that not the correct reading, the partitive in Latin is normally expressed by either genitive case or the preposition *ex*, not by *de*.

10. The absence of the reading in (72) means that *two books* has an 'exactly' reading only; if the 'at least' reading were available, then a single plural individual could have been both 'at least two books' and 'at least twenty books' at the same time. See also chapter 2 on this point.

11. An alternative derivation would be to first pluralize each of the conjoined NPs, then apply the set-product operation to the result; the outcome would be exactly the same as in (82).

12. Another case to consider is (i), which does not involve repetition of the same modifier and yet lacks the intersective reading.

(i) those twelve professors and two committees

Why shouldn't twelve professors also form two committees? We propose that the answer lies in the difference between plural individuals and collections: *committee* does not denote a set of plural individuals and therefore its intersection with *twelve professors* is empty.

13. Additive readings are also available to modified NPs (e.g., *the tall giraffes and large elephants*), which are generally assumed not to denote kinds (unlike NPs containing relational adjectives: McNally and Boleda 2004). This suggests that what we really require is not a kind but rather a concept (Krifka 1995; Bouchard 2002, 2005), since concepts can be formed on the basis of modified NPs. Ultimately, whether we treat *man and woman* as coordination of kinds or as coordination of concepts does not affect our proposal.

14. We are concerned here only with atomic sets, as we hypothesize that it is only atomic sets that compose with cardinals.

15. This analysis means that a conjunction of two cardinal-containing NPs, such as *two students and five professors*, which we have shown to be derived by the set-product operation (section 5.8.2), also involves the conjunction of kinds. We stress that type shifting from predicates to kinds with *two students* is a purely formal operation, and does not entail that *two students* is a natural kind. Indeed, cardinal-containing NPs do not behave like kinds on any test of kind reference (cf. Krifka et al. 1995; Chierchia 1998): they cannot combine with kind predicates (compare *#Five bears are widespread here* to *Bears are widespread here*); they do not obtain kind readings in the object position of *invent*-type verbs (*Babbage invented two computers* can only mean that Babbage invented two kinds of computers, *not* that he invented the two-computers kind); and they cannot denote kinds in episodic sentences (compare *Rats reached Australia in 1770*, which is about the rat kind, to *Five rats reached Australia in 1770*, which can only mean that five specific rats reached Australia, not that the five-rats kind did). We leave open the question of why cardinal-containing NPs do not behave like kinds yet can be shifted to kinds for the sake of coordination. We note that any analysis which treats cardinal-containing NPs as having type $\langle e,t \rangle$ (Link 1987; Verkuyl 1997; Landman 2003) has to explain why *two students and five professors* has an additive rather than an intersective reading.

Chapter 6

1. The cardinal *sto* 'hundred' is unique in having two paucal forms: the dual (*dve-sti* 'two hundred') and the nondual paucal (*tri-sta* 'three hundred', *pol-tora-sta* 'one hundred fifty', *četyre-sta* 'four hundred'). See also section 6.7.

2. Just as in English, in Russian these cardinals can occur in the plural even in the absence of another cardinal:

(i) Na ploščad' vyšli   tyšjači/milliony   studentov.
    on square   came out.PL thousand.PL.NOM/million.PL.NOM student.PL.GEN
    'Thousands/millions of students came out onto the square.'

The interpretation that obtains here, just as in the English equivalent, is the "plural of abundance" interpretation (see section 4.1.2). While it is not clear whether we are dealing here with cardinals or with the corresponding numerical nouns, the difference in behavior between 'thousand/million' and the lower cardinals suggests that they have a more nominal nature.

3. Only singular nouns of the first (-*a*) declension and feminine singular adjectives have a dedicated accusative form, with the exponents -*u*- and -*uju*- respectively. For all other nouns and adjectives (including demonstratives, possessives, and quite a few quantifiers; see Halle and Matushansky 2006) the underlying accusative is spelled out as genitive on animate NPs (and all personal pronouns) and as nominative otherwise.

4. The noun *para* 'pair' would seem to be an exception to this claim, since examples like (i) are well attested.

(i) èti   para časov
    this.PL pair hour.PL.GEN
    'this couple of hours'

Para can, however, be argued to be a vague numeral rather than a noun, based on both its interpretation ('couple') and its ability to trigger default agreement on the verb (ii). While it does not exhibit the homogeneous pattern of case assignment in oblique cases (iii), it can oddly surface in the subject position in what looks like the accusative case (iv).

(ii) Prišlo   para   posylok.
     arrived.N.SG pair.NOM parcel.PL.GEN
     'A couple of parcels arrived.'

(iii) s   paroj   čelovek/*čelovekami/*ljud'mi
      with pair.INS person.PL.GEN/person.PL.INS/people.INS
      'with a couple of people'

(iv) Bylo   paru   čelovek,   s   obščimi   dokladami.
     was.N.SG pair.ACC person.PL.GEN with general.PL.INS presentation.PL.INS
     'There was a couple of people, with general presentations.'
     (Letuchiy and Xolodilova 2011)

Other nouns naming imprecise quantities (e.g., *djužina* 'dozen', *polovina* 'half', etc.) may show the same behavior. See Letuchiy and Xolodilova 2011 for some discussion.

5. It is most likely that such gender agreement involves not proper cardinals but rather null-derived collective nouns (cf. *sto* 'hundred' vs. *sotnja* 'group of one hundred').

6. Mel'čuk adds to these two properties a third: the ability to combine with the count form (see section 6.6.2). However, as (18c) shows, collective nouns denoting quantities can also combine with the count form.

7. Mel'čuk (1985, 309–311), Babko-Malaya (1998), and Timberlake (2004, 197) note that in the direct cases an agreeing form is available for the quantifiers *mnogo* 'many' and *nemnogo* 'not many', which has the strong (proportional) reading. Rakhlin (2003) argues that the agreeing forms appearing in oblique cases are in fact proportional rather than cardinal and that the cardinals *mnogo* 'many' and *nemnogo* 'not many' do not in fact have oblique forms. We consider this analysis incorrect: on the one hand, the agreeing forms for *neskol'ko* 'several', *skol'ko* 'how many', and *stol'ko* 'so many' cannot appear in direct contexts, showing that the pattern is in principle available, and on the other hand, in oblique contexts cardinal readings of *mnogo* 'many' and *nemnogo* 'not many' are possible. We set this issue aside.

8. Russian adjectives fall into two declensional patterns, which Halle and Matushansky (2006) call *regular* (those using the theme suffix *-oj-* in singular cells of the paradigm and the theme suffix *-yj-* in plural cells) and *irregular* (those lacking these suffixes in the direct cases). Functional or semifunctional items, such as possessives, determiners, and quantifiers, and lexical adjectives can be found in both groups (Halle and Matushansky 2006), but cardinals only show the irregular declensional pattern.

9. The dual Q *poltorá/poltorý* 'one and a half.M/F' has a deficient paradigm (oblique *polútora*), with the gender distinction neutralized in the oblique case. The morphology of the direct case is compatible both with treating it as agreeing in number and with failing to do so. We will not decide between these two options here.

10. The remaining question is why interpretable number specification on the higher cardinals does not lead to plural morphology, irrespective of whether they are specified for gender (the case of the cardinals higher than 'thousand') or not. Intuitively, the lexically specified plurality is a purely formal feature for pluralia tantum nouns but forms part of the lexical meaning of cardinals. Given that the two are treated differently for agreement processes (Corbett 1979, 1983, 1991, 2006; Wechsler and Zlatić 2000, 2003, 2012; etc.; see Landau 2015a for a recent Minimalist treatment), it is not implausible that a formal distinction between being valued and being interpretable must be drawn, as suggested by Pesetsky and Torrego (2001). Alternatively, it is not impossible for the same lexical item to be simultaneously specified for interpretable and uninterpretable instances of the same feature.

11. Surnames derived from possessive adjectives decline like regular adjectives in oblique cases and like nouns in direct cases (the *irregular declension pattern* of Halle and Matushansky 2006), which means that their nominative-plural ending is identical to the genitive-singular ending of feminine nouns. This is why Franks (1995) interprets the behavior of such surnames as evidence that the paucal form corresponds to a particular case marking and that this case is direct rather than oblique. We will return to this issue in section 6.3.6.

12. As discussed by Mel'čuk (1985, 432), the genitive-singular form of the nouns in (29) is also available with paucal cardinals, depending on the context. According to our informants, the genitive singular is possible for all but one of these nouns (*čas* 'hour'). There would appear to be semantic/pragmatic factors behind the variation between the paucal and genitive-singular forms, but we cannot determine them here.

13. Pesetsky (2008) proposes that the lexical NP appearing in the paucal form is in fact marked the genitive singular. However, Xiang et al. (2011) provide data from processing that strongly suggest that the paucal form has an underlying representation different from the surface genitive singular.

14. Zaliznjak (1967) proposes that the paucal form lies outside the normal case-number paradigm and is therefore unspecified for either case or number. Since this hypothesis does not appear to predict the distribution of the paucal form, we set it aside here.

15. The syncretism between the genitive singular and the nominative plural appears to be peripheral to the issue, as shown by the fact that in Ukrainian and Belarusian (Mayer 1973; Akiner 1983) both nominative- and genitive-plural marking can appear on the attributive adjectives in cardinal-containing NPs, despite the lack of such syncretism in certain cases (see section 6.3.4). Likewise, in nineteenth-century Russian

(Vinogradov 1952, 372–377), masculine NP complements of a paucal cardinal allowed both nominative and genitive plural on the attributive adjectives, while for feminine NPs the genitive plural on the attributive adjective was dispreferred compared to the nominative plural. Furthermore, Vinogradov (1952) provides examples where the nominative plural and the genitive singular of a masculine noun have different realizations:

(i) ... uvidel   čërnuju   borodu   i   dva   sverkajuščie
    see.M.PAST black.ACC beard.ACC and two.M.ACC/NOM shining.PL.ACC/NOM
    gláza.
    eye.SG.GEN/PAUC
    '... saw a black beard and two shining eyes.'

(ii) ... Fifi sdelala   dva   nerešitel'nye   šaga ...
      Fifi make.F.PAST two.M.ACC/NOM hesitant.PL.ACC/NOM step.SG.GEN
      '... Fifi made two hesistant steps ...'

Moreover, first-declension masculine nouns (e.g., *mužčina* 'a man (male human)') disallow nominative on the adjective, even though their nominative-plural and genitive-singular forms are also generally identical.

The diachronic change occurring here seems straightforward: from the optionality of realizing the paucal form of the adjective as either genitive or nominative (probably due to the uncertainty of speakers as to whether it is a case, in which case it is a subspecies of genitive, or a number, in which case the entire NP is analyzed as nominative), through its realization as nominative only in cases of systematic nominal syncretism (the current stage), towards its systematic realization as genitive plural due to the reanalysis of paucal as a deficient variant of genitive.

16. Pesetsky (2008) notes that, contrary to the accepted wisdom, corpus data from Google show the acceptability of the genitive-plural form of feminine proper names (assimilating them to attributive adjectives), but not of the genitive-singular form.

17. Null derivation is less frequent in Russian than it is in English, but nonetheless is found in, for instance, shop names (*buločnaja* 'bakery') and room names (*vannaja* 'bathroom'); in addition, Russian has a number of lexical items that behave like nouns and never demonstrate adjectival syntax or semantics yet decline like adjectives (*portnoj* 'tailor', *vselennaja* 'universe', etc.; see Halle and Matushansky 2006, 379).

18. The cardinal *pol-* is a phonological clitic and is generally spelled as one word with the following noun. The cardinal *poltora* 'one and a half' reflects this convention.

19. Mel'čuk (1985, 282–288) argues that the "oblique case form" *polu-* is in reality a separate morpheme with certain peculiar properties, such as the absence of direct-case forms, since its use is restricted to a closed set of nouns, which don't allow adjectival modification (unlike *pol-*). However, he also notes (287) that alongside these forms colloquial Russian also allows *pol-* in the oblique cases, including the genitive (i) and the locative (ii).

(i) A   ob   vodke ni   pol-slova.
    and about vodka NEG half.ACC/NOM-word.SG.GEN
    'And not a word about vodka.'

(ii) V pol-djužine   tjurem   proizošli   ser'ëznye
      in half.ACC/NOM-dozen.LOC jail.PL.GEN happen.PL.PAST serious
      volnenija.
      disturbance.PL.NOM
      'In half a dozen jails, there were serious disturbances.'

We will assume here that the alternation between *pol-* and *polu-* is a matter of phonology conditioned by linear adjacency.

20. Garde (1998, 242) claims that *poltora* shows gender distinctions not only in the direct cases but also in the oblique ones, but Vinogradov (1952, 383–384) marks these as archaic. Vinogradov also notes that in modern Russian the lexical NP appears in the plural in oblique cases (unlike with the cardinal *pol-*), but such was not the case in the nineteenth century.

21. Unlike *pol-* 'half' and *poltora* 'one and a half', *četvert'* 'quarter' allows *čas* to show up in either the genitive singular or the paucal (count) form (ia). At the same time, *četvert'* requires genitive singular for other nouns (ib), and shows the heterogeneous case pattern in oblique cases (ic). Given that in all other respects *četvert'* behaves like *polovina* 'half', we have no explanation for the paucal form that surfaces with *čas*.

(i) a. za četvert'   čása/časá
        for quarter.ACC/NOM hour.SG.GEN/PAUC
        'quarter of an hour'
    b. četvert'   šára/*šará
        quarter.ACC/NOM balloon.SG.GEN/PAUC
        'quarter of a balloon'
    c. četvertju   *časá/čása/*časom   ranee
        quarter.INS hour.PAUC/SG.GEN/INS earlier
        'a quarter of an hour earlier'

22. While singular marking on the predicate is the default, it is possible to construct a context in which plural marking is allowed, for example, when the years in (59a) are definite, referential, or partitive. The same is true for event-oriented readings. This is consistent with the account of Pereltsvaig (2006b), on which definiteness/specificity correlate with plural agreement.

23. It is not possible to provide examples in which the plural suppletive form occurs in the genitive singular, because most of these suppletive forms are pluralia tantum and do not have a paradigm cell (genitive singular) that could be used with the paucal cardinals.

24. We are indebted to Roumyana Pancheva for the Bulgarian examples.

25. For the cardinal 'two', two variants are available: the more archaic *dvu-* and the modern *dvux-*, which is identical to the genitive form. For some compound-based derivations both options are available (e.g., *dvuletnij/dvuxletnij* 'two-year'), whereas for others only one is possible (e.g., *dvuličnyj* 'duplicitous', *dvuxtomnyj* 'two-volume'). For a discussion and some regularities see Rozental, Džandžakova, and Kabanova 1998, 232–233.

26. Es'kova (2011, 279) notes that the genitive forms are colloquially possible in ordinals. This option is excluded from standard literary Russian.

27. To capture the varying properties of different cardinals in Polish, Klockmann (2017) adopts a ϕ-deficiency approach in which different cardinals are differently specified for ϕ-features such as number and gender. We will not go into such details here, but see Matushansky and Ionin, to appear.

28. All the examples in section 6.8, except where indicated otherwise, are taken from Miechowicz-Mathiasen 2011.

29. The cardinal 'two' also has a form *dwu*, which does not distinguish for gender at all (Brooks 1975, 316).

30. Russian has a similar phenomenon whereby collective rather than cardinal lower numerals must be used with first-declension masculine nouns, as shown in (ia); second-declension masculine nouns can combine with either form (ib).

(i) a. Dvoe   muzščin   / *dva   muzščiny prišli.
       two.COLL man.PL.GEN two.NOM/ACC man.PAUC arrive.PL.PAST
       'Two men arrived.'
    b. Dvoe   malčikov / dva   malčika prišli.
       two.COLL boy.PL.GEN two.NOM/ACC boy.PAUC arrive.PL.PAST
       'Two boys arrived.'

With regard to the Polish data, Decaux (1964) claims that the use of the nominative form is only possible for a specific numeral-NP subject, whereas the genitive form correlates with its nonspecific interpretation. If correct, this datum provides further evidence in favor of not treating the nominative form as a cardinal.

31. We note that (101c) with *tych* is acceptable on a particular reading, where *tych* is taken as a modifier of 'letters' that has been fronted: 'a kilogram of these letters', not 'this kilogram of letters'. This is not relevant for our purposes.

Chapter 7

1. All examples in this section are taken from Hiraiwa 2013, except where indicated otherwise.

2. While Hiraiwa (2013) regards the forms in (3a–b) and (3c–d) as singular and plural, respectively, Bodomo and Marfo (2006) argue that the situation is more complex and that the singular–plural opposition in Dagaare is expressed by inverse marking (see also Grimm 2010, 2012). We leave this complication aside as not directly relevant to our analysis.

3. We have no explanation for the fact that, crosslinguistically, head-initial languages typically have deletion of the NP in the leftmost conjunct and head-final languages typically have deletion of the NP in the rightmost conjunct. Exceptions do exist, for example in Scottish Gaelic, Biblical Welsh, and Biblical Hebrew, as discussed in chapter 5.

4. There is much variation in the phonological (and correspondingly, orthographic) form of each cardinal word, depending on the environment; see Hurford 1975 for the details.

5. As discussed by Nurmio and Willis (2015), in Modern Welsh this apparent plural is in fact a special numerative form, distinct from the true plural of 'year', *blynyddoedd*. We abstract away from this complication here. We also note that, according to Nurmio and Willis, this noun is special in yet another way, in that it has a different allomorph, *blwydd*, when 'years of age' is intended (e.g., *tair blwydd* 'three years of age').

6. While Hurford glosses *onid* and *namyn* as 'minus', Pughe (1832) translates them, respectively, as 'if not, unless, except' (389) and 'except, but, save' (368), which indicates that they are in fact caritative prepositions and potentially clausal connectors.

7. Williams (1980, 41) notes that at an earlier stage of Welsh, plural nouns could (or had to) follow cardinals, as in (i). In Modern Welsh (ii), this option is no longer available.

(i) pedair   merched   (ii) pedair merch
    four.F   girl.PL   four.F girl.SG
    'four girls' (Acts 21:9)   *'four girls'*

See note 5 discussing the hypothesis of Nurmio and Willis (2015) that the surface plural is a special numerative form.

8. For the Modern Welsh examples and judgments in this section that are not attributed to other sources we are indebted to Peredur Webb-Davies.

9. The choice between the two allomorphs in (26) depends upon what follows the cardinal: if nothing does, or if an 'of' PP is used, the longer form is selected. We will return to the matter in section 7.3.2.

10. The cardinal *deg* 'ten' changes in such examples to *deng*. King (2003, 113) claims that the context for the change is a measure noun starting with [m], as well as the nouns *blynedd* 'year' and *diwrnod* 'day'. It must also be noted that *blynedd* is a special numerative form, which appears in all cardinal-containing NPs denoting pluralities (see Borsley, Tallerman, and Willis 2007, 166, for evidence that it patterns with the plural form with respect to mutation); the singular form *blwyddyn* 'year' is used in all singular contexts, as well as with the cardinal 'one' (Ball and Fife 1993, 312). No other noun in Welsh distinguishes between the numerative and the singular.

11. The older singular pattern also correlates with the vigesimal system used for cardinal formation and is obligatory in time measurement (see below), whereas the newer plural pattern correlates with the decimal system.

12. We are grateful to Norah Boneh and Yael Gertner for the Modern Hebrew judgments in section 7.4.

13. As noted by Faust (2014), the bound form of nouns is not restricted to the construct state, it also occurs in word formation. To explain this, Faust analyzes the distribution of the bound form syntactically (the bound form is used in the absence of a phase boundary between the host and the next word), which is compatible with our proposal.

14. The nominal cardinal *me'a* 'hundred', being a canonical feminine noun, should have the bound form *me'at*, which is in fact available and is optionally used as a simplex

cardinal (Glinert 1989, 83; however, Glinert does not describe its distribution). Conversely, *elef* 'thousand' in the singular is not expected to have a morphologically distinct bound form, due to its status as a canonical masculine noun; the same is the case for *meot* 'hundreds', as a canonical feminine plural.

15. In spoken Hebrew, just as it is possible to place the definite article before the totality of the construct state (42b), so it is possible to combine the definite article with the entire cardinal-containing NP (Shlonsky 2004; Borer 2005, 213fn.), as in (i). The same holds for complex cardinals involving multiplication, as in (ii).

(i)   %ha-hamišim sfarim
         the-fifty   book.M.PL
         'the fifty books'

(ii) %ha-šlošet   alafim   sfarim
         the-three.M.BOUND thousand.M.PL book.M.PL
         'the three thousand books'

The ability of the definite article to appear before the cardinal is a property of spoken Hebrew only. We note that this placement of the definite article is also what we find with measure nouns, for which no other option is possible, as shown in (iii).

(iii) a. ha-kilo kemax
             the-kilo flour
             'the kilo of flour'
       b. *kilo ha-kemax
             kilo the-flour

16. Modern Hebrew does not have productive dual formation, although a closed class of nouns (certain lexical duals, such as 'pants', and some time specifications, such as 'months') occurs in the dual; see Glinert 1989.

17. We note that the structure in (55) is in principle also applicable to the cardinals *thirteen* through *nineteen* in English, if *teen* is given the same special semantics as dependent 'ten' in Hebrew. However, since English cardinals do not exhibit gender agreement, we have no independent motivation for treating *thirteen* through *nineteen* as built in the syntax rather than the lexicon (see also discussion in chapter 5).

18. At this point, the authors wish to impart the following joke to the reader. A mathematician presenting a problem to the class declares at some point, "The proof of this theorem is quite obvious." One of the students asks for details. The mathematician ponders for a while, then proceeds to write on the board for the next two hours, and finally declares, "See, I told you it was obvious."

19. We note that there is one other environment, apart from complex cardinals, where bound forms are impossible, namely NP ellipsis. This is readily seen for 'two' in (ia). For other cardinals, the required contrast only arises in definite NPs (ib).

(i) a. Ra'iti   šloša   sfarim,   aval kaniti   rak šnayim/*šney.
             saw.1SG three.M book.M.PL but bought.1SG only two.M/two.M.BOUND
             'I saw three books, but I bought only two.'
       b. Ra'iti   šloša   sfarim,   ve kaniti   et kol ha-šloša/*šlošet.
             saw.1SG three.M book.M.PL and bought.1SG ACC all three.M/three.M.BOUND
             'I saw three books, and I bought all three.'

20. In Biblical Hebrew, though, the superlative is formed by the construct state of the adjective (D. W. Thomas 1953), as in (i), suggesting that the head of the construct state does not necessarily determine the category of the entire phrase.

(i) ktan   banav
     small son.PL.3SG.M.POSS
     'the smallest of his sons' (1 Samuel 16:11)

21. Examples from Modern Standard Arabic in section 7.5 not attributed to other sources are due to Abelaadim Bidaoui.

22. Alqarni (2015), like us, argues that 'one' and 'two' in Arabic are adjectival whereas 'twenty' through 'thousand' are nominal. As for 'three' through 'ten', which do not behave straightforwardly like either nouns or adjectives, Alqarni analyzes them as quantifiers. For us, on the other hand, cardinals are on a continuum from the most adjectival to the most nominal, with 'three' through 'ten' in the middle of the continuum.

23. The fact that 'one' in Standard Arabic exhibits normal agreement might suggest an alternative explanation: the chiastic-agreement pattern exhibited by 'three' through 'ten' may be a form of number agreement. This view is supported by the fact that in closely related Cushitic languages, including Somali, plurals of masculine nouns are feminine and plurals of feminine nouns are masculine (see Lecarme 2002 for data and references). Taking this view on board would mean that all lower cardinals in Arabic languages agree not only for gender but also for number. Higher cardinals, on the other hand, do not show number or gender agreement.

A formal account of these facts is hampered by the lack of appropriate machinery. As the examples in (67b,d) show, the cardinal 'two' is formally marked for (dual) number, but it also, per our theory, provides the source for the valuation of the uninterpretable number feature on the lexical NP. One possibility is that 'two' bears two instances of the number feature, an interpretable one that determines number marking on the lexical NP (which is semantically singular, per our theory) and an uninterpretable one that determines number marking on the cardinal itself. The alternative is that 'two' bears only the interpretable number feature, and its uninterpretable counterpart, which is realized as the dual morphology on the cardinal, is borne by a higher functional head to which the cardinal 'two' moves via head movement. Yet a third possibility is that both the lexical NP and the cardinal bear the uninterpretable number feature, and their interpretable counterpart reflecting the semantic plurality of the entire NP is located on a functional head higher in the NP. We cannot at present decide among these three possibilities.

24. The agreement patterns of cardinal-containing NPs in Palestinian Arabic are nearly identical to those of Lebanese Arabic (Elias Shakkour, pers. comm.). As in Lebanese Arabic, lower cardinals combine with a plural lexical NP and trigger obligatory plural agreement on verbs and adjectives, while higher cardinals combine with a singular lexical NP and may fail to trigger agreement on verbs and adjectives, with the same consequences for collective versus distributive readings. A notable feature of Palestinian Arabic is that this optionality appears to be restricted to [+human] NPs; [−human] NPs obligatorily trigger plural agreement on verbs and adjectives, although there may be some exceptions. We do not have an explanation for this phenomenon.

Chapter 8

1. As discussed in chapter 2, we follow Matushansky and Zwarts (2017) in treating measure nouns on a par with regular nouns as denoting in the domain of individuals rather than degrees. Such unified treatment has the advantage of allowing us to provide a single analysis for both measure and nonmeasure nouns inside the modified-cardinal construction.

2. Bylinina, Dotlačil, and Klockmann (2016) try to take into account the contribution of the NP by relativizing *mere* and *amazing* to the context. However, this does not solve the problem for examples like (4), where the context must be the same for both instances of 'ten'.

3. Honda (1984) and Ellsworth, Lee-Goldman, and Rhodes (2008) observe that the modified-cardinal construction is compatible with a range denotation (i), albeit with certain restrictions. We leave this issue for future research. We also note that the modified-cardinal construction is ungrammatical with prepositional numerals, as in (ii) (thanks to an anonymous reviewer for pointing this out) and in (iii). This is compatible with our analysis of prepositional numerals as PPs rather than NPs (see chapter 9). As PPs, prepositional numerals cannot be modified by an adjective nor appear with an article.

(i) a. a whopping three/thirty/three hundred to four/forty/four hundred admirals
     b. a mind-boggling million to billion star clusters

(ii) a. *a whopping from three/thirty/three hundred to forty/forty/four hundred admirals
     b. *a whopping between three/thirty/three hundred and four/forty/four hundred admirals
     c. *a mind-boggling from (a) million to (a) billion star clusters
     d. *a mind-boggling between (a) million and (a) billion star clusters

(iii) *an amazing over one hundred customers

4. Indefinite-article insertion in plural cardinal-containing NPs cannot reflect any kind of formal [singular] specification. We suggest therefore that the English indefinite article might be better viewed as encoding the formal property of being quantized.

Chapter 9

1. We have nothing to say about the semantics of superlative numerals, but we point the reader to a number of recent analyses (Krifka 1999; Hackl 2000; Geurts and Nouwen 2007; Büring 2008; Geurts 2010; and Cummins and Katsos 2010, among others) which have independently challenged the complex-determiner analysis of superlative numerals (1b) on semantic grounds. We note that some of these analyses, in particular Krifka's (1999) and Geurts and Nouwen's (2007), take up the uses of superlative modifiers with expressions other than numerals, as in (i). On our analysis, such uses are expected, and the correct semantic analysis of *at most/at least* should allow this modifier to combine with a variety of expressions, including cardinal-containing NPs.

(i) a. Julie took at least the allowed number of classes.
  b. Robin talked with at least John.
  c. At least it is not raining.
  d. Julie is at least an associate professor.

2. There are in principle two different 'much' readings: the additive reading (five sandwiches + something else, as in (19b)) and the replacement reading (something more substantial than five sandwiches, as in *I ate more than five sandwiches—I ate a whole dinner instead!*). Results from a survey of native speakers that we report in Ionin and Matushansky 2013a indicate that in English, replacement readings are not accepted as readily as additive readings (for Russian clausal *bol'še* comparatives, discussed below, both readings are accepted). We leave this issue aside.

Chapter 10

1. Some types of vague measure partitives (*a number of cats*) as well as regular measure partitives (*two bottles of wine*) have been divided into *real partitives* and *pseudopartitives* (Selkirk 1977). We set this distinction aside here.

2. The examples in this section come primarily from Ionin, Matushansky, and Ruys 2006.

3. If, as argued by Van Geenhoven (1996), bare plurals denote properties, the ungrammaticality of (11) follows, but the grammaticality of bare plurals in (vague) measure partitives doesn't.

4. The contrast between partitives and all other PPs disappears if a plausible antecedent for NP ellipsis can be found in the context. We hypothesize that this effect also explains why superlatives and ordinals combine with PPs much more easily when the resulting DPs denote humans.

5. One exception, *every man jack of them*, would not seem to disprove the point. Furthermore, though the examples in (i) might have been taken to provide evidence that the first NP_1 can be overt when NP_2 is elided, (ia) is not equivalent to (ib): in (ia), *two books* is preferentially interpreted as specific, unlike (ib). That is, on the most natural reading of (ia), the speaker is referring to two particular books (*two books* scopes over *plan*), whereas there is no such preference in (ib).

(i) a. I plan to read two books of the three that you gave me.
  b. I plan to read two of the three books that you gave me.

6. Implicit in all of these definitions is the extension of the notion of *partition* to material parts. We have not been able to detect any adverse consequences of such an extension.

7. The question arises whether the fixed gender of fractions in many languages results from the presence of a null noun meaning 'part' or from nominalization. While in many languages an overt noun 'part' can be used with a fractional meaning, and in Russian (ia) the gender of that noun and the gender of an isolated fraction coincide, in Spanish (ib) they don't: while fractions are masculine, the word *parte* 'part' is feminine.

(i) a. Tret'ja čast' nasledstva   ušla na nalogi.   Russian
   third part inheritance.GEN left on taxes
   'One-third of the inheritance was spent on taxes.'

    b. El área a ser descontaminada equivale a una séptima parte de   Spanish
   the area to be decontaminated equals   to one seventh.F part.F of
   la provincia.
   the province
   'The area to be decontaminated is equal to one-seventh of the province.'

We leave the matter unresolved here, since nothing crucial hinges on this decision.

8. The blueberry-pie example (43b) is grammatical if interpreted as a token of a type, especially in a restaurant setting, where *two of the blueberry pie* means 'two tokens of the type blueberry pie'.

    An anonymous reviewer points out that *one of the blueberry pie/every donation* is just as ungrammatical as (43b), even though it does have a meaning (it means simply *the blueberry pie/every donation*). Ionin, Matushansky, and Ruys (2006) explain this via pragmatics: use of *one of the blueberry pie* is infelicitous when the same meaning can be expressed more simply as *the blueberry pie*. Support for this analysis comes from the fact that this pragmatic constraint can be overridden in the presence of *only*, as in (i).

(i) a. There is only one of me.
    b. You have only one of each parent, so you should value us.

References

Abbott, Barbara. 1976. "Right Node Raising as a test for constituenthood." *Linguistic Inquiry* 7: 639–642.

Abbott, Barbara. 1996. "Doing without a partitive constraint." In *Partitives*, ed. Jacob Hoeksema, 25–56. Berlin: Mouton de Gruyter.

Abney, Steven. 1987. "The English noun phrase in its sentential aspect." PhD dissertation, Massachusetts Institute of Technology.

Acquaviva, Paolo. 2005. "The morphosemantics of transnumeral nouns." In *Morphology and Linguistic Typology: On-Line Proceedings of the Fourth Mediterranean Morphology Meeting*, ed. Geert Booij, Emiliano Guevara, Angela Ralli, Salvatore Sgroi, and Sergio Scalise. Bologna: University of Bologna.

Acquaviva, Paolo. 2006. "Goidelic inherent plurals and the morphosemantics of number." *Lingua* 116 (11): 1860–1887.

Adger, David. 2010. "Gaelic syntax." In *The Edinburgh Companion to the Gaelic Language*, ed. Moray Watson and Michelle Macleod, 304–351. Edinburgh: Edinburgh University Press.

Akiner, Shirin. 1983. "The syntax of the numeral in Byelorussian, compared with Ukrainian, Russian and Polish." *Slavonic and East European Review* 61 (1): 55–68.

Akmajian, Adrian, and Adrienne Lehrer. 1976. "NP-like quantifiers and the problem of determining the head of an NP." *Linguistic Analysis* 2 (4): 395–413.

Alegre, Maria A., and Peter Gordon. 1996. "Red rats eater exposes recursion in children's word formation." *Cognition* 60 (1): 65–82.

Alexandropoulou, Stavroula. 2015. "Testing the nature of variation effects with modified numerals." In *Proceedings of Sinn und Bedeutung 19*, ed. Eva Csipak and Hedde Zeijlstra, 36–53. Göttingen, Germany: University of Göttingen.

Alexandropoulou, Stavroula, Jakub Dotlacil, Yaron McNabb, and Rick Nouwen. 2016. "Pragmatic inferences with numeral modifiers: Novel experimental data." *Proceedings of SALT* 25: 533–549.

Allegranza, Valerio. 1998. "Determiners as functors: NP structure in Italian." In *Romance in HPSG*, ed. Sergio Balari and Luca Dini, 55–108. Stanford, CA: Center for the Study of Language and Information.

Allen, Joseph Henry, James Bradstreet Greenough, George Lyman Kittredge, Albert Andrew Howard, and Benjamin Leonard D'Ooge. 1903. *Allen and Greenough's New Latin Grammar for Schools and Colleges, Founded on Comparative Grammar.* London: Ginn.

Alqarni, Muteb. 2015. "The morphosyntax of numeral-noun constructions in Modern Standard Arabic." PhD dissertation, University of Florida.

Andersen, Henning. 2005. "The plasticity of Universal Grammar." In *Convergence: Interdisciplinary Communications 2004/2005*, ed. Willy Østreng, 21–26. Oslo: Centre for Advanced Study.

Aoun, Joseph, Elabbas Benmamoun, and Dominique Sportiche. 1994. "Agreement, word order, and conjunction in some varieties of Arabic." *Linguistic Inquiry* 25: 195–220.

Arkadiev, Peter M. 2014. "Case and word order in Lithuanian infinitival clauses revisited." In *Grammatical Relations and Their Non-canonical Encoding in Baltic*, ed. Axel Holvoet and Nicole Nau, 43–95. Amsterdam: John Benjamins.

Arregi, Karlos. 2010. "The syntax of comparative numerals." In *Proceedings of NELS 40*, ed. Seda Kan, Claire Moore-Cantwell, and Robert Staubs, 45–58. Amherst, MA: Graduate Linguistic Student Association.

Babby, Leonard H. 1985. "Noun phrase internal case agreement in Russian." *Russian Linguistics* 9: 1–15.

Babby, Leonard H. 1987. "Case, pre-quantifiers, and discontinuous agreement in Russian." *Natural Language and Linguistic Theory* 5 (1): 91–138.

Babko-Malaya, Olga. 1998. "Context-dependent quantifiers restricted by focus." In *Proceedings of the Workshop on Focus*, ed. Elena E. Benedicto, Maribel Romero, and Satoshi Tomioka, 1–18. Amherst, MA: Graduate Linguistic Student Association.

Bachrach, Asaf, and Roni Katzir. 2009. "Right-Node Raising and delayed spellout." In *InterPhases: Phase-Theoretic Investigations of Linguistic Interfaces*, ed. Kleanthes Grohmann, 283–316. Oxford: Oxford University Press.

Backhouse, Anthony E. 2004. "Inflected and uninflected adjectives in Japanese." In *Adjective Classes: A Cross-Linguistic Typology*, ed. R. M. W. Dixon and Alexandra Y. Aikhenvald, 50–73. Oxford: Oxford University Press.

Bailyn, John. 2001. "The syntax of Slavic predicate Case." *ZAS Papers in Linguistics* 22: 1–26.

Bailyn, John, and Andrew Ira Nevins. 2008. "Russian genitive plurals are impostors." In *Inflectional Identity*, ed. Asaf Bachrach and Andrew Ira Nevins, 237–270. Oxford: Oxford University Press.

Baker, Mark C. 2003. "Verbal adjectives as adjectives without phi-features." In *Proceedings of the Fourth Tokyo Conference on Psycholinguistics*, ed. Yukio Otsu, 1–22. Tokyo: Keio University.

Bale, Alan, Michaël Gagnon, and Hrayr Khanjian. 2011. "Cross-linguistic representations of numerals and number marking." In *Semantics and Linguistic Theory (SALT) 20*, ed. Nan Li and David Lutz, 582–598. eLanguage.

Ball, Martin, and James Fife. 1993. *The Celtic Languages*. London: Routledge.

Barbiers, Sjef. 2005. "Variation in the morphosyntax of ONE." *Journal of Comparative Germanic Linguistics* 8 (3): 159–183.

Barker, Chris. 1998. "Partitives, double genitives and anti-uniqueness." *Natural Language and Linguistic Theory* 16: 679–717.

Barker, Chris. 1999. "Temporary accommodation: I'm the oldest of my siblings." *Studies in the Linguistic Sciences* 29 (1): 3–12.

Barlow, Michael. 1992. *A Situated Theory of Agreement.* New York: Garland.

Barwise, Jon, and Robin Cooper. 1981. "Generalized quantifiers and natural language." *Linguistics and Philosophy* 4: 159–219.

Batchelor, John. 1905. *Ainu-English-Japanese Dictionary (Including a Grammar of the Ainu Language).* Tokyo: Methodist Publishing House.

Bennett, Michael R. 1974. "Some extensions of a Montague fragment of English." PhD dissertation, University of California, Los Angeles.

Bennis, Hans, Norbert Corver, and Marcel den Dikken. 1998. "Predication in nominal phrases." *Journal of Comparative Germanic Linguistics* 1: 85–117.

Berent, Iris, and Steven Pinker. 2007. "The dislike of regular plurals in compounds." *Mental Lexicon* 2 (2): 129–181.

Bielec, Dana. 1999. *Basic Polish: A Grammar and Workbook.* New York: Routledge.

Billings, Loren Allen. 1995. "Approximation in Russian and the single-word constraint." PhD dissertation, Princeton University.

Blevins, Juliette. 2004. "The morphology of Yurok numerals." Paper presented at the Workshop on Numerals in the World's Languages, Max Planck Institute for Evolutionary Anthropology, Leipzig, March 29–30.

Blok, Dominique. 2013. "Directional prepositions as numeral modifiers." Ms., Utrecht University.

Blok, Dominique. 2016a. "Directional numeral modifiers: An implicature-based account." In *Proceedings of NELS 46*, ed. Christopher Hammerly and Brandon Prickett, 123–136. Amherst, MA: Graduate Linguistic Student Association.

Blok, Dominique. 2016b. "The semantics and pragmatics of directional numeral modifiers." *Proceedings of SALT* 25: 471–490.

Bobaljik, Jonathan David. 2012. *Universals in Comparative Morphology.* Cambridge, MA: MIT Press.

Bock, Kathryn, and Carol A. Miller. 1991. "Broken agreement." *Cognitive Psychology* 23 (1): 45–93.

Bodomo, Adams. 1997. *The Structure of Dàgáárè.* Stanford, CA: CSLI Publications.

Bodomo, Adams. 2000. *Dàgáárè.* Munich: Lincom.

Bodomo, Adams, and Charles Marfo. 2006. "The morphophonology of noun classes in Dagaare and Akan." *Studi Linguistici e Filologici Online* 4 (2): 205–244.

Borer, Hagit. 1996. "The construct in review." In *Studies in Afro-asiatic Languages*, ed. Jacqueline Lecarme, Jean Lowenstamm, and Ur Shlonsky, 30–61. The Hague: Holland Academic Graphics.

Borer, Hagit. 1999. "Deconstructing the construct." In *Beyond Principles and Parameters*, ed. Kyle Johnson and Ian Roberts, 43–90. Dordrecht, the Netherlands: Kluwer.

Borer, Hagit. 2005. *Structuring Sense*, vol. 1, In Name Only. Oxford: Oxford University Press.

Borer, Hagit. 2011. "Afro-Asiatic, Semitic: Hebrew." In *The Oxford Handbook of Compounding*, ed. Rochelle Lieber and Pavol Štekauer, 491–511. Oxford: Oxford University Press.

Borer, Hagit. 2013. "The syntactic domain of content." In *Generative Linguistics and Acquisition: Studies in Honor of Nina M. Hyams*, ed. Misha Becker, John Grinstead, and Jason Rothman, 205–248. Amsterdam: John Benjamins.

Borsley, Robert D. 1983. "A note on the generalized Left Branch Condition." *Linguistic Inquiry* 14: 169–174.

Borsley, Robert D., Maggie Tallerman, and David Willis. 2007. *The Syntax of Welsh*. Cambridge: Cambridge University Press.

Bouchard, Denis. 2002. *Adjectives, Number and Interfaces: Why Languages Vary*. Oxford: Elsevier Science.

Bouchard, Denis. 2005. "Sériation des adjectifs dans le SN et formation de concepts" [Serialization of adjectives in the NP and formation of concepts]. In *L'adjectif*, ed. Patricia Cabredo Hofherr and Ora Matushansky, 125–142. Saint-Denis, France: Presses Universitaires de Vincennes.

Breheny, Richard. 2008. "A new look at the semantics and pragmatics of numerically quantified noun phrases." *Journal of Semantics* 25 (2): 93–140.

Bresnan, Joan. 1973. "Syntax of the comparative clause construction in English." *Linguistic Inquiry* 4: 275–343.

Broekhuis, Hans, and Marcel den Dikken. 2012. *The Syntax of Dutch*, vol. 2, Nouns and Noun Phrases. Amsterdam: Amsterdam University Press.

Brooks, Maria Zagórska. 1975. *Polish Reference Grammar*. Berlin: Mouton.

Buccola, Brian, and Benjamin Spector. 2016. "Modified numerals and maximality." *Linguistics and Philosophy* 39 (3): 151–199.

Büring, Daniel. 2008. "The least *at least* can do." In *Proceedings of the 26th West Coast Conference on Formal Linguistics*, ed. Charles B. Chang and Hannah J. Haynie, 114–120. Somerville, MA: Cascadilla Proceedings Project.

Büring, Daniel. 2009. "*At least* and *at most*: The logic of bounds and insecurity." Paper presented at the Massachusetts Institute of Technology, April 24.

Bylinina, Lisa, Jakub Dotlačil, and Heidi Klockmann. 2016. "Adjectival modification of numerals." Paper presented at the LogiCon workshop, Utrecht, the Netherlands, September 19–20.

Bylinina, Lisa, Natalia Ivlieva, Alexander Podobryaev, and Yasutada Sudo. 2015. "An in situ semantics for ordinals." In *NELS 45: Proceedings of the 45th Meeting of the North East Linguistic Society*, ed. Thuy Bui and Deniz Özyıldız. Amherst, MA: Graduate Linguistic Student Association.

Bylinina, Lisa, and Rick Nouwen. 2017. "On 'zero' and semantic plurality." Ms., Leiden University and Utrecht University.

Caldwell, Robert. 1913. *A Comparative Grammar of the Dravidian or South-Indian Family of Languages*. New Delhi: Asian Educational Services.

Cappelletti, Marinella, Brian Butterworth, and Michael D. Kopelman. 2001. "Spared numerical abilities in a case of semantic dementia." *Neuropsychologia* 39: 1224–1239.

Cappelletti, Marinella, Michael D. Kopelman, John Morton, and Brian Butterworth. 2005. "Dissociations in numerical abilities revealed by progressive cognitive decline in a patient with semantic dementia." *Cognitive Neuropsychology* 22 (7): 771–793.

Cardinaletti, Anna, and Giuliana Giusti. 1992. "Partitive *ne* and the QP-hypothesis: A case study." In *Proceedings of the XVII Meeting of Generative Grammar*, ed. Elisabetta Fava, 122–141. Turin: Rosenberg and Sellier.

Cardinaletti, Anna, and Giuliana Giusti. 2005. "The syntax of quantified phrases and quantitative clitics." In *The Blackwell Companion to Syntax*, ed. Henk van Riemsdijk and Martin Everaert, 23–93. Oxford: Blackwell.

Carlson, Greg N. 1977a. "Reference to kinds in English." PhD dissertation, University of Massachusetts, Amherst.

Carlson, Greg N. 1977b. "A unified analysis of the English bare plural." *Linguistics and Philosophy* 1: 413–457.

Carpenter, Bob. 1998. *Type-Logical Semantics*. Cambridge, MA: MIT Press.

Carston, Robyn. 1998. "Informativeness, relevance and scalar implicature." In *Relevance Theory: Applications and Implications*, ed. Robyn Carston and Seiji Uchida, 179–236. Amsterdam: John Benjamins.

Champollion, Lucas. 2016. "Ten men and women got married today: Noun coordination and the intersective theory of conjunction." *Journal of Semantics* 33 (3): 561–622.

Chandralal, Dileep. 2010. *Sinhala*. Amsterdam: John Benjamins.

Cheng, Lisa L.-S., and Rint Sybesma. 1999. "Bare and not-so-bare nouns and the structure of NP." *Linguistic Inquiry* 30: 509–542.

Chierchia, Gennaro. 1984. "Topics in the syntax and semantics of infinitives and gerunds." PhD dissertation, University of Massachusetts, Amherst.

Chierchia, Gennaro. 1985. "Formal semantics and the grammar of predication." *Linguistic Inquiry* 16: 417–443.

Chierchia, Gennaro. 1998. "Reference to kinds across languages." *Natural Language Semantics* 6: 339–405.

Chierchia, Gennaro. 2004. "Numerals and 'formal' vs. 'substantive' features of mass and count." Paper presented at Linguistic Perspectives on Numerical Expressions, Utrecht, the Netherlands, June 10–11.

Chierchia, Gennaro. 2010. "Mass nouns, vagueness and semantic variation." *Synthèse* 174: 99–149.

Childers, R. C. 1876. "Notes on the Sinhalese language, no. II: Proofs of the Sanscritic origins of Sinhalese." *Journal of the Royal Asiatic Society* 8: 131–156.

Chomsky, Noam. 1986. *Knowledge of Language*. New York: Praeger.

Chung, Sandra. 2013. "The syntactic relations behind agreement." In *Diagnosing Syntax*, ed. Lisa L.-S. Cheng and Norbert Corver, 251–270. Oxford: Oxford University Press.

Comrie, Bernard S. 2005a. "Endangered numeral systems." In *Bedrohte Vielfalt: Aspekte des Sprach(en)tods*, ed. Jan Wohlgemuth and Tyko Dirksmeyer, 203–230. Berlin: Weißensee.

Comrie, Bernard S. 2005b. "Numeral bases." In *The World Atlas of Language Structures*, ed. Martin Haspelmath, Matthew S. Dryer, David Gil, and Bernard S. Comrie, 530–533. Oxford: Oxford University Press.

Coppock, Elizabeth, and Thomas Brochhagen. 2013. "Raising and resolving issues with scalar modifiers." *Semantics and Pragmatics* 6 (3): 1–57.

Corbett, Greville G. 1978. "Universals in the syntax of cardinal numbers." *Lingua* 46: 355–368.

Corbett, Greville G. 1979. "The agreement hierarchy." *Journal of Linguistics* 15: 203–224.

Corbett, Greville G. 1983. *Hierarchies, Targets and Controllers: Agreement Patterns in Slavic*. University Park, PA: Pennsylvania State University Press.

Corbett, Greville G. 1991. *Gender*. Cambridge: Cambridge University Press.

Corbett, Greville G. 1993. "The head of Russian numeral expressions." In *Heads in Grammatical Theory*, ed. Greville G. Corbett, Norman M. Fraser, and Scott McGlashan, 11–35. Cambridge: Cambridge University Press.

Corbett, Greville G. 2000. *Number*. Cambridge: Cambridge University Press.

Corbett, Greville G. 2006. *Agreement*. Cambridge: Cambridge University Press.

Corver, Norbert, and Joost Zwarts. 2006. "Prepositional numerals." *Lingua* 116 (6): 811–835.

Creissels, Denis, and Anna Jurievna Urmančieva. 2017. "Soninke jazyk" [The Soninke language]. In *Jazyki mande*, ed. Valentin Vydrin, Julia Viktorovna Mazurova, Andrei Aleksandrovich Kibrik, and Elena Borisovna Markus, 248–283. St. Petersburg: Nestor-Istorija.

Crockett, Dina B. 1976. *Agreement in Contemporary Standard Russian*. Bloomington, IN: Slavica.

Cummins, Chris, and Napoleon Katsos. 2010. "Comparative and superlative quantifiers: Pragmatic effects of comparison type." *Journal of Semantics* 27: 271–305.

Dal Pozzo, Lena. 2007. "The Finnish noun phrase." Ms., Università Ca' Foscari di Venezia.

Danon, Gabi. 2012. "Two structures for numeral-noun constructions." *Lingua* 122 (12): 1282–1307.

Davies, Mark. 2008. *The Corpus of Contemporary American English*. http://corpus .byu.edu/coca/.

Decaux, Étienne. 1964. "L'expression de la détermination au pluriel numérique en polonais." *Revue des études slaves* 40: 61–72.

Dehaene, Stanislas. 1997. *The Number Sense*. Oxford: Oxford University Press.

de Hoop, Helen. 1997. "A semantic reanalysis of the Partitive Constraint." *Lingua* 103: 151–174.

de Hoop, Helen. 1998. "Partitivity." *GLOT International* 3 (2): 3–10.

den Dikken, Marcel. 2001. "'Pluringulars,' pronouns and quirky agreement." *Linguistic Review* 18: 19–41.

Derbyshire, William W. 1993. *A Basic Reference Grammar of Slovene*. Columbus, OH: Slavica.

Dixon, R. M. W. 1977. "Where have all the adjectives gone?" *Studies in Language* 1: 19–80.

Djemai, Salem. 2008. "L'expression de la qualité en berbère: Étude de quelques aspects morphologiques, sémantiques et syntaxiques de l'adjectif en kabyle" [Expression of quality in Berber: A study of some morphological, semantic, and syntactic aspects of the adjective in Kabyle]. MA thesis, Institut National des Langues et Civilisations Orientales, Paris.

Doetjes, Jenny. 1997. "Quantifiers and selection: On the distribution of quantifying expressions in French, Dutch, and English." PhD dissertation, Leiden University.

Domahs, Frank, Lisa Bartha, Aliette Lochy, Thomas Benke, and Margarete Delazer. 2005. "Number words are special: Evidence from a case of primary progressive aphasia." *Journal of Neurolinguistics* 19 (1): 1–37.

Donabédian, Anaïd. 1993. "Le pluriel en arménien moderne" [The plural in modern Armenian]. *Faits de Langues* 2: 179–188.

Downing, Pamela. 1984. "Japanese numeral classifiers: A syntactic, semantic and functional profile." PhD dissertation, University of California, Berkeley.

Downing, Pamela. 1996. *Numeral Classifier Systems: The Case of Japanese*. Philadelphia: John Benjamins.

Dowty, David R. 1986. "A note on collective predicates, distributive predicates and *all*." In *Proceedings of the Third Eastern States Conference on Linguistics (ESCOL '86)*, ed. Fred Marshall, Ann Miller, and Zheng-sheng Zhang, 97–115. Columbus, OH: Ohio State University.

Drellishak, Scott. 2004. "A survey of coordination strategies in the world's languages." MA thesis, University of Washington.

Dvořák, Boštjan, and Uli Sauerland. 2005. "The semantics of the Slovenian dual." In *Formal Approaches to Slavic Linguistics 14: The Princeton Meeting 2005*, ed. James Lavine, Steven Franks, Mila Tasseva-Kurktchieva, and Hana Filip, 98–112. Ann Arbor, MI: Michigan Slavic Publications.

Dyła, Stefan. 1984. "Across-the-board dependencies and case in Polish." *Linguistic Inquiry* 15: 701–705.

Eberhard, Kathleen M., J. Cooper Cutting, and Kathryn Bock. 2005. "Making syntax of sense: Number agreement in sentence production." *Psychological Review* 112 (3): 531–559.

Elbourne, Paul. 2002. "Situations and individuals." PhD dissertation, Massachusetts Institute of Technology.

Ellsworth, Michael, Russell Lee-Goldman, and Russell Rhodes. 2008. "Two paradoxes of English determination: The construction of complex number expressions." In *Proceedings of the 38th Western Conference on Linguistics*, ed. Michael Grosvald and Dionne Soares, 24–33. Davis, CA: Department of Linguistics, University of California, Davis.

Elšik, Viktor, and Yaron Matras. 2006. *Markedness and Language Change: The Romani Sample*. Berlin: Mouton de Gruyter.

Embick, David. 1997a. "Voice and interfaces of syntax." PhD dissertation, University of Pennsylvania.

Embick, David. 1997b. "Voice systems and the syntax/morphology interface." In *Papers from the UPenn/MIT Roundtable on Argument Structure and Aspect*, ed. Heidi Harley, 41–72. Cambridge, MA: MIT Working Papers in Linguistics.

Eschenbach, Carola. 1993. "Semantics of number." *Journal of Semantics* 10 (1): 1–31.

Es'kova, N. A. 2011. *Izbrannye raboty po rusistike: Fonologija, morfonologija, morfologija, orfografija, leksikografija*. Moscow: *Jazyki Slavjanskix Kul'tur*.

Etxeberria, Urtzi, and Ricardo Etxepare. 2008. "Number agreement with weak quantifiers in Basque." In *Proceedings of the 27th West Coast Conference on Formal Linguistics*, ed. Natasha Abner and Jason Bishop, 159–167. Somerville, MA: Cascadilla Proceedings Project.

Everett, Daniel L. 2005. "Cultural constraints on grammar and cognition in Piraha: Another look at the design features of human language." *Current Anthropology* 46: 621–646.

Fabricius-Hansen, Cathrine, Peter Gallmann, Peter Eisenberg, Reinhard Fiehler, and Jörg Peters. 2009. *Duden: Die Grammatik*. 8th ed. Mannheim, Germany: Dudenverlag.

Farkas, Donka. 1981. "Quantifier scope and syntactic islands." In *Proceedings of the 17th Annual Meeting of the Chicago Linguistics Society*, ed. Roberta Hendrick, Carrie Masek, and Mary Frances Miller, 59–66. Chicago: Chicago Linguistic Society.

Farkas, Donka, and Henriëtte de Swart. 2003. *The Semantics of Incorporation: From Argument Structure to Discourse Transparency*. Stanford, CA: CSLI Publications.

Farkas, Donka, and Henriëtte de Swart. 2010. "The semantics and pragmatics of plurals." *Semantics and Pragmatics* 3 (6): 1–54.

Faust, Noam. 2014. "Where it's [at]: A phonological effect of phasal boundaries in the construct state of Modern Hebrew." *Lingua* 150: 315–331.

Feldstein, Ronald F. 2001. *A Concise Polish Grammar*. Durham, NC: Slavic and Eurasian Language Resource Center.

Fodor, Janet Dean, and Ivan Sag. 1982. "Referential and quantificational indefinites." *Linguistics and Philosophy* 5: 355–398.

Fowler, George. 1987. "The syntax of the genitive case in Russian." PhD dissertation, University of Chicago.

Frank, Michael C., Daniel L. Everett, Evelina Fedorenko, and Edward Gibson. 2008. "Number as a cognitive technology: Evidence from Pirahã language and cognition." *Cognition* 108: 819–824.

Franks, Steven. 1993. "On parallelism in across-the-board dependencies." *Linguistic Inquiry* 24 (3): 509–529.

Franks, Steven. 1994. "Parametric properties of numeral phrases in Slavic." *Natural Language and Linguistic Theory* 12: 597–674.

Franks, Steven. 1995. *Parameters of Slavic Morphosyntax*. Oxford: Oxford University Press.

Franks, Steven. 2002. "A Jakobsonian feature-based analysis of the Slavic numeric quantifier genitive." *Journal of Slavic Linguistics* 10: 141–181.

Frege, Gottlöb. 1884. *Die Grundlagen der Arithmetik: Eine logisch mathematische Untersuchung über den Begriff der Zahl [The foundations of arithmetic: A logico-mathematical investigation of the concept of number]*. Breslau, Germany: Wilhelm Koebner.

Garde, Paul. 1998. *Grammaire russe: Phonologie et morphologie [Russian grammar: Phonology and morphology]*. 2nd ed. Paris: Institut d'Études Slaves.

Gärtner, Hans-Martin. 2004. "Naming and economy." *Empirical Issues in Formal Syntax and Semantics* 5: 63–73.

Gawron, Jean Mark. 2002. "Two kinds of quantizers in DP." Paper presented at the Annual Meeting of the Linguistic Society of America, San Francisco, January 3–6.

Gawron, Jean Mark, and Andrew Kehler. 2002. "The semantics of the adjective *respective*." In *WCCFL 21 Proceedings*, ed. Line Hove Mikkelsen and Christopher Potts, 85–98. Somerville, MA: Cascadilla Press.

Geniušienė, Emma. 1987. *The Typology of Reflexives*. Berlin: Mouton de Gruyter.

Geurts, Bart. 2006. "Take 'five': The meaning and use of a number word." In *Non-definiteness and Plurality*, ed. Liliane Tasmowski and Svetlana Vogeleer, 311–329. Amsterdam: John Benjamins.

Geurts, Bart. 2010. *Quantity Implicatures*. Cambridge: Cambridge University Press.

Geurts, Bart, Napoleon Katsos, Chris Cummins, Jonas Moons, and Leo Noordman. 2010. "Scalar quantifiers: Logic, acquisition and processing." *Language and Cognitive Processes* 25: 130–148.

Geurts, Bart, and Rick Nouwen. 2007. "At least *et al.*: The semantics of scalar modifiers." *Language* 83: 533–559.

Gillon, Brandon. 1984. "The logical form of quantification and plurality in natural language." PhD dissertation, Massachusetts Institute of Technology.

Gillon, Brandon. 1987. "The readings of plural noun phrases in English." *Linguistics and Philosophy* 10: 199–219.

Gilmore, Camilla, Shannon McCarthy, and Elizabeth Spelke. 2007. "Symbolic arithmetic knowledge without instruction." *Nature* 447: 589–591.

Giusti, Giuliana. 1991. "The categorial status of quantified nominals." *Linguistische Berichte* 136: 438–452.

Giusti, Giuliana. 1997. "The categorial status of determiners." In *The New Comparative Syntax*, ed. Liliane Haegeman, 94–113. Cambridge: Cambridge University Press.

Giusti, Giuliana, and Giuseppina Turano. 2004. "Case-assignment in the pseudo-partitives of Standard Albanian and Arbëresh: A case for micro-variation." *University of Venice Working Papers in Linguistics* 14: 173–194.

Glinert, Lewis. 1989. *The Grammar of Modern Hebrew*. New York: Cambridge University Press.

Görgülü, Emrah. 2010. "Nominals and the count/mass distinction in Turkish." Paper presented at the 26th Northwest Linguistics Conference, Vancouver, May 8–9.

Görgülü, Emrah. 2012. "Nominal semantics and number in Turkish." In *Proceedings of the 40th Western Conference on Linguistics*, ed. Christina Galeano, Emrah Görgülü, and Irina Presnyakova, 236–247. Fresno, CA: Department of Linguistics, California State University.

Graudina, Ljudmila Karlovna, Viktor Aleksandrovič Ickovič, and Lia Pavlovna Katlinskaja. 1976. *Grammatičeskaja pravil'nost' russkoj reči: Stilističeskij slovar' variantov*. Moscow: Nauka.

Greenberg, Joseph H. 1978. "Generalizations about numeral systems." In *Universals of Human Language*, 249–295. Stanford, CA: Stanford University Press.

Greenberg, Joseph H. 2000. "Numeral." In *Morphology: An International Handbook on Inflection and Word-Formation*, ed. Geert Booij, Christian Lehmann, and Joachim Mugdan, 770–783. Berlin: De Gruyter.

Greene, David. 1992. "Celtic." In *Indo-European Numerals*, ed. Jadranka Gvozdanović, 497–554. Berlin: Mouton de Gruyter.

Grestenberger, Laura. 2015. "Number marking in German measure phrases and the structure of pseudo-partitives." *Journal of Comparative Germanic Linguistics* 18 (2): 93–138.

Grice, Paul. 1975. "Logic and conversation." In *Speech Acts*, ed. Peter Cole and Jerry L. Morgan, 41–58. New York: Academic Press.

Grimm, Scott. 2010. "Number and markedness: A view from Dagaare." In *Proceedings of Sinn und Bedeutung 14*, ed. Martin Prinzhorn, Viola Schmitt, and Sarah Zobel, 168–184. Vienna: University of Vienna.

Grimm, Scott. 2012. "Individuation and inverse number marking in Dagaare." In *Count and Mass across Languages*, ed. Diane Massam, 75–98. Oxford: Oxford University Press.

Grosu, Alexander. 1976. "A note on subject raising to object and right node raising." *Linguistic Inquiry* 7: 642–645.

Hackl, Martin. 2000. "Comparative quantifiers." PhD dissertation, Massachusetts Institute of Technology.

Hackl, Martin. 2002. "The ingredients of essentially plural predicates." In *Proceedings of NELS 32*, ed. Masako Hirotani. Amherst, MA: Graduate Linguistic Student Association.

Haegeman, Liliane, and Jacqueline Guéron. 1999. *English Grammar: A Generative Perspective*. Oxford: Blackwell.

Halle, Morris. 1990. "An approach to morphology." In *Proceedings of NELS 20*, ed. Juli Carter, Rose-Marie Déchaine, Bill Philip, and Tim Sherer, 150–184. Amherst, MA: Graduate Linguistic Student Association.

Halle, Morris. 1994. "The morphology of numeral phrases." In *Annual Workshop of Formal Approaches to Slavic Linguistics: The MIT Meeting*, ed. Sergey Avrutin, Steven Franks, and Ljiljana Progovac, 178–215. Ann Arbor, MI: Michigan Slavic Publications.

Halle, Morris, and Ora Matushansky. 2006. "The morphophonology of Russian adjectival inflection." *Linguistic Inquiry* 37 (3): 351–404.

Hankamer, Jorge, and Line Hove Mikkelsen. 2008. "Definiteness marking and the structure of Danish pseudopartitives." *Journal of Linguistics* 44 (2): 317–346.

Hanke, Thomas. 2010. "Additional rarities in the typology of numerals." In *Rethinking Universals: How Rarities Affect Linguistic Theory*, ed. Jan Wohlgemuth and Michael Cysouw, 61–90. Boston: De Gruyter Mouton.

Harbour, Daniel. 2008. *Morphosemantic Number: From Kiowa Noun Classes to UG Number Features*. Dordrecht, the Netherlands: Springer.

Hartmann, Katharina. 2000. *Right Node Raising and Gapping: Interface Conditions on Prosodic Deletion*. Amsterdam: John Benjamins.

Haspelmath, Martin. 1987. *Transitivity Alternations of the Unaccusative Type*. Cologne, Germany: Institut für Sprachwissenschaft, Universität zu Köln.

Haspelmath, Martin. 1990. "The grammaticization of passive morphology." *Studies in Language* 14 (1): 25–72.

Haspelmath, Martin. 1993. "More on the typology of inchoative/causative verb alternations." In *Causatives and Transitivity*, ed. Bernard S. Comrie and Maria Polinsky, 87–120. Amsterdam: John Benjamins.

Hazout, Ilan. 2000. "Adjectival genitive constructions in Modern Hebrew: A case study in coanalysis." *Linguistic Review* 9: 29–52.

Heath, Jeffrey. 2005. *A Grammar of Tamashek (Tuareg of Mali)*. Berlin: De Gruyter.

Heim, Irene. 1982. "The semantics of definite and indefinite noun phrases." PhD dissertation, University of Massachusetts, Amherst.

Heim, Irene. 1985. "Notes on comparatives and related matters." Ms., University of Texas, Austin.

Heim, Irene. 1991. "Artikel und Definitheit" [Articles and definiteness]. In *Semantics: An International Handbook of Contemporary Research*, ed. Arnim von Stechow and Dieter Wunderlich, 487–535. Berlin: Walter de Gruyter.

Heim, Irene. 1995. "Notes on superlatives." Ms., Massachusetts Institute of Technology.

Heim, Irene. 2000. "Degree operators and scope." In *Proceedings of Semantics and Linguistic Theory (SALT) 10*, ed. Brendan Jackson and Tanya Matthews, 40–64. Ithaca, NY: CLC Publications.

Heim, Irene, and Angelika Kratzer. 1998. *Semantics in Generative Grammar*. Malden, MA: Blackwell.

Hetzron, Robert. 1967. "Agaw numerals and incongruence in Semitic." *Journal of Semitic Studies* 12 (2): 169–193.

Heycock, Caroline, and Roberto Zamparelli. 2000. "Plurality and NP-coordination." In *Proceedings of NELS 30*, ed. Masako Hirotani, Andries Coetzee, Nancy Hall, and Ji-yung Kim, 341–352. Amherst, MA: Graduate Linguistic Student Association.

Heycock, Caroline, and Roberto Zamparelli. 2003. "Coordinated bare definites." *Linguistic Inquiry* 34: 443–469.

Heycock, Caroline, and Roberto Zamparelli. 2005. "Friends and colleagues: Plurality, coordination, and the structure of DP." *Natural Language Semantics* 17 (3): 201–270.

Higginbotham, James. 1981. "Reciprocal interpretations." *Journal of Linguistic Research* 1 (3): 97–117.

Higginbotham, James. 1987. "Indefiniteness and predication." In *The Representation of (In)definiteness*, ed. Eric Reuland and Alice ter Meulen, 43–70. Cambridge, MA: MIT Press.

Hiraiwa, Ken. 2013. "The numeral system and its arithmetic structure in Dagaare." Ms., Meiji Gakuin University.

Hoeksema, Jack. 1984. "Partitives." Ms., University of Groningen.

Hoeksema, Jack. 1988. "The semantics of non-Boolean 'and'." *Journal of Semantics* 6 (1): 19–40.

Hoeksema, Jack. 1996a. *Introduction to Partitives: Studies on the Syntax and Semantics of Partitive and Related Constructions*, 1–24. Berlin: Mouton de Gruyter.

Hoeksema, Jack, ed. 1996b. *Partitives: Studies on the Syntax and Semantics of Partitive and Related Constructions*. Berlin: Mouton de Gruyter.

Hofweber, Thomas. 2005. "Number determiners, numbers, and arithmetic." *Philosophical Review* 114 (2): 179–225.

Honda, Masaru. 1984. "On the syntax and semantics of numerals in English." *Journal of Osaka Jogakuin 2-Year College* 14: 97–115.

Horn, Laurence. 1972. "On the semantic properties of logical operators in English." PhD dissertation, University of California, Los Angeles.

Hornstein, Norbert. 1999. "Movement and control." *Linguistic Inquiry* 30 (1): 69–96.

Hornstein, Norbert. 2001. *Move! A Minimalist Theory of Construal*. Oxford: Blackwell.

Horton, Alonso E. 1949. *A Grammar of Luvale. Johanesburg*: Witwatersrand University Press.

Huang, Yi Ting, and Jesse Snedeker. 2009. "Online interpretation of scalar quantifiers: Insight into the semantics-pragmatics interface." *Cognitive Psychology* 58 (3): 376–415.

Hurford, Jim. 1975. *The Linguistic Theory of Numerals*. Cambridge: Cambridge University Press.

Hurford, Jim. 1987. *Language and Number: The Emergence of a Cognitive System*. Oxford: Blackwell.

Hurford, Jim. 2001. "Numeral systems." In *International Encyclopedia of the Social and Behavioral Sciences*, ed. N. J. Smelser and P. B. Baltes, 10756–10761. Amsterdam: Pergamon.

Hurford, Jim. 2003. "The interaction between numerals and nouns." In *Noun Phrase Structure in the Languages of Europe*, ed. Frans Plank, 561–620. The Hague: Mouton de Gruyter.

Hyman, Larry M. 2003. "Basaa A.43." In *The Bantu Languages*, ed. Derek Nurse and Gérard Philippson, 257–282. London: Routledge.

Ionin, Tania, and Ora Matushansky. 2004. "A singular plural." In *WCCFL 23: Proceedings of the 23rd West Coast Conference on Formal Linguistics*, ed. Benjamin Schmeiser, Vineeta Chand, Ann Kelleher, and Angelo Rodriguez, 399–412. Stanford, CA: Center for the Study of Language and Information.

Ionin, Tania, and Ora Matushansky. 2006. "The composition of complex cardinals." *Journal of Semantics* 23 (4): 315–360.

Ionin, Tania, and Ora Matushansky. 2013a. "More than one comparative in more than one Slavic language: An experimental investigation." In *Formal Approaches to Slavic Linguistics #21: The Third Indiana Meeting*, ed. Steven Franks, Markus Dickinson, George Fowler, Melissa Whitcombe, and Ksenia Zenon, 91–107. Ann Arbor, MI: Michigan Slavic Publications.

Ionin, Tania, and Ora Matushansky. 2013b. *Numerals*. Oxford Bibliographies Online. Oxford University Press. http://www.oxfordbibliographies.com/view/document/obo-9780199772810/obo-9780199772810-0131.xml.

Ionin, Tania, Ora Matushansky, and E. G. Ruys. 2006. "Parts of speech: Toward a unified semantics for partitives." In *Proceedings of NELS 36*, ed. Christopher Davis, Amy Rose Deal, and Youri Zabbal, 357–370. Amherst, MA: Graduate Linguistic Student Association.

Isakadze, N. V. 1998. "Otraženie morfologii i referencial'noj semantiki imennoj gruppy v formal'nom sintaksise." PhD dissertation, Moscow State University.

Izard, Véronique, Coralie Sann, Elizabeth Spelke, and Arlette Streri. 2009. "Newborn infants perceive abstract numbers." *Proceedings of the National Academy of Sciences* 106 (25): 10382–10385.

Jackendoff, Ray. 1968. "Possessives in English." In *Studies in Transformational Grammar and Related Topics*, ed. Stephen R. Anderson, Ray Jackendoff, and Samuel Jay Keyser, 25–51. Waltham, MA: Brandeis University Press.

Jackendoff, Ray. 1977. *X-Bar Syntax: A Study of Phrase Structure*. Cambridge, MA: MIT Press.

Jenks, Peter. 2011. "The hidden structure of Thai noun phrases." PhD dissertation, Harvard University.

Kadmon, Nirit. 1992. *On Unique and Non-unique Reference and Asymmetric Quantification*. New York: Garland.

Kamp, Hans, and Uwe Reyle. 1993. *From Discourse to Logic*. Dordrecht, the Netherlands: Kluwer.

Karlsson, Fred. 2002. *Finnish: An Essential Grammar*. New York: Routledge.

Karttunen, Lauri. 2006. "Numbers and Finnish numerals." In *A Man of Measure: Festschrift in Honour of Fred Karlsson on his 60th Birthday*, ed. Mickael Suominen, Antti Arppe, Anu Airola, Orvokki Heinämäki, Matti Miestamo, Urho Määttä, Jussi Niemi, Kari K. Pitkänen, and Kaius Sinnemäki, 407–421. Turku, Finland: Linguistic Association of Finland.

Kayne, Richard S. 2003. "Silent years, silent hours." In *Grammar in Focus: Festschrift for Christer Platzack*, ed. Lars-Olof Delsing, Cecilia Falk, Gunlög Josefsson, and Halldór Á. Sigurðsson, 209–226. Lund, Sweden: Wallin and Dalholm.

Keenan, Caitlin. 2013. "'A pleasant three days in Philadelphia': Arguments for a pseudopartitive analysis." *University of Pennsylvania Working Papers in Linguistics* 19 (1): 86–96.

Keenan, Edward L., and Jonathan Stavi. 1986. "A semantic characterization of natural language determiners." *Linguistics and Philosophy* 9: 253–326.

Kenesei, István, Robert Michael Vágó, and Anna Fenyvesi. 1998. *Hungarian*. London: Routledge.

Kennedy, Christopher. 2013. "A scalar semantics for scalar readings of number words." In *From Grammar to Meaning: The Spontaneous Logicality of Language*, ed. Ivano Caponigro and Carlo Cecchetto, 172–200. Cambridge: Cambridge University Press.

Kennedy, Christopher. 2015. "A 'de-Fregean' semantics (and a neo-Gricean pragmatics) for modified and unmodified numerals." *Semantics and Pragmatics* 8 (10): 1–44.

Kester, Ellen Petra. 1996. "The nature of adjectival inflection." PhD dissertation, Utrecht Institute of Linguistics, Utrecht University.

Khachaturyan, Maria. 2015. "Grammaire du mano." *Mandenkan* 54: 1–252.

Kim, Ji-yung. 2002. "Adjectives in construct." In *Sinn and Bedeutung VI: Proceedings of the 6th Annual Meeting of the Gesellschaft für Semantik*, ed. Graham Katz, Sabine Reinhard, and Philip Reuter, 185–200. Osnabrück, Germany: Institute of Cognitive Science, University of Osnabrück.

Kim, Jong-Bok. 2011. "Floating numeral classifiers in Korean: A thematic-structure perspective." In *Proceedings of the HPSG 2011 Conference*, ed. Stefan Müller, 302–313. Stanford, CA: CSLI Publications.

King, Gareth. 2003. *Modern Welsh: A Comprehensive Grammar*. London: Routledge.

Kiparsky, Paul. 1982. "Lexical morphology and phonology." In *Linguistics in the Morning Calm*, ed. In-Seok Yang, 3–91. Seoul: Hanshin.

Kiparsky, Paul, and Judith Tonhauser. 2012. "Semantics of inflection." In *Semantics: An International Handbook of Natural Language Meaning*, ed. Klaus von Heusinger, Claudia Maienborn, and Paul Portner, 2070–2097. Berlin: Mouton de Gruyter.

Kiss, Katalin É. 2002. *The Syntax of Hungarian*. Cambridge: Cambridge University Press.

Klockmann, Heidi. 2012. "Polish numerals and quantifiers: A syntactic analysis of subject-verb agreement mismatches." MA thesis, Utrecht University.

Klockmann, Heidi. 2013. "Phi-defective numerals in Polish: Bleeding and default agreement." Paper presented at the thirty-fifth annual conference of the German Linguistic Society (DGfS), Potsdam, Germany, March 12–15.

Klockmann, Heidi. 2017. "The design of semi-lexicality: Evidence from case and agreement in the nominal domain." PhD dissertation, Utrecht University.

Klockmann, Heidi, and Anja Šarić. 2015. "The categorial status of numerals: Evidence from Polish and Serbian." Paper presented at the forty-eighth annual meeting of the Societas Linguistica Europaea, Leiden, the Netherlands, September 2–5.

Klooster, Wim. 1972. *The Structure Underlying Measure Phrase Sentences*. Dordrecht, the Netherlands: Reidel.

Kobuchi-Philip, Mana. 2003. "On the syntax and semantics of the Japanese numeral quantifier." PhD dissertation, City University of New York.

Koenig, Jean-Pierre. 1991. "Scalar predicates and negation: Punctual semantics and interval interpretations." In *Papers from the 27th Regional Meeting of the Chicago Linguistic Society*, vol. 2, The Parasession on Negation, ed. Lise M. Dobrin, 140–155. Chicago: Chicago Linguistic Society.

Koptjevskaja-Tamm, Maria. 2001. "*A piece of the cake* and *a cup of tea*: Partitive and pseudo-partitive nominal constructions in the Circum-Baltic languages." In *Circum-Baltic Languages*, ed. Östen Dahl and Maria Koptjevskaja-Tamm, 523–568. Amsterdam: John Benjamins.

Koptjevskaja-Tamm, Maria, and Bernhard Wälchli. 2001. "The Circum-Baltic languages: An areal-typological approach." In *Circum-Baltic Languages*, ed. Östen Dahl and Maria Koptjevskaja-Tamm, 615–750. Amsterdam: John Benjamins.

Korunec, I. V. 2003. *Porivnjal'na tipologija anglijs'koï ta ukraïns'koï mov: Navčal'nij posibnik*. Vinnicja, Ukraine: Nova Kniga.

Kratzer, Angelika. 1989. "An investigation of the lumps of thought." *Linguistics and Philosophy* 12: 607–653.

Kratzer, Angelika. 1998a. "More structural analogies between pronouns and tenses." Paper presented at Semantics and Linguistic Theory 8, Ithaca, NY, May 8–10.

Kratzer, Angelika. 1998b. "Scope or pseudoscope? Are there wide scope indefinites?" In *Events and Grammar*, ed. Susan Rothstein, 163–196. Dordrecht, the Netherlands: Kluwer.

Krifka, Manfred. 1986. "Nominalreferenz und Zeitkonstitution: Zur Semantik von Massentermen, Pluraltermen und Aspektklassen." PhD dissertation, University of Munich.

Krifka, Manfred. 1990a. "Boolean and non-Boolean *and*." In *Papers from the Second Symposium on Logic and Language*, ed. Lászlo Kálman and Lászlo Polos, 161–188. Budapest: Akadémiai Kiadó.

Krifka, Manfred. 1990b. "Four thousand ships passed through the lock: Object-induced measure functions on events." *Linguistics and Philosophy* 13: 487–520.

Krifka, Manfred. 1995. "Common nouns: A contrastive analysis of Chinese and English." In *The Generic Book*, ed. Greg N. Carlson and Francis Jeffry Pelletier, 398–411. Chicago: University of Chicago Press.

Krifka, Manfred. 1999. "At least some determiners aren't determiners." In *The Semantics/Pragmatics Interface from Different Points of View*, ed. Ken Turner, 257–291. Amsterdam: Elsevier Science.

Krifka, Manfred, Francis Jeffry Pelletier, Greg N. Carlson, Alice ter Meulen, Gennaro Chierchia, and Godehard Link. 1995. "Genericity: An introduction." In *The Generic Book*, ed. Greg N. Carlson and Francis Jeffry Pelletier, 1–124. Chicago: University of Chicago Press.

Krifka, Manfred, and Sabine Zerbian. 2008. "Quantification across Bantu languages." In *Quantification: A Cross-Linguistic Perspective*, ed. Lisa Matthewson, 383–414. Bingley, UK: Emerald.

Kripke, Saul. 1980. *Naming and Necessity*. Oxford: Blackwell.

Kubo, Miori. 1992. "Japanese syntactic structures and their constructional meaning." PhD dissertation, Massachusetts Institute of Technology.

Labelle, Marie. 2008. "The French reflexive and reciprocal *se*." *Natural Language and Linguistic Theory* 26: 833–876.

Ladusaw, William. 1982. "Semantic constraints on the English partitive constructions." In *Proceedings of WCCFL 1*, ed. Daniel P. Flickinger, Marlys Macken, and Nancy Wiegand, 231–242. Stanford, CA: Stanford Linguistics Association.

Landau, Idan. 2015a. "DP-internal semantic agreement: A configurational analysis." *Natural Language and Linguistic Theory* 34: 975–1020.

Landau, Idan. 2015b. *A Two-Tiered Theory of Control*. Cambridge, MA: MIT Press.

Landman, Fred. 2000. *Events and Plurality: The Jerusalem Lectures*. Dordrecht, the Netherlands: Kluwer.

Landman, Fred. 2003. "Predicate-argument mismatches and the adjectival theory of indefinites." In *From NP to DP*, vol. 1, The Syntax and Semantics of Noun Phrases, ed. Martine Coene and Yves D'hulst, 211–237. Amsterdam: John Benjamins.

Langendoen, D. Terence. 1978. "The logic of reciprocity." *Linguistic Inquiry* 9 (2): 177–197.

Lasersohn, Peter. 1995. *Plurality, Conjunction, and Events*. Dordrecht, the Netherlands: Kluwer.

Leafgren, John. 2011. *A Concise Bulgarian Grammar*. Durham, NC: Slavic and Eurasian Language Resource Center. http://www.seelrc.org:8080/grammar/pdf/stand _alone_bulgarian.pdf.

Le Bruyn, Bert, and Henriëtte de Swart. 2014. "Bare coordination: The semantic shift." *Natural Language and Linguistic Theory* 32 (1): 205–246.

Le Bruyn, Bert, Henriëtte de Swart, and Joost Zwarts. 2013. "*Have, with* and *without*." *Proceedings of SALT* 23: 535–548.

Lecarme, Jacqueline. 2002. "Gender 'polarity': Theoretical aspects of Somali nominal morphology." In *Many Morphologies*, ed. Paul Boucher and Marc Plénat, 109–141. Somerville, MA: Cascadilla Press.

Lechner, Winfried. 1998. "Comparatives and DP-structure." PhD dissertation, University of Massachusetts, Amherst.

Letuchiy, Alexander. 2009. "Slavic systems of grammatical and lexical markers of reflexivity and reciprocality in a typological perspective." Paper presented at the eighth European conference on Formal Description of Slavic Languages, Potsdam, Germany, December 2–5.

Letuchiy, Alexander, and Marija Xolodilova. 2011. "*Bylo paru čelovek*: Ob odnoj količestvennoj konstrukcii v russkom jazyke." Paper presented at Russkij Jazyk: Konstrukcionnye i Leksiko-semantičeskie Podxody, St. Petersburg, March 24.

Levinson, Stephen C. 1983. *Pragmatics*. Cambridge: Cambridge University Press.

Li, Yen-hui Audrey. 1999. "Plurality in a classifier language." *Journal of East Asian Linguistics* 8: 75–99.

Link, Godehard. 1983. "The logical analysis of plurals and mass terms: A lattice theoretical approach." In *Meaning, Use, and the Interpretation of Language*, ed. Rainer Bauerle, Christoph Schwarze, and Arnim von Stechow, 302–323. Berlin: De Gruyter.

Link, Godehard. 1984. "Hydras: On the logic of relative clause constructions with multiple heads." In *Varieties of Formal Semantics*, ed. Fred Landman and Frank Veltman, 245–257. Dordrecht, the Netherlands: Foris.

Link, Godehard. 1987. "Generalized quantifiers and plurals." In *Generalized Quantifiers*, ed. P. Gärdenfors, 151–180. Dordrecht, the Netherlands: Reidel.

Löbner, Sebastian. 1985. "Definites." *Journal of Semantics* 4: 279–326.

Longobardi, Giuseppe. 1994. "Reference and proper names." *Linguistic Inquiry* 25 (4): 609–665.

Longobardi, Giuseppe. 1999. "Some reflections on proper names." Ms., University of Trieste.

Ludlow, Peter, and Stephen Neale. 1991. "Indefinite descriptions: In defense of Russell." *Linguistics and Philosophy* 14: 171–202.

Maekawa, Takafumi. 2013. "An HPSG analysis of 'a beautiful two weeks'." *Linguistic Research* 30 (3): 407–433.

Martí Girbau, Núria. 2010. "The syntax of partitives." PhD dissertation, Universitat Autònoma de Barcelona.

Marty, Paul, Emmanuel Chemla, and Benjamin Spector. 2015. "Phantom readings: The case of modified numerals." *Language, Cognition, and Neuroscience* 30 (4): 462–477.

Matushansky, Ora. 2002. "Movement of degree/degree of movement." PhD dissertation, Massachusetts Institute of Technology.

Matushansky, Ora. 2006. "Head-movement in linguistic theory." *Linguistic Inquiry* 37 (1): 69–109.

Matushansky, Ora. 2008a. "On the attributive nature of superlatives." *Syntax* 11 (1): 26–90.

Matushansky, Ora. 2008b. "A case study of predication." In *Studies in Formal Slavic Linguistics: Contributions from Formal Description of Slavic Languages 6.5*, ed. Franc Marušič and Rok Žaucer, 213–239. Frankfurt: Peter Lang.

Matushansky, Ora. 2008c. "On the linguistic complexity of proper names." *Linguistics and Philosophy* 31 (5): 573–627.

Matushansky, Ora. 2010. "Russian predicate case, *encore*." In *Proceedings of FDSL 7.5*, ed. Gerhild Zybatow, Philip Dudchuk, Serge Minor, and Ekaterina Pshehotskaya, 117–135. Frankfurt: Peter Lang.

Matushansky, Ora. 2012. "On the internal structure of case in Finno-Ugric small clauses." *Finno-Ugric Languages and Linguistics* 1 (1–2): 3–43.

Matushansky, Ora. 2013. "Gender confusion." In *Diagnosing Syntax*, ed. Lisa L.-S. Cheng and Norbert Corver, 271–294. Oxford: Oxford University Press.

Matushansky, Ora. 2015. "On Russian approximative inversion." In *Slavic Grammar from a Formal Perspective: The 10th Anniversary FDSL Conference*, ed. Gerhild

Zybatow, Petr Biskup, Marcel Guhl, Claudia Hurtig, Olav Mueller-Reichau, and Maria Yastrebova, 303–316. Frankfurt: Peter Lang.

Matushansky, Ora, and Tania Ionin. 2014. "More than one solution." In *Proceedings of CLS 47*, ed. Carissa Abrego-Collier, Arum Kang, Martina Martinovic, and Chieu Nguyen, 231–245. Chicago: Chicago Linguistic Society.

Matushansky, Ora, and Tania Ionin. To appear. "Polish numeral NP agreement as a function of surface morphology." In *Proceedings of FASL 25*, ed. Wayles Browne, Miloje Despic, Naomi Enzinna, Robin Karlin, Simone De Lemos, and Draga Zec. Ann Arbor, MI: Michigan Slavic Publications.

Matushansky, Ora, and E. G. Ruys. 2014. "On the syntax of measure." Paper presented at TIN-dag 2014, Utrecht, the Netherlands, February 1.

Matushansky, Ora, and E. G. Ruys. 2015a. "4000 measure NPs: Another pass through the *иллюз*." In *Proceedings of FASL 23*, ed. Małgorzata Szajbel-Keck, Roslyn Burns, and Darya Kavitskaya, 184–205. Ann Arbor, MI: Michigan Slavic Publications.

Matushansky, Ora, and E. G. Ruys. 2015b. "Measure for measure." In *Slavic Grammar from a Formal Perspective: The 10th Anniversary FDSL Conference*, ed. Gerhild Zybatow, Petr Biskup, Marcel Guhl, Claudia Hurtig, Olav Mueller-Reichau, and Maria Yastrebova, 317–330. Frankfurt: Peter Lang.

Matushansky, Ora, E. G. Ruys, and Joost Zwarts. 2017. "On the structure and composition of pseudo-partitives." Paper presented at the Séminaire LaGraM, Unité Mixte de Recherche 7023, Paris, January 16.

Matushansky, Ora, and Joost Zwarts. 2017. "Making space for measures." In *NELS 47: Proceedings of the Forty-Seventh Annual Meeting of the North East Linguistic Society*, ed. Andrew Lamont and Katerina Tetzlo, 261–274. Amherst, MA: Graduate Linguistic Student Association.

Mayer, Gerald L. 1973. "Common tendencies in the syntactic development of 'two,' 'three,' and 'four' in Slavic." *Slavic and East European Journal* 17 (3): 308–314.

McNally, Louise, and Gemma Boleda. 2004. "Relational adjectives as properties of kinds." *Empirical Issues in Formal Syntax and Semantics* 5: 179–196. http://www.cssp.cnrs.fr/eiss5.

Mel'čuk, Igor. 1980. "O padeže čislovogo vyraženija v russkix slovosočetani'x tipa *(bol'še) na dva mal'čika* ili *po troe bol'nyx*." *Russian Linguistics* 5 (1): 55–74.

Mel'čuk, Igor. 1985. *Poverxnostnyj sintaksis russkix čislitel'nyx vyraženij*. Vienna: Institut für Slawistik der Universität Wien.

Miechowicz-Mathiasen, Katarzyna. 2011. "The syntax of Polish cardinal numerals." Ms., Adam Mickiewicz University.

Miechowicz-Mathiasen, Katarzyna. 2012. "Licensing Polish higher numerals: An account of the Accusative Hypothesis." *Generative Linguistics in Wrocław* 2: 57–75.

Miljan, Merilin, and Ronnie Cann. 2013. "Rethinking case marking and case alternation in Estonian." *Nordic Journal of Linguistics* 36 (3): 333.

Milner, Jean-Claude. 1978. *De la syntaxe à l'interprétation [From syntax to interpretation]*. Paris: Seuil.

Mittendorf, Ingo, and Louisa Sadler. 2005. "Numerals, nouns and number in Welsh NPs." In *Proceedings of the LFG05 Conference*, ed. Miriam Butt and Tracy Holloway King, 294–312. Stanford, CA: CSLI Publications.

Miyagawa, Shigeru. 1987. "Lexical categories in Japanese." *Lingua* 73: 29–51.

Moltmann, Friederike. 2012a. *Abstract Objects and the Semantics of Natural Language*. Oxford: Oxford University Press.

Moltmann, Friederike. 2012b. "Reference to numbers in natural language." *Philosophical Studies* 162 (3): 499–536.

Moltmann, Friederike. 2017. "Number word as number names." *Linguistics and Philosophy* 40 (4): 331–345.

Montague, Richard. 1974. *Formal Philosophy: Selected Papers of Richard Montague*, ed. Richmond Thomason. New Haven, CT: Yale University Press.

Moser, Mary B., and Stephen A. Marlett. 1994. "Los números en seri" [Numbers in Seri]. In *II Encuentro de Lingüística en el Noroeste: Memorias*, ed. Zarina Estrada Fernández, 63–79. Hermosillo, Mexico: Departamento de Letras y Lingüística, División de Humanidades y Bellas Artes, Universidad de Sonora.

Munn, Alan. 1987. "Coordinate structure and X-bar theory." *McGill Working Papers in Linguistics* 4 (1): 121–140.

Munn, Alan. 1993. "Topics in the syntax and semantics of coordinate structures." PhD dissertation, University of Maryland, College Park.

Munn, Alan. 1999. "First conjunct agreement: Against a clausal analysis." *Linguistic Inquiry* 30 (4): 643–668.

Muromatsu, Keiko. 1998. "On the syntax of classifiers." PhD dissertation, University of Maryland, College Park.

Neidle, Carol. 1988. *The Role of Case in Russian Syntax*. Dordrecht, the Netherlands: Kluwer.

Nelson, Diane, and Ida Toivonen. 2000. "Counting and the grammar: Case and numerals in Inari Sami." *Leeds Working Papers in Linguistics* 8: 179–192.

Nishiyama, Kunio. 1999. "Adjectives and the copulas in Japanese." *Journal of East Asian Linguistics* 8: 183–222.

Norris, Mark. 2015. "Case-marking in Estonian pseudopartitives." In *Proceedings of the 41st Annual Meeting of the Berkeley Linguistics Society*, ed. Anna E. Jurgensen, Hannah Sande, Spencer Lamoureux, Kenny Baclawski, and Alison Zerbe, 371–395. Berkeley, CA: Berkeley Linguistics Society.

Nouwen, Rick. 2008. "Directionality in modified numerals: The case of *up to*." In *Proceedings of Semantics and Linguistic Theory (SALT) 18*, ed. Tova Friedman and Satoshi Ito, 569–582. eLanguage.

Nouwen, Rick. 2010. "Two kinds of modified numerals." *Semantics and Pragmatics* 3: 1–41.

Nouwen, Rick. 2015. "Modified numerals: The epistemic effect." In *Epistemic Indefinites*, ed. Luis Alonso-Ovalle and Paula Menéndez-Benito, 244–266. Oxford: Oxford University Press.

Noyer, Rolf. 1992. "Features, positions, and affixes in autonomous morphological structure." PhD dissertation, Massachusetts Institute of Technology.

Nunes, Jairo. 1995. "The copy theory of movement and linearization of chains in the Minimalist Program." PhD dissertation, University of Maryland, College Park.

Nurmio, Silva, and David Willis. 2015. "The rise and fall of a minor number: The case of the Welsh numerative." Ms., University of Cambridge and University of Dublin.

Odijk, J. 1992. "Uninflected adjectives in Dutch." In *Linguistics in the Netherlands 1992*, ed. R. Bok-Bennema and R. van Hout, 197–208. Amsterdam: John Benjamins.

Ohna, Tsutomu. 2003. "*A beautiful two weeks*: Its syntactic structure and the semantic relations of the adjective to the numeral and head noun." In *Empirical and Theoretical Investigations into Language: A Festschrift for Masaru Kajita,* ed. Shuji Chiba, 23–40. Tokyo: Kaitakusha.

Õispuu, Jaan. 1999. *Spravočnik po èstonskomu jazyku.* Tallinn, Estonia: Koolibri.

Ó Maolalaigh, Roibeard. 2013. "*Corpas na Gàidhlig* and singular nouns with the numerals 'three' to 'ten' in Scottish Gaelic." In *Language in Scotland: Corpus-Based Studies,* ed. Wendy Anderson, 113–142. Amsterdam: Rodopi.

Oomen, Stanly. 2012. "Adjectives as a distinct class in Eastern Riffian (Berber, Afroasiatic)." Paper presented at the Syntax Interface Lectures, Utrecht, the Netherlands, April 16.

Ouwayda, Sarah. 2014. "Where number lies: Plural marking, numerals, and the collective-distributive distinction." PhD dissertation, University of Southern California.

Ouwayda, Sarah, and Ur Shlonsky. 2015. "Order in the DP! On word order and the structure of the DP." *LSA Annual Meeting Extended Abstracts* 6. https://journals.linguisticsociety.org/proceedings/index.php/ExtendedAbs/article/view/3019.

Panagiotidis, E. Phoevos. 2002. *Pronouns, Clitics and Empty Nouns: "Pronominality" and Licensing in Syntax.* Amsterdam: John Benjamins.

Pancheva, Roumyana. 2006. "Phrasal and clausal comparatives in Slavic." In *Proceedings of FASL 14: The Princeton Meeting,* ed. James Lavine, Steven Franks, Mila Tasseva-Kurktchieva, and Hana Filip, 236–257. Ann Arbor, MI: Michigan Slavic Publications.

Pancheva, Roumyana. 2010. "More students attended FASL than CONSOLE." In *Formal Approaches to Slavic Linguistics 18: The Cornell Meeting, 2009,* ed. Wayles Browne, Adam Cooper, Alison Fisher, Esra Kesici, Nikola Predolac, and Draga Zec, 383–400. Ann Arbor, MI: Slavica.

Panizza, Daniele, Gennaro Chierchia, and Charles Clifton. 2009. "On the role of entailing patterns in the interpretation and processing of numerals and scalar quantifiers." *Journal of Memory and Language* 61 (4): 503–518.

Partee, Barbara H. 1986. "Noun phrase interpretation and type-shifting principles." In *Studies in Discourse Representation Theory and the Theory of Generalized Quantifiers,* ed. Jeroen Groenendijk, Dick de Jongh, and Martin Stokhof, 115–143. Dordrecht, the Netherlands: Foris.

Partee, Barbara H. 1989. "Binding implicit variables in quantified contexts." In *Papers from CLS 25,* ed. Caroline R. Wiltshire, Bradley Music, and Randolph Graczyk, 342–365. Chicago: Chicago Linguistic Society.

Partee, Barbara H. 1999. "Weak NP's in HAVE sentences." In *JFAK: Essays Dedicated to Johan van Benthem on the Occasion of His 50th Birthday*, ed. Jelle Gerbrandy, Maarten Marx, Maarten de Rijke, and Yde Venema. Amsterdam: University of Amsterdam. http://festschriften.illc.uva.nl/j50/.

Partee, Barbara H., and Mats Rooth. 1983. "Generalized conjunction and type ambiguity." In *Meaning, Use and Interpretation of Language*, ed. Rainer Bauerle, Christoph Schwarze, and Arnim von Stechow, 361–383. Berlin: Mouton de Gruyter.

Paterson, John. 1968. *Gaelic Made Easy: A Gaelic Guide for Beginners*. Glasgow: Dionnasg Gaidhlig na h-Alba (The Gaelic League of Scotland).

Payne, John R. 1985. "Complex phrases and complex sentences." In *Language Typology and Syntactic Description*, ed. Timothy Shopen, 3–41. Cambridge: Cambridge University Press.

Pereltsvaig, Asya. 2006a. "Passing by cardinals: In support of head movement in nominals." In *Proceedings of FASL 14: The Princeton Meeting*, ed. James Lavine, Steven Franks, Mila Tasseva-Kurktchieva, and Hana Filip, 277–292. Ann Arbor, MI: Michigan Slavic Publications.

Pereltsvaig, Asya. 2006b. "Small nominals." *Natural Language and Linguistic Theory* 24 (2): 433–500.

Pesetsky, David. 1982. "Paths and categories." PhD dissertation, Massachusetts Institute of Technology.

Pesetsky, David. 2008. "Russian case morphology and the syntactic categories." Ms., Massachusetts Institute of Technology.

Pesetsky, David, and Esther Torrego. 2001. "T-to-C movement: Causes and consequences." In *Ken Hale: A Life in Language*, ed. Michael Kenstowicz, 355–426. Cambridge, MA: MIT Press.

Pica, Pierre, Cathy Lemer, Véronique Izard, and Stanislas Dehaene. 2004. "Exact and approximate arithmetic in an Amazonian indigene group." *Science* 306: 499–503.

Plank, Frans. 2004. "From local adpositions to approximative adnumerals, in German and wherever." *Studies in Language* 28 (1): 165–201.

Pollard, Carl, and Ivan A. Sag. 1994. *Head-Driven Phrase Structure Grammar*. Chicago: CSLI.

Postal, Paul. 1969. "On so-called 'pronouns' in English." In *Modern Studies in English: Readings in Transformational Grammar*, ed. David A. Reibel and Sanford A. Schane, 201–224. Englewood Cliffs, NJ: Prentice-Hall.

Postal, Paul. 1974. *On Raising*. Cambridge, MA: MIT Press.

Potet, Jean-Paul G. 1992. "Numeral expressions in Tagalog." *Archipel* 44: 167–181.

Preminger, Omer. 2011. "Agreement as a fallible operation." PhD dissertation, Massachusetts Institute of Technology.

Progovac, Ljiljana. 1998. "Structure for coordination (part 1)." *GLOT International* 3 (7): 3–6.

Przepiórkowski, Adam. 1997. "Case assignment in Polish: Towards an HPSG analysis." In *Formale Slavistik*, ed. Uwe Junghanns and Gerhild Zybatow, 307–319. Frankfurt: Vervuert.

Przepiórkowski, Adam, and Agnieszka Patejuk. 2012. "The puzzle of case agreement between numeral phrases and predicative adjectives in Polish." In *The Proceedings of the LFG' 12 Conference*, ed. Miriam Butt and Tracy Holloway King, 490–502. Stanford, CA: CSLI Publications.

Przepiórkowski, Adam, and Alexandr Rosen. 2008. "On the case of predicative complements in Czech infinitival clauses." In *Formal Description of Slavic Languages: The Fifth Conference*, ed. Gerhild Zybatow, Uwe Junghanns, Roland Meyer, and Luka Szucsich, 478–492. Frankfurt: Peter Lang.

Pughe, W. Owen. 1832. *Dictionary of the Welsh Language: Explained in English*. Denbigh, Wales, UK: Thomas Gee.

Rakhlin, Natalia. 2003. "Genitive of quantification in Russian: What morphology can tell us about syntax." In *Proceedings of CONSOLE XI*, ed. Marjo van Koppen and Mark de Vos. Leiden, the Netherlands: Student Organization of Linguistics in Europe. http://www.hum2.leidenuniv.nl/pdf/lucl/sole/console11/console11-rakhlin.pdf.

Rappaport, Gilbert C. 2002. "Numeral phrases in Russian: A minimalist approach." *Journal of Slavic Linguistics* 10: 329–342.

Rappaport, Gilbert C. 2003a. "Case syncretism, features, and morphosyntax of Polish numeral phrases." In *Generative Linguistics in Poland*, ed. Piotr Banski and Adam Przepiórkowski, 123–137. Warsaw: Academy of Sciences.

Rappaport, Gilbert C. 2003b. "The grammatical role of animacy in a formal model of Slavic morphology." In *American Contributions to the Thirteenth International Congress of Slavists (Ljubljana, 2003)*, ed. Robert A. Maguire and Alan Timberlake, 149–166. Bloomington, IN: Slavica.

Rappaport, Gilbert C. To appear. "The Slavic noun phrase in comparative perspective." In *Comparative Slavic Morphosyntax*, ed. Stephanie Harves and James Lavine. Bloomington, IN: Slavica.

Reed, Ann M. 1991. "On interpreting partitives." In *Bridges between Psychology and Linguistics: A Swarthmore Festschrift for Lila Gleitman*, ed. Donna Jo Napoli and Judy Anne Kegl, 207–223. Hillsdale, NJ: Lawrence Erlbaum Associates.

Reinhart, Tanya. 1997. "Quantifier scope: How labor is divided between QR and choice functions." *Linguistics and Philosophy* 20: 335–397.

Reinhart, Tanya, and Tal Siloni. 2004. "Against the unaccusative analysis of reflexives." In *The Unaccusativity Puzzle*, ed. Artemis Alexiadou, Elena Anagnostopoulou, and Martin Everaert, 288–331. Oxford: Oxford University Press.

Rendahl, Anne-Charlotte. 2001. "Swedish dialects around the Baltic Sea." In *Circum-Baltic Languages*, ed. Östen Dahl and Maria Koptjevskaja-Tamm, 137–177. Amsterdam: John Benjamins.

Reuland, Eric. 2005. "Binding conditions: How are they derived?" In *HPSG05: Proceedings of the 12th International Conference on Head-Driven Phrase Structure Grammar*, ed. Stefan Müller, 578–593. Stanford, CA: CLSI Publications.

Reuland, Eric, and Yoad Winter. 2009. "Binding without identity: Towards a unified semantics for bound and exempt anaphors." In *Anaphora Processing and Applications*, ed. Antonio Branco, Sobha Lalitha Devi, and Ruslan Mitkov, 69–79. Berlin: Springer.

Ritter, Elisabeth. 1987. "NSO noun phrase in Modern Hebrew." In *Proceedings of NELS 17*, ed. Joyce McDonough and Bernadette Plunkett, 521–537. Amherst, MA: Graduate Linguistic Student Association.

Ritter, Elisabeth. 1988. "A head-movement approach to construct-state noun phrases." *Linguistics* 26: 909–929.

Ritter, Elisabeth. 1991. "Two functional categories in noun phrases: Evidence from Modern Hebrew." In *Perspectives on Phrase Structure, Syntax and Semantics*, 37–62. New York: Academic Press.

Roberts, Craige. 1987. "Modal subordination, anaphora, and distributivity." PhD dissertation, University of Massachusetts, Amherst.

Roberts, Gareth. 2000. "Bilingualism and number in Wales." *International Journal of Bilingual Education and Bilingualism* 3 (1): 44–56.

Rosén, Haiim B. 1977. *Contemporary Hebrew*. The Hague: Mouton.

Rosenthal, Franz. 2006. *A Grammar of Biblical Aramaic*. Wiesbaden, Germany: Harrassowitz.

Ross, John R. 1967. "Constraints on variables in syntax." PhD dissertation, Massachusetts Institute of Technology.

Ross, John R. 1972. "The category squish: Endstation hauptwork." In *Papers from the Eighth Regional Meeting of the Chicago Linguistic Society*, ed. Paul M. Peranteau, Judith N. Levi, and Gloria C. Phares, 316–318. Chicago: University of Chicago.

Rothstein, Susan. 2009. "Individuating and measure readings of classified constructions: Evidence from modern Hebrew." *Brill's Annual of Afroasiatic Languages and Linguistics* 1 (1): 106–145.

Rothstein, Susan. 2011a. "Counting, measuring, and the semantics of classifiers." In *The Baltic International Yearbook of Condition, Logic and Communication*, ed. Barbara Partee, Michael Glanzberg, and Jurģis Šķilters, 1–42. Manhattan, KS: New Prairie Press.

Rothstein, Susan. 2011b. "Numbers: Counting, measuring, and classifying." Paper presented at Sinn und Bedeutung 16, Utrecht University, September 6–9.

Rothstein, Susan. 2013. "A Fregean semantics for number words." In *Proceedings of the 19th Amsterdam Colloquium*, ed. Maria Aloni, Michael Franke, and Floris Roelofsen, 179–186. Amsterdam: Institute for Logic, Language, and Computation.

Rothstein, Susan. 2016. "Counting and measuring: A theoretical and crosslinguistic account." *Baltic International Yearbook of Cognition, Logic, and Communication* 11: 1–49.

Rothstein, Susan. 2017. *The Semantics of Counting and Measuring*. Cambridge: Cambridge University Press.

Rothstein, Susan, and Keren Khrizman. 2015. "Approximative inversion in Russian as a measure construction." In *Slavic Grammar from a Formal Perspective: The 10th Anniversary FDSL Conference*, ed. Gerhild Zybatow and Petr Biskup, 259–272. Frankfurt: Peter Lang.

Rozental, D. E., E. V. Džandžakova, and N. P. Kabanova. 1998. *Spravočnik po pravopisaniju, proiznošeniju, literaturnomu redaktirovaniju*. Moscow: ČeRo.

Rullmann, Hotze. 2003. "Bound-variable pronouns and the semantics of number." In *Proceedings of the Western Conference on Linguistics (WECOL) 2002*, ed. Brian Agbayani, Paivi Koskinen, and Vida Samiian, 243–254. Fresno, CA: California State University.

Rutkowski, Paweł. 2002. "The syntax of quantifier phrases and the inherent vs. structural case distinction." *Linguistic Research* 7 (1): 43–74.

Rutkowski, Paweł. 2007. "Hipoteza frazy przedimkowej jako narzędzie opisu składniowego polskich grup imiennych." PhD dissertation, Warsaw University.

Rutkowski, Paweł, and Hanna Maliszewska. 2007. "On prepositional phrases inside numeral expressions in Polish." *Lingua* 117 (5): 784–813.

Rutkowski, Paweł, and Kamil Szczegot. 2001. "On the syntax of functional elements: Numerals, pronouns, and expressions of approximation." In *Generative Linguistics in Poland: Syntax and Morphosyntax*, ed. Adam Przepiórkowski and Piotr Bański, 187–196. Warsaw: Instytut Podstaw Informatyki PAN.

Ruys, E. G. 1992. "The scope of indefinites." PhD dissertation, Utrecht University.

Ruys, E. G. 2017. "Two Dutch *many*'s and the structure of pseudo-partitives." *Glossa* 2 (1): 7–33.

Ryding, Karin. 2005. *A Reference Grammar of Modern Standard Arabic*. Cambridge: Cambridge University Press.

Sabbagh, Joseph. 2003. "Ordering and linearizing rightward movement." In *Proceedings of WCCFL 22*, ed. Gina Garding and Mimu Tsujimura, 436–449. Somerville, MA: Cascadilla Press.

Sadler, Louisa. 2000. "Noun phrase structure in Welsh." In *Argument Realization*, ed. Miriam Butt and Tracy Holloway King, 73–110. Stanford, CA: CSLI Publications.

Sadler, Louisa. 2003. "Coordination and asymmetric agreement in Welsh." In *Nominals: Inside and Out*, ed. Miriam Butt and Tracy Holloway King, 85–118. Stanford, CA: CSLI Publications.

Sadler, Louisa. 2010. "Arabic numeral noun constructions: Some preliminary thoughts." Ms., University of Essex.

Saeed, John. 1999. *Somali*. Amsterdam: John Benjamins.

Safir, Ken. 2004. *The Syntax of (In)dependence*. Cambridge, MA: MIT Press.

Sauerland, Uli. 2003. "A new semantics for number." In *Proceedings of Semantics and Linguistic Theory (SALT) 13*, ed. Robert B. Young and Yuping Zhou, 258–275. Ithaca, NY: CLC Publications.

Sauerland, Uli. 2004. "A comprehensive semantics for agreement." Ms., Leibniz-Zentrum Allgemeine Sprachwissenschaft.

Sauerland, Uli, Jan Anderssen, and Kazuko Yatsushiro. 2005. "The plural is semantically unmarked." In *Linguistic Evidence: Empirical, Theoretical, and Computational Perspectives*, ed. Stephan Kepser and Marga Reis, 413–434. Berlin: Mouton de Gruyter.

Sauerland, Uli, and Paul Elbourne. 2002. "Total reconstruction, PF movement, and derivational order." *Linguistic Inquiry* 33 (2): 283–319.

Savu, Carmen. 2012. "On different analyses for additive and prepositional numeral noun constructions." Ms., University of Bucharest.

Scatton, Ernest A. 1975. *Bulgarian Phonology*. Cambridge, MA: Slavica.

Scatton, Ernest A. 1984. *A Reference Grammar of Modern Bulgarian*. Columbus, OH: Slavica.

Scha, Remko. 1981. "Distributive, collective, and cumulative quantification." In *Formal Methods in the Study of Language*, ed. Jeroen Groenendijk, Martin Stokhof, and Theo M. V. Janssen, 483–512. Amsterdam: Mathematisch Centrum, University of Amsterdam.

Schenker, Alexander M. 1971. "Some remarks on Polish quantifiers." *The Slavic and East European Journal* 15 (1): 54–60.

Schroeder, Christoph. 1999. *The Turkish Nominal Phrase in Spoken Discourse*. Wiesbaden, Germany: Harrassowitz.

Schuh, Russell G. 1989. "Number and gender in Miya." In *Current Progress in Chadic Linguistics: Proceedings of the International Symposium on Chadic Linguistics*, ed. Zygmunt Frajzyngier, 171–181. Amsterdam: John Benjamins.

Schuh, Russell G. 1998. *A Grammar of Miya*. Berkeley: University of California Press.

Schuh, Russell G., and Vaziya Ciroma Tilde Miya. 2010. "Miya-English-Hausa dictionary." Ms., University of California, Los Angeles.

Schwarz, Bernhard. 2001. "Two kinds of long-distance indefinites." In *Proceedings of the Thirteenth Amsterdam Colloquium*, ed. Robert van Rooy and Martin Stokhof, 192–197. Amsterdam: Institute for Logic, Language, and Computation/Department of Philosophy, University of Amsterdam.

Schwarz, Bernhard. 2016a. "*At least* and ignorance: A reply to Coppock and Brochhagen (2013)." *Semantics and Pragmatics* 9 (10): 1–17.

Schwarz, Bernhard. 2016b. "Consistency preservation in Quantity implicature: the case of *at least*." *Semantics and Pragmatics* 9 (1): 1–47.

Schwarz, Bernhard, Brian Buccola, and Michael Hamilton. 2012. "Two types of class B numeral modifiers: A reply to Nouwen 2010." *Semantics and Pragmatics* 5 (1): 1–25.

Schwarzschild, Roger. 1994. "Plurals, presuppositions, and the sources of distributivity." *Natural Language Semantics* 2: 201–248.

Schwarzschild, Roger. 1996. *Pluralities*. Dordrecht, the Netherlands: Kluwer.

Schwarzschild, Roger. 2002a. "The grammar of measurement." Ms., Rutgers University.

Schwarzschild, Roger. 2002b. "Singleton indefinites." *Journal of Semantics* 19 (3): 289–314.

Schwarzschild, Roger. 2005. "Measure phrases as modifiers of adjectives." In *L'adjectif*, ed. Patricia Cabredo Hofherr and Ora Matushansky, 207–228. Saint-Denis, France: Presses Universitaires de Vincennes.

Schwarzschild, Roger. 2011. "Stubborn distributivity, multiparticipant nouns, and the count/mass distinction." In *Proceedings of the Thirty-Ninth Annual Meeting of the*

North East Linguistic Society (NELS 39), ed. Suzi Lima, Kevin Mullin and Brian Smith, 661–678. Amherst, MA: Graduate Linguistic Student Association.

Scontras, Gregory. 2013. "A unified semantics for number marking, numerals, and nominal structure." In *Proceedings of Sinn und Bedeutung 17*, ed. Emmanuel Chemla, Vincent Homer, and Grégoire Winterstein, 545–562. Semantics Archive. https://semanticsarchive.net/sub2012/.

Scontras, Gregory. 2014. "The semantics of measurement." PhD dissertation, Harvard University.

Selkirk, Elisabeth. 1977. "Some remarks on noun phrase structure." In *Formal Syntax*, ed. Peter W. Culicover, Thomas Wasow, and Adrian Akmajian, 285–316. London: Academic Press.

Selkirk, Elisabeth. 1982. *The Syntax of Words*. Cambridge, MA: MIT Press.

Shlonsky, Ur. 2004. "The form of Semitic noun phrases." *Lingua* 114 (12): 1465–1526.

Sigler, Michele. 1992. "Number agreement and specificity in Armenian." In *Papers from the 28th Regional Meeting of the Chicago Linguistic Society*, ed. Costas P. Canakis, Grace P. Chan, and Jeanette Marshall Denton, 499–514. Chicago: Chicago Linguistic Society.

Sigler, Michele. 1996. "Specificity and agreement in standard Western Armenian." PhD dissertation, Massachusetts Institute of Technology.

Siloni, Tal. 2001. "Adjectival constructs and inalienable constructions." In *Themes and Issues in the Syntax of Arabic and Hebrew*, ed. Jamal Ouhalla and Ur Shlonsky, 161–187. Dordrecht, the Netherlands: Kluwer.

Simpson, Andrew. 2005. "Classifiers and DP structure in Southeast Asia." In *The Oxford Handbook of Comparative Syntax*, ed. Guglielmo Cinque and Richard S. Kayne, 806–838. Oxford: Oxford University Press.

Sneed, Elisa. 2002. "The acceptability of regular plurals in compounds." In *Papers from the 38th Annual Meeting of the Chicago Linguistic Society*, ed. Mary Andronis, Erin Debenport, Anne Pycha, and Keiko Yoshimura, 617–631. Chicago: Chicago Linguistic Society.

Solt, Stephanie. 2007. "Two types of modified cardinals." Paper presented at the Colloque International sur les Adjectifs, Université Lille 3, September 13–15.

Spelke, Elizabeth S. 2011. "Natural number and natural geometry." In *Space, Time and Number in the Brain: Searching for the Foundations of Mathematical Thought*, ed. Elizabeth Brannon and Stanislas Dehaene, 287–317. Oxford: Oxford University Press.

Staroverov, Peter. 2007. "Relational nouns and reciprocal plurality." In *Proceedings of SALT 17*, ed. Tova Gibson and Masayuki Friedman, 300–316. Ithaca, NY: Cornell University.

Stassen, Leon. 2000. "AND-languages and WITH-languages." *Linguistic Typology* 4: 1–54.

Stateva, Penka. 2002. "How different are different degree donstructions?" PhD dissertation, University of Connecticut.

Stavrou, Melita. 2003. "Semi-lexical nouns, classifiers, and the interpretation(s) of the pseudopartitive construction." In *From NP to DP*, vol. 1, The Syntax and Semantics of

Noun Phrases, ed. Martine Coene and Yves D'hulst, 329–353. Amsterdam: John Benjamins.

Stepanov, Arthur. 2001. "Late adjunction and minimalist phrase structure." *Syntax* 4 (1): 94–125.

Stolz, Thomas, and Ljuba N. Veselinova. 2005. "Ordinal numerals." In *The World Atlas of Language Structures*, ed. Martin Haspelmath, Matthew S. Dryer, David Gil, and Bernard S. Comrie, 218–221. Oxford: Oxford University Press.

Stump, Gregory. 2010. "The derivation of compound ordinal numerals: Implications for morphological theory." *Word Structure* 3 (2): 205–233.

Swan, Oscar E. 2002. *A Grammar of Contemporary Polish*. Bloomington, IN: Slavica.

Szabolcsi, Anna. 1997. "Strategies for scope taking." In *Ways of Scope Taking*, ed. Anna Szabolcsi, 109–154. Dordrecht, the Netherlands: Kluwer.

Tennant, Neil. 2015. *Introducing Philosophy: God, Mind, World, and Logic*. London: Taylor and Francis.

Thomas, D. Winton. 1953. "A consideration of some unusual ways of expressing the superlative in Hebrew." *Vetus Testamentum* 3 (3): 209–224.

Thomas, Peter Wyn. 1996. *Gramadeg y Gymraeg*. Caerdydd, Wales, UK: Gwasg Prifysgol Cymru.

Thorne, David. 1993. *A Comprehensive Welsh Grammar*. Oxford: Basil Blackwell.

Timberlake, Alan. 2004. *A Reference Grammar of Russian*. Cambridge: Cambridge University Press.

van der Does, Jaap. 1992. "Applied quantifier logics." PhD dissertation, University of Amsterdam.

van der Does, Jaap. 1993. "Sums and quantifiers." *Linguistics and Philosophy* 16: 509–550.

van Geenhoven, Veerle. 1996. "Semantic incorporation and indefinite descriptions: Semantic and syntactic aspects of noun incorporation in West Greenlandic." PhD dissertation, Universität Tuebingen.

Vasmer, Max. 1986. *Ètimologičeskij slovar' russkogo jazyka*. Moscow: Progress.

Verkuyl, Henk. 1993. *A Theory of Aspectuality: The Interaction between Temporal and Atemporal Structure*. Cambridge: Cambridge University Press.

Verkuyl, Henk J. 1997. "Some issues in the analysis of multiple quantification with plural NPs." In *Plurality and Quantification*, ed. Fritz Hamm and Erhard Hinrichs, 283–319. Dordrecht, the Netherlands: Kluwer.

Verkuyl, Henk J., and Jaap van der Does. 1991. "The semantics of plural noun phrases." In *Quantifiers, Logic, and Language*, ed. Jaap van der Does and Jan van Eyck, 337–374. Stanford, CA: Center for the Study of Language and Information.

Veselinova, Ljuba. 2004. "Cross-linguistic distribution of derived numerals." Paper presented at the Workshop on Numerals in the World's Languages, Max Planck Institute for Evolutionary Anthropology, Leipzig, Germany, March 29–30.

Veselovská, Ludmila. 2001. "Agreement patterns of Czech group nouns and quantifiers." In *Semi-lexical Categories: The Function of Content Words and the Content of*

Function Words, ed. Norbert Corver and Henk van Riemsdijk, 273–320. Berlin: Mouton de Gruyter.

Vinogradov, V. V., ed. 1952. *Grammatika russkogo jazyka*. Moscow: Soviet Academy of Sciences.

von Stechow, Arnim. 2003. "Feature deletion under semantic binding: Tense, person, and mood under verbal quantifiers." In *Proceedings of NELS 33*, ed. Makoto Kadowaki and Shigeto Kawahara, 379–403. Amherst, MA: Graduate Linguistic Student Association.

Vos, Riet. 1999. "A grammar of partitive constructions." PhD dissertation, Tilburg University.

Wągiel, Marcin. 2015. "Sums, groups, genders, and Polish numerals." In *Slavic Grammar from a Formal Perspective: The 10th Anniversary FDSL Conference*, ed. Gerhild Zybatow, Petr Biskup, Marcel Guhl, Claudia Hurtig, Olav Mueller-Reichau, and Maria Yastrebova, 495–513. Frankfurt: Peter Lang.

Wechsler, Stephen. 2011. "Mixed agreement, the person feature, and the index/concord distinction." *Natural Language and Linguistic Theory* 29 (4): 999–1031.

Wechsler, Stephen, and Larisa Zlatić. 2000. "A theory of agreement and its application to Serbo-Croatian." *Language* 76 (4): 799–832.

Wechsler, Stephen, and Larisa Zlatić. 2003. *The Many Faces of Agreement*. Stanford, CA: Center for the Study of Language and Information.

Wechsler, Stephen, and Larisa Zlatić. 2012. "The wrong two faces." *Language* 88 (2): 380–387.

Westera, Matthijs, and Adrian Brasoveanu. 2014. "Ignorance in context: The interaction of modified numerals and QUDs." *Proceedings of SALT* 24: 414–431.

Wiese, Heike. 2003. *Numbers, Language, and the Human Mind*. Cambridge: Cambridge University Press.

Wiese, Heike. 2007. "The co-evolution of number concepts and counting words." *Lingua* 117 (5): 758–772.

Wilder, Chris. 1997. "Some properties of ellipsis in coordination." In *Studies on Universal Grammar and Typological Variation*, ed. Artemis Alexiadou and Tracy Hall, 59–107. Amsterdam: John Benjamins.

Wilder, Chris. 1999. "Right-Node Raising and the LCA." In *WCCFL 18: Proceedings of the 18th West Coast Conference on Formal Linguistics*, ed. Sonya Bird, Andrew Carnie, Jason D. Haugen, and Peter Norquest, 586–598. Somerville, MA: Cascadilla Press.

Williams, Stephen Joseph. 1980. *A Welsh Grammar. Cardiff, Wales*, UK: University of Wales Press.

Winter, Yoad. 1995. "Syncategorematic conjunction and structured meanings." In *Proceedings of Semantics and Linguistic Theory (SALT) 5*, ed. Mandy Simons and Teresa Galloway, 387–404. Ithaca, NY: CLC Publications.

Winter, Yoad. 1996. "A unified semantic treatment of singular NP coordination." *Linguistics and Philosophy* 19: 337–391.

Winter, Yoad. 1997. "Choice functions and the scopal semantics of indefinites." *Linguistics and Philosophy* 20: 399–467.

Winter, Yoad. 1998. "Flexible Boolean semantics: Coordination, plurality, and scope in natural language." PhD dissertation, Utrecht University.

Winter, Yoad. 2001a. *Flexibility Principles in Boolean Semantics: Coordination, Plurality and Scope in Natural Language*. Cambridge, MA: MIT Press.

Winter, Yoad. 2001b. "Plural predication and the Strongest Meaning Hypothesis." *Journal of Semantics* 18: 333–365.

Winter, Yoad. 2005. "On some problems of (in)definiteness in flexible semantics." *Lingua* 115: 767–786.

Witkoś, Jacek. 2008. "Control and agreement with predicative adjectives in Polish." In *Elements of Slavic and Germanic Grammars: A Comparative View*, ed. Gisbert Fanselow and Jacek Witkoś, 255–277. Frankfurt: Peter Lang.

Woolford, Ellen. 1987. "An ECP account of constraints on across-the-board extraction." *Linguistic Inquiry* 18 (1): 166–171.

Worth, Dean S. 1959. "Grammatical and lexical quantification in the syntax of the Russian numeral." *International Journal of Slavic Linguistics and Poetics* 1–2: 117–132.

Wunderlich, Dieter. 1991. "How do prepositional phrases fit into compositional syntax and semantics?" *Linguistics* 29 (4): 591–622.

Wynn, Karen. 1992. "Addition and subtraction by human infants." *Nature* 358: 749–750.

Xalipov, S. G. 1995. *Kratkaja grammatika vallijskogo jazyka*. St. Petersburg: DEAN+ADIA-M.

Xiang, Ming, Boris Harizanov, Maria Polinsky, and Ekaterina Kravtchenko. 2011. "Processing morphological ambiguity: An experimental investigation of Russian numerical phrases." *Lingua* 121 (3): 548–560.

Xu, Fei, and Elizabeth Spelke. 2000. "Large number discrimination in 6-month-old infants." *Cognition* 74 (1): B1–B11.

Yadroff, Michael. 1999. "Formal properties of functional categories: The minimalist syntax of Russian nominal and prepositional expressions." PhD dissertation, Indiana University.

Yadroff, Michael, and Loren Billings. 1998. "The syntax of approximative inversion in Russian (and the general architecture of nominal expressions)." In *Proceedings of the 6th Annual Workshop on Formal Approaches to Slavic Linguistics: The Connecticut Meeting 1997*, ed. Željko Bošković, Steven Franks, and William Snyder, 319–338. Ann Arbor, MI: Michigan Slavic Publications.

Zabbal, Youri. 2002. "The semantics of number in the Arabic noun phrase." MA thesis, University of Calgary.

Zabbal, Youri. 2005. "The syntax of numeral expressions." Ms., University of Massachusetts, Amherst.

Zaliznjak, A. A. 1967. *Russkoe imennoe slovoizmenenie*. Moscow: Nauka.

Zalta, Edward N. 1999. "Natural numbers and natural cardinals as abstract objects: A partial reconstruction of Frege's *Grundgesetze* in object theory." *Journal of Philosophical Logic* 28 (6): 617–658.

Zamparelli, Roberto. 1995. "Layers in the Determiner Phrase." PhD dissertation, University of Rochester.

Zamparelli, Roberto. 1998. "A theory of kinds, partitives, and *of/z* possessives." In *Possessors, Predicates and Movement in the Determiner Phrase*, ed. Chris Wilder and Artemis Alexiadou, 259–301. Amsterdam: John Benjamins.

Zamparelli, Roberto. 2002a. "Definite and bare kind-denoting noun phrases." In *Romance Languages and Linguistic Theory 2000: Selected Papers from Going Romance 2000*, ed. Claire Beyssade, Reineke Bok-Bennema, Frank Drijkoningen, and Paola Monachesi, 305–342. Amsterdam: John Benjamins.

Zamparelli, Roberto. 2002b. "Dei ex machina." Ms., Università di Bergamo.

Zamparelli, Roberto. 2004. "Every two days." *Snippets* 9: 19–20.

Zaroukian, Erin. 2012. "Approximative inversion revisited." In *Formal Approaches to Slavic Linguistics 19: The College Park Meeting*, ed. John Bailyn, Ewan Dunbar, Yakov Konrad, and Chris LaTerza, 146–160. Ann Arbor, MI: Michigan Slavic Publications.

Zhang, Niina Ning. 2012. "Countability and numeral classifiers in Mandarin Chinese." In *A Cross-Linguistic Exploration of the Count-Mass Distinction*, ed. Diane Massam, 220–237. Oxford: Oxford University Press.

Zoerner, Ed. 1995. "Coordination: The syntax of &P." PhD dissertation, University of California, Irvine.

Zuckermann, Ghil'ad. 2006. "A new vision for 'Israeli Hebrew': Theoretical and practical implications of analyzing Israel's main language as a semi-engineered Semito-European hybrid language." *Journal of Modern Jewish Studies* 5 (1): 57–71.

Zwarts, Joost, and Yoad Winter. 2000. "Vector space semantics: A model-theoretic analysis of locative prepositions." *Journal of Logic, Language, and Information* 9: 169–211.

Zweig, Eytan. 2006. "Nouns and adjectives in numeral NPs." In *Proceedings of NELS 35*, ed. Leah Bateman and Cherlon Ussery, 663–679. Amherst, MA: Graduate Linguistic Student Association.

Author Index

Language Index

Ainu, 144–145
Albanian, 341n
Andoke, 120
Arabic, 7, 9, 62, 89, 106, 108, 121, 223–224, 257–263, 265, 267–268, 336. *See also* Lebanese Arabic; Palestinian Arabic; Standard Arabic
Armenian, 341n; *See also* Western Armenian

Bantu, 2, 68, 122. *See also* Basaa; Chinyanja; Luvale; Shona
Basaa, 68
Basque, 2, 7, 10, 292–293, 295, 299, 304
Bayso, 88–89
Belarusian, 9, 162, 183–184, 260
Berber. *See* Eastern Riffian; Kabyle; Tamashek
Biblical Hebrew, 9, 122, 243
Biblical Welsh, 6–7, 15, 116, 122–123, 125, 137, 142–144, 223, 227–231, 233–234, 240–241
British English, 110, 113, 173
Bulgarian, 7, 9, 162, 195–196, 199–201, 203, 208, 221

Celtic, 15. *See also* Scottish Gaelic; Welsh
Chinese, 7, 76. *See also* Mandarin
Chinyanja, 68
Classical Welsh, 227
Czech, 296

Dagaare, 7, 9, 223–227, 268
Danish, 15, 102
Dutch, 8, 10, 20, 31, 33–34, 37, 50, 84–85, 87, 94, 98, 101, 109, 112, 116, 121, 126, 145, 280, 304, 309–310, 313–317

Eastern Riffian, 181
English, 8, 10, 15, 20, 30–31, 33–34, 47, 50, 55, 58–59, 61–62, 74, 76, 78–79, 81, 83–87, 93–94, 110–113, 116, 120–122,

125–126, 135–136, 138, 140–141, 173, 184, 189, 231, 251, 269–270, 280–281, 283–285, 288–289, 296, 298, 300–301, 307, 309, 315, 319, 326. *See also* British English
Estonian, 141, 341n
Estonian Swedish, 108

Finnish, 7–8, 15, 49, 52, 62–63, 78–80, 83, 87–88, 90–93, 127–129, 135–137, 260
Finno-Ugric, 2. *See also* Estonian; Finnish; Hungarian; Inari Sami
French, 15, 30, 33, 43, 51, 54, 139, 189, 270, 284, 288, 300–303, 325, 328
Fula, 89

German, 102, 121, 134–135, 141, 300, 315

Hebrew, 3, 6–7, 9–10, 36–37, 62, 68, 121–122, 134–136, 223, 235, 241–253, 255–258, 260–262, 268, 292–293, 295, 299, 304. *See also* Biblical Hebrew; Modern Hebrew
Hungarian, 7–8, 15, 71, 78, 83, 87–92, 102

Inari Sami, 7, 9, 51–52, 78, 115, 127–129, 135–137, 260, 281
Italian, 280, 326

Japanese, 2, 7, 76, 121, 181

Kabyle, 181
Korean, 2

Latin, 7–8, 49, 53, 61–66, 75–76, 93–94, 117, 144–145
Lebanese Arabic, 7, 9, 223, 257, 263, 265, 267
Lithuanian, 68
Luvale, 68, 122

Subject Index

Linguistic Inquiry Monographs

Samuel Jay Keyser, general editor